热分析应用手册
Application Handbook
Thermal Analysis

热固性树脂
Thermosets

Rudolf Riesen 著

陆立明 译

本应用手册提供各种精选的应用实例。实验是由瑞士梅特勒-托利多热分析实验室采用在每个应用实例中描述的特定仪器完成的,并以最新知识为依据对实验结果进行评估。
然而,这并非意味着读者无需用自己的适合样品的方法、仪器和用途进行亲自测试。由于对实例的效仿和应用是无法控制的,所以我们当然无法承担任何责任。
使用化学品、溶剂和气体时,必须遵循常规安全规范和制造商或供应商提供的使用指南。
This application handbook presents selected application examples. The experiments were conducted with the utmost care using the instruments specified in the description of each application at METTLER TOLEDO Themal Analysis Lab in Switzerland. The results have been evaluated according to the current state of our knowledge.
This does not however absolve you from personally testing the suitability of the examples for your own methods, instruments and purposes. Since the transfer and use of an application is beyond our control, we cannot of course accept any responsibility.
When chemicals, solvents and gases are used, general safety rules and the instructions given by the manufacturer or supplier must be observed.

图书在版编目(CIP)数据

热固性树脂/(瑞士)里森(Riesen,R.)著;陆立明译.
—上海:东华大学出版社,2008.11
(热分析应用手册系列丛书)
ISBN 978－7－81111－458－4

Ⅰ.热... Ⅱ.①里...②陆... Ⅲ.热固性树脂
Ⅳ.TQ323

中国版本图书馆 CIP 数据核字(2008)第 188203 号

责任编辑 竺海娟
封面设计 蔡顺兴

热固性树脂

东华大学出版社出版
上海市延安西路1882号
邮政编码:200051 电话:(021)62193056
新华书店上海发行所发行 苏州望电印刷有限公司印刷
开本:889×1194 1/16 印张:21.5 字数:688千字
2009年3月第1版 2019年12月第4次印刷
ISBN 978－7－81111－458－4
定价:105.00元

序

热分析是仪器分析的一个重要分支,它对物质的表征发挥着不可替代的作用。热分析历经百年的悠悠岁月,从矿物、金属的热分析兴起,近几十年在高分子科学和药物分析等方面唤起了勃勃生机。

我国在20世纪50～60年代,科研单位、高校和产业部门为满足科研、教学和生产的需要,经历了从原理出发自行设计研制热分析仪器的艰苦创业阶段;30～40年前,先进的热分析仪器还只是在少数科研单位的测试中心才拥有,而随着我国综合国力的增强和对科研支持力度的加大,现已逐渐成为许多实验室的通用仪器,在科研和生产中起着越加重要的作用。广大相关专业的科技人员为了更好地利用这些设备,迫切需要深入掌握热分析仪器及相关的基础和应用方面的知识。这套热分析应用系列丛书就是在这样的形势下应蕴而生的。

本书的基础数据主要是由瑞士的梅特勒—托利多(Mettler-Toledo)公司提供的,该公司是全球著名的精密仪器制造和经销商。早在1945年,就曾以首台单秤盘替代法天平而闻名于世。随后,又将其与加热炉结合,在1964年推出了世界上第一台商品化TGA/DTA热分析仪器。1968年又有TGA/MS联用仪和差示扫描量热仪(DSC)相继问世。40余年来,梅特勒—托利多一直是全球热分析仪器的主要供应商之一。现今具有包括DSC、TGA/DSC、TMA、DMA等完备的现代热分析仪器。近年取得的新进展有如:多星型热电堆DSC 2006年荣获美国R&D100奖,该奖项是每年颁发给当年在全球技术领域具有代表性新产品的开发者;2005年开发的随机多频温度调制DSC技术TOPEM™,能在一次实验中测定准稳态比热容,由热流与升温速率的相关性分析分别得到可逆、不可逆热流量和总热流量,以及反应(转变)过程与频率的关系。

《热分析应用手册》是系统介绍热分析在诸多领域应用的一项系统工程,《热固性树脂》是其中的一个分册。这套丛书汇集梅特勒—托利多公司瑞士总部和梅特勒—托利多(中国)公司科技人员的智慧而潜心编著的。

该分册的主要作者Rudolf Riesen博士,1978年在瑞士理工学院(Swiss Institute of Technology)获得化学工程博士学位,随即加入梅特勒—托利多公司,一直从事高分子(尤其热固性聚合物)的热分析研究,积累了丰富的经验。

译者陆立明先生1985年在华东理工大学获得聚合物材料工学硕士,后在上海市合成树脂研究所从事聚合物研究开发工作12年(其中3年在德国柏林技术大学进修高分子物理)。加入梅特勒—托利多(中国)公司以来一直从事热分析的技术应用和管理工作。

本书的酝酿和出版正值我国改革开放30年,我国在世界的影响全面提升。据称这一时期我国科技人员在国际期刊发表的论文数量提高近35倍(从1981年的约2千篇到2006年的约7万篇)。本书的一个明显特点是以中英文对照的形式出版,这就为熟悉英语论文的写作方法提供了一种借鉴。

相信这套丛书的出版,将会对我国热分析技术的普及与提高起到重要的推动作用。

刘振海

2009年1月20日 于长春

著者序

为了能深入地了解热固性树脂，本应用手册给出了大量的实例。用于样品测试的主要技术为差示扫描量热法（DSC）、热重分析（TGA）、热机械分析（TMA）和动态热机械分析（DMA）。在特别情况下，也使用如 SDTA、图像监测和逸出气体分析这样的在线联用技术。

热固性树脂是经历了称为固化即交联的永久性化学反应而形成交联网状结构的热固性聚合物。它们是坚硬的、典型不溶的机械强度和温度稳定性高的材料。与热塑性塑料不同，热固性树脂在固化后不能再熔融或再成型为另一个形状。热固性材料包括大范围的在化学上不同的化合物。当今，为了开发新的材料性能，越来越多的不同类型的聚合物被结合在一起。因而有时难以区分热固性树脂、热塑性塑料和弹性体。由于这个原因，本手册的重心放在可被清晰确定为热固性树脂的相对简单的体系。

第一部分为全面的评述和对常用于热固性树脂表征的分析技术的扼要说明。第二部分论述各个热固性树脂的化学性能和讨论这些材料的用途。这部分是供热固性聚合物领域的新人和期望学习更多热固性树脂性能和应用的人们使用的。第三部分讨论可用不同热分析技术研究的性能和效应。通常，为了利于比较，用相同的树脂体系来测试。

第四至第九部分集中于实际例子。按照树脂体系类型被细分。应用实例描述了在热固性树脂的生命周期中可被研究、测试或只是检查的不同性能。

我希望本书中描述的应用将为这个相当复杂但十分有意义的领域的专家和新手找到广泛的兴趣并激励新思想。

我对我的梅特勒－托利多同事们为本书提供的许多贡献表示非常感谢。我感谢倪菁女士、Jürgen Schawe 博士、Markus Schubnell 博士、Matthias Wagner 博士、Georg Widmann 和 Marco Zappa。特别应该提及 Myrta Pfister 女士，她为本出版物做了大量的样品测试。

最后，我要感谢我所有的同事，特别是 Jürgen Schawe 博士和 Georg Widmann 有价值的讨论和校对，以及 Dudley May 博士将德文原稿翻译为英文。

Rudolf Riesen 博士

Preface

This applications handbook provides an insight into the thermal analysis of thermosets and presents a large number of practical examples. The main techniques used for sample measurement are differential scanning calorimetry (DSC), thermogravimetric analysis (TGA), thermomechanical analysis (TMA) and dynamic mechanical analysis (DMA). In special cases, combined on-line techniques such as SDTA, visual monitoring, and evolved gas analysis have also been employed.

Thermosets are thermosetting polymers that have undergone a permanent chemical reaction known as curing or crosslinking to form a giant crosslinked network structure. They are rigid, typically insoluble materials of high mechanical strength and high temperature stability. In contrast to thermoplastics, thermosets cannot be remelted or remolded into another shape after curing. Thermosetting materials include a wide range of chemically different compounds. Nowadays, more and more different types of polymers are being combined in order to develop new material properties. This makes it sometimes difficult to distinguish between thermosets, thermoplastics, and elastomers. For this reason, the focus in this handbook is on relatively simple systems that can be clearly identified as thermosets.

The first part presents an overview and brief description of the analytical techniques commonly used to characterize thermosets. The second part deals with the chemistry of individual thermosets and discusses the use of these materials. This section is intended for readers who are new to the field of thermosetting polymers and who wish to learn more about the properties and applications of thermosets. The third part discusses the properties and effects that can be investigated using different thermoanalytical techniques. In general, the same resin systems were used for the measurements in order to facilitate comparison.

Part four to nine concentrate on practical examples. These have been subdivided according to the type of resin system. The applications describe the different properties that can be investigated, measured, or simply checked during the lifecycle of a thermoset.

I hope that the applications described in this book will find wide interest and stimulate new ideas both for experts and for newcomers to this rather complex but immensely interesting field.

I am very grateful for the many contributions in this book supplied by my colleagues at METTLER TOLEDO. My thanks go to Mrs. Ni Jing, Dr. Jürgen Schawe, Dr. Markus Schubnell, Dr. Matthias Wagner, Georg Widmann and Marco Zappa. Mrs. Myrta Pfister deserves special mention for the large number of sample measurements she performed for this publication.

Finally, I thank all my colleagues, especially Dr. Jürgen Schawe and Georg Widmann for valuable discussions and proofreading, and Dr. Dudley May for translating the original German manuscript into English.

Dr. Rudolf Riesen

出版前言

《热分析应用手册系列丛书》是由梅特勒-托利多瑞士热分析实验室专家撰写的系列手册，包括《热分析基础》、《热塑性聚合物》、《热固性树脂》、《弹性体》、《食品》、《药物》、《无机物》、《化学品》和《热重-逸出气体分析》等分册。

本套书既注重实用性，又注重学术性。它们可以作为应用手册查询，也可以作为实验指南，如帮助选择合适的热分析测试技术和方法、制备和处理样品、设定实验参数等。手册中的所有应用实例都经过认真挑选，实验方法经精心设计，测试曲线重复可靠，数据处理严格谨慎，对实验结果的解释和对实验结论的推导科学合理。

本套手册面向所有用到热分析和对热分析感兴趣的教授、科学家、工程师和学生（特别是研究生）及其他科技工作者，适合所有热分析仪器的直接使用者。

本书是《热分析应用手册系列丛书》之《热固性树脂》分册。

热固性树脂是经交联固化反应生成的具有巨大网状结构的热固性聚合物，具有出色的力学性能和热稳定性。本分册通过大量实例全面深入地介绍和讨论了热分析在热固性树脂方面的应用。主要内容：热分析技术 DSC、TMDSC、TGA、TMA 和 DMA 等简介；热固性树脂的结构、性能和应用；热固性树脂的基本热效应；环氧树脂、不饱和聚酯树脂、酚醛树脂、丙烯酸类树脂、聚氨酯树脂等的热分析—固化反应（等温固化、光固化、后固化、反应动力学等）、玻璃化转变（T_g 与固化度、T_g 的各种测试法、固化反应中的玻璃化、凝胶化、时间—温度转换图等）、填料和增强纤维等的影响、印制线路板分析（T_g、分层、老化）、缩聚、加聚、层压板、黏合剂……

与其他分册一样，本书以中英文对照方式出版，读者可以阅读中文，同时可对照原著。无论对热分析工作者，还是热分析学习者，应该都有帮助和裨益。

这里要特别感谢刘振海教授，他仔细审阅了本书全部书稿，提出了宝贵的意见并亲自进行修改，使本书的质量得到了很大提高。

我的同事唐远旺和蔡艺参加了本书初稿的部分翻译，在此表示感谢。对于东华大学出版社编辑的辛勤工作一并致谢。

译文甚至原著中，有错误之处，恳望读者指正，以便能在再版时改正，不胜感谢。

陆立明
2009 年 1 月，上海

目 录

应用一览表(第一至第三章) Application list (Chapter 1 to 3) VII
应用一览表(第四至第九章) Application list (Chapter 4 to 9) IX

1. 热分析概论 Introduction to Thermal Analysis 1
    1.1 差示扫描量热法(DSC) Differential Scanning Calorimetry 1
        1.1.1 常规 DSC Conventional DSC 1
        1.1.2 温度调制 DSC(TMDSC) Temperature-modulated DSC 3
            1.1.2.1 ADSC 3
            1.1.2.2 IsoStep 4
            1.1.2.3 TOPEM™ 5
    1.2 热重分析(TGA) Thermogravimetric Analysis 6
    1.3 热机械分析(TMA) Thermomechanical Analysis 7
    1.4 动态热机械分析(DMA) Dynamic Mechanical Analysis 9
    1.5 与 TGA 的同步测量 Simultaneous Measurements with TGA 11
        1.5.1 同步 DSC 和差热分析(DTA,SDTA) Simultaneous DSC and Differential Thermal Analysis 11
        1.5.2 逸出气体分析(EGA) Evolved Gas Analysis 12
            1.5.2.1 TGA—MS 12
            1.5.2.2 TGA—FTIR 13

2. 热固性树脂的结构、性能和应用 Structure, Properties and Applications of Thermosets 15
    2.1 概述 Introduction 15
    2.2 热固性树脂的化学结构 Chemical Structure of Thermosets 16
        2.2.1 大分子 Macromolecules 16
        2.2.2 热固性树脂概述 A general overview of thermosets 17
        2.2.3 树脂 Resins 19
            2.2.3.1 环氧树脂 Epoxy Resins,EP 20
            2.2.3.2 酚醛树脂 Phenolic Resins, PF 23
            2.2.3.3 氨基树脂 Amino Resins, UF, MF 26
            2.2.3.4 醇酸树脂,不饱和聚酯树脂 Alkyds, Unsaturated polyester resins, UP 27
            2.2.3.5 乙烯基酯树脂 Vinyl ester resins, VE 29
            2.2.3.6 烯丙基、DAP 模塑料 Allylics, DAP molding compounds 29
            2.2.3.7 聚丙烯酸酯 Polyacrylate, PAK 30
            2.2.3.8 聚氨酯体系 Polyurethane systems, PUR 31
            2.2.3.9 二氰酸酯树脂 Dicyanate resins 32
            2.2.3.10 聚酰亚胺、双马来酰亚胺树脂 Polyimides (PI), Bis maleimide (BMI) resins 32
            2.2.3.11 硅树脂 Silicone resins, SI 33
    2.3 固化反应 The curing reaction 33

2.3.1	交联步骤 Crosslinking steps	33
2.3.2	TTT 图 TTT diagram	35
2.3.3	固化动力学 Curing Kinetics	37

2.4 热固性树脂的应用 Applications of Thermosets 41
 2.4.1 热固性树脂的性能 Properties of Thermosets 41
 2.4.2 加工 Processing 42
 2.4.3 各种树脂的应用领域和性能 Areas of application and properties of individual resins 43
  2.4.3.1 环氧树脂 Epoxy Resins，EP 43
  2.4.3.2 酚醛树脂 Phenol-formaldehyde resins，PF 44
  2.4.3.3 氨基树脂 Amino resins，UF/MF 45
  2.4.3.4 聚酯树脂 Polyester resins，UP 46
  2.4.3.5 乙烯基酯树脂 Vinyl ester resins，VE 47
  2.4.3.6 苯二酸二烯丙酯模塑料 DAP molding compounds 47
  2.4.3.7 丙烯酸酯树脂 Acrylate 48
  2.4.3.8 聚氨酯 Polyurethane，PUR 48
  2.4.3.9 聚酰亚胺 PI 和 BMI Polyimide，PI and BMI 49
  2.4.3.10 硅树脂 Silicone resines 49
  2.4.3.11 使用范围和应用概述 Overview of the areas of use and application 50

2.5 热固性树脂的表征方法 Characterization methods for thermosets 52
 2.5.1 所需信息的概述 Overview of information required 52
 2.5.2 表征热固性树脂的热分析技术 TA techniques for the characterization of thermosets 54
 2.5.3 玻璃化转变 Glass transition 56
  2.5.3.1 玻璃化转变和松弛:热学和动态玻璃化转变 Glass transition and relaxation：thermal and dynamic glass transition 56
  2.5.3.2 玻璃化转变温度的测定 Determination of the glass transition temperature 59
 2.5.4 热固性树脂分析的标准方法 Standard methods for thermoset analysis 61

3. 热固性树脂的基本热效应 Basic thermal effects of thermosets 69
3.1 热效应的 DSC 测量 Measurement effects with DSC 69
 3.1.1 玻璃化转变的测定 Determination of the glass transition 69
  3.1.1.1 玻璃化转变温度的 DSC 测量 Measurement of the glass transition temperature by DSC 70
  3.1.1.2 用 DSC 计算玻璃化转变的方法 Evaluation possibilities for the glass transition by DSC 72
  3.1.1.3 样品预处理对玻璃化转变的影响 Influence of sample pretreatment on the glass transition 78
  3.1.1.4 玻璃化转变的 ADSC 测量 Measurement of the glass transition by ADSC 81
 3.1.2 比热容测定 Determination of the specific heat capacity 83

3.1.3 用DSC测试的固化反应 The curing reaction measured by DSC ······ 85
   3.1.3.1 动态固化：第一次和第二次升温测量 Dynamic curing: first and second heating measurements ······ 86
   3.1.3.2 等温固化的DSC测量 Isothermal curing by DSC ······ 89
   3.1.3.3 后固化和固化度的DSC测量 Postcuring and degree of cure by DSC ······ 91
   3.1.3.4 玻璃化转变与转化率的关系 Glass transition as a function of the conversion ······ 94
   3.1.3.5 固化速率和动力学的等温测量 Rate of cure and kinetics, isothermal measurements ······ 97
   3.1.3.6 固化速率的动态测量 Curing rate, dynamic measurements ······ 101
   3.1.3.7 动力学计算和预测 Kinetic evaluations and predictions ······ 102
3.1.4 玻璃化转变和后固化的分离（TOPEM™法）Separation of the glass transition and postcuring (TOPEM™) ······ 105
3.1.5 紫外光固化的DSC测量 UV curing measured by DSC ······ 107

3.2 效应的TGA测量 Measurement effects with TGA ······ 111
 3.2.1 热固性树脂升温时的质量变化 Mass changes on heating a thermoset ······ 111
 3.2.2 含量测定：水分、填料和树脂含量 Content determination: moisture, filler and resin content ······ 113
 3.2.3 苯酚—甲醛缩合反应的TGA分析 TGA analysis of a phenol-formaldehyde condensation reaction ······ 115

3.3 效应的TMA测量 Measurement effects with TMA ······ 116
 3.3.1 线膨胀系数的测定 Determination of the linear expansion coefficient ······ 117
 3.3.2 玻璃化转变的TMA测量 Measurement of the glass transition by TMA ······ 120
   3.3.2.1 测定玻璃化转变的膨胀曲线 Determination of the glass transition by means of the expansion curve ······ 120
   3.3.2.2 薄涂层软化温度的测定 Determination of the softening temperature of thin coatings ······ 122
   3.3.2.3 由弯曲测试测定玻璃化转变 Determination of the glass transition from bending measurements ······ 123
 3.3.3 固化反应的TMA测量 Measurement of the curing reaction by TMA ······ 126
   3.3.3.1 固化反应的弯曲测量研究 Investigation of the curing reaction using bending measurements ······ 127
   3.3.3.2 凝胶时间的DLTMA测定 Determination of the gelation time by DLTMA ······ 130

3.4 效应的DMA测量 Measurement effects with DMA ······ 131
 3.4.1 玻璃化转变的DMA测量 Determination of the glass transition by DMA ······ 132
 3.4.2 玻璃化转变的频率依赖性 The frequency dependence of the glass transition ······ 135
 3.4.3 动态玻璃化转变 The dynamic glass transition ······ 138
 3.4.4 等温频率扫描 Isothermal frequency sweeps ······ 140
 3.4.5 主曲线绘制和力学松弛频率谱 Master curve construction and mechanical relaxation spectrum ······ 142

3.4.6 固化的 DMA 测量 Curing measured by DMA ……144
3.5 玻璃化转变 DSC、TMA 和 DMA 测量的比较 A comparison of the glass transition measured by DSC, TMA and DMA ……146

4. 环氧树脂 Epoxy resins ……150
4.1 影响固化反应的因素 Factors affecting curing reactions ……150
4.1.1 固化条件（温度、时间）的影响 Influence of curing conditions (temperature, time) ……150
4.1.2 组分混合比例的影响 Influence of the mixing ratio of the components ……152
4.1.3 促进剂类型的影响 Influence of the type of accelerator ……155
4.1.4 促进剂含量对固化反应的影响 Influence of accelerator content on the curing reaction ……157
4.1.5 环氧树脂：转化率行为的预测和验证 EP: Prediction of conversion behavior and verification ……160
4.1.6 环氧树脂固化的 DMA 测量 Curing of an EP resin measured by DMA ……163
4.1.7 预浸料固化的 DMA 测量 Curing of a prepreg measured by DMA ……166
4.1.8 粉末涂层的固化 Curing of a powder coating ……168

4.2 影响玻璃化转变的因素 Influences affecting the glass transition ……170
4.2.1 重复后固化对玻璃化转变的影响 Effect of repeated postcuring on the glass transition ……170
4.2.2 化学计量对固化和最终玻璃化转变温度的影响 The effect of stoichiometry on curing and the resulting glass transition temperature ……173
4.2.3 活性稀释剂对最终玻璃化转变温度的影响 Influence of reactive diluents on the resulting glass transition temperature ……176
4.2.4 玻璃化 Vitrification ……181
4.2.4.1 玻璃化转变温度与转化率关系的测定 Determination of the dependence of the glass transition temperature on conversion ……181
4.2.4.2 等温固化反应中化学引发玻璃化转变的温度调制 DSC 测量 Chemically induced glass transition in an isothermal curing reaction measured by temperature-modulated DSC ……185
4.2.4.3 非模型动力学和固化过程中的玻璃化 Model free kinetics and vitrification during curing ……187
4.2.4.4 固化过程中玻璃化的测量 Measurement of vitrification during curing ……191
4.2.5 TTT 图的测定 Determination of a TTT diagram ……193
4.2.5.1 TTT 图：由后固化实验测定 TTT diagram: Determination from postcuring experiments ……193
4.2.5.2 TTT 图：温度调制 DSC 的应用 TTT diagram: Application of temperature-modulated DSC ……195
4.2.5.3 玻璃化和非模型动力学 Vitrification and model free kinetics ……197
4.2.6 等温固化的凝胶点和力学玻璃化转变 Gel point and mechanical glass transition during

　　　　　　isothermal curing ……………………………………………………………… 203
　　　　4.2.6.1　固化反应中剪切模量的变化 Change of the shear modulus during the curing reaction
　　　　　　…………………………………………………………………………………… 203
　　　　4.2.6.2　固化反应中剪切模量的频率依赖性 Frequency dependence of the shear modulus during
　　　　　　a curing reaction …………………………………………………………………… 206
4.3　贮存效应 Storage effects ………………………………………………………………… 209
　　4.3.1　贮存后的后固化 Postcuring after storage ……………………………………… 209
　　4.3.2　环氧树脂-碳纤维：贮存对预浸料的影响 EP-CF：Influence of storage on preprgs … 211
4.4　填料和增强纤维 Fillers and reinforcement fibers …………………………………… 213
　　4.4.1　玻璃化转变温度和"固化因子"按照 IPC-TM-650 的 DSC 测定 Glass transition temperature
　　　　and "Cure Factor" by DSC according to IPC-TM-650 ………………………… 213
　　4.4.2　玻璃化转变温度和 z-轴热膨胀按照 IPC-TM-650 的 TMA 测定 Glass transition temperature
　　　　and z-axis thermal expansion by TMA according to IPC-TM-650 …………… 214
　　4.4.3　印制线路板,纤维取向对膨胀行为的影响 Printed circuit boards, influence of fiber
　　　　orientation on expansion behavior ……………………………………………… 216
　　4.4.4　碳纤维增强树脂玻璃化转变的测定 Determination of the glass transition of CF-reinforced
　　　　resins ………………………………………………………………………………… 218
　　4.4.5　复合材料纤维含量的热重分析测定 Determination of the fiber content of composites by
　　　　thermogravimetric analysis ………………………………………………………… 221
　　4.4.6　预浸料中的碳纤维含量 Carbon fiber content in prepregs …………………… 224
4.5　材料性能的检测 Checking material properties ……………………………………… 226
　　4.5.1　印制线路板生产中的质量保证 Quality assurance in the production of printed circuit boards
　　　　……………………………………………………………………………………… 226
　　4.5.2　碳纤维增强热固性树脂的玻璃化转变测定 Determination of the glass transition of carbon
　　　　fiber reinforced thermosets ………………………………………………………… 229
　　4.5.3　按照 ASTM 标准 E1641 和 E1877 求解分解动力学和长期稳定性 Decomposition kinetics
　　　　and long-term stability according to ASTM standards E1641and E1877 ……… 233
　　4.5.4　印制线路板的老化 Aging of printed circuit boards …………………………… 235
　　4.5.5　分解产物的 TGA-MS 分析 Analysis of decomposition products by TGA-MS ……… 238
　　4.5.6　印制线路板分层的 TMA-EGA 测量 Delamination of printed circuit boards by TMA-EGA
　　　　……………………………………………………………………………………… 240
　　4.5.7　印制线路板分层时间按照 IPC-TM-650 的 TMA 测定 Time to delamination of printed
　　　　circuit board by TMA according to IPC-TM-650 ……………………………… 242
　　4.5.8　质量保证,黏结层的失效分析 Quality assurance, failure analysis of adhesive bonds ……
　　　　……………………………………………………………………………………… 244
　　4.5.9　油与增强环氧树脂管的相互作用 Interaction of oil with a reinforced EP resin pipe … 246

5. 不饱和聚酯树脂 Unsaturated polyester resins ………………………………………… 249
　　5.1　进货控制:固化特性和玻璃化转变 Incoming goods control：curing characteristics and glass
　　　　transition ……………………………………………………………………………… 249

5.2 不饱和聚酯:促进剂含量的影响 UP:Influence of the accelerator content 250
5.3 不饱和聚酯:硬化剂含量的影响 UP:Influence of the hardener content 252
5.4 抑制剂对等温固化的影响 Influence of the inhibitor on isothermal curing 254
5.5 不饱和聚酯:贮存后的固化行为 UP:Curing behavior after storage 255
5.6 乙烯基酯树脂:由促进剂引起的固化温度的移动 VE:Shift of curing temperature due to the accelerator 257
5.7 乙烯基酯－玻璃纤维:使用后管材的固化度 VE-GF:Degree of cure of a pipe after use 258
5.8 粉末涂料的紫外光固化 Curing of powder coatings using UV light 260
5.9 加工片状模塑料的模塑时间 Molding times for processing SMC 266

6. 甲醛树脂 Formaldehyde resins 269

6.1 酚醛树脂:测试条件的影响 PF:Influence of measurement conditions 269
6.2 酚醛树脂:用 TMA 区别完全和部分固化的酚醛树脂 PF:Differentiation between completely and partially cured phenolic resins by TMA 270
6.3 酚醛树脂:树脂的软化行为 PF:Softening behavior of resins 272
6.4 两种不同的填充三聚氰胺甲醛/酚醛树脂模塑料 Two different filled MF/PF molding compounds 276
6.5 酚醛树脂:胶合板的纸预浸料 PF:Paper prepregs for plywood 278
6.6 酚醛树脂:缩聚反应的 TGA/SDTA 研究 PF:Condensation reaction investigated by TGA/SDTA 280
6.7 酚醛树脂:可溶性酚醛树脂的固化动力学 PF:Curing kinetics of Resol resins 284
6.8 脲醛树脂模塑料:加工(模塑)的影响 UF molding compounds:Influence of processing (molding) 286
6.9 脲醛树脂:模塑料固化动力学 UF:Curing kinetics of molding compounds 288
6.10 酚醛树脂:热导率的测定 PF:Determination of thermal conductivity 291

7. 甲基丙烯酸类树脂 Methacrylate/Acrylic resins (MMA) 296

7.1 牙科复合材料的光固化 Light curing of a dental composite 296

8. 聚氨酯体系 PUR systems 298

8.1 聚氨酯:含溶剂的双组分体系 PUR:Two-component system with solvent 298
8.2 聚氨酯:在不同温度下的加成聚合 PUR:Polyaddition at different temperatures 299
8.3 聚氨酯漆涂层的软化温度 Softening temperature of PUR lacquer coatings 301
8.4 聚氨酯模塑料:作为质量标准的玻璃化转变 PUR casting compounds:Glass transition as a quality criterion 303

9. 其它树脂体系 Other resin systems 307

9.1 双马来酰亚胺树脂－碳纤维:贮存温度对预浸料黏性的影响 BMI-CF:Influence of storage temperature on tackiness of prepregs 307
9.2 黏合剂的光固化 Light curing of adhesives 309

附录:缩写和首字母缩拼词 Appendix: Abbreviations and acronyms 316
与热固性树脂有关的所用术语 Terms used in connection with thermosets 318
文献 Literature 322

应用一览表（第一至第三章）Application list (Chapter 1 to 3)

标题 Title	主题 Topics					方法 Methods				页码 Page
	玻璃化转变 Glass transition	Physical properties (C_p, GTE, 模量) 物理性能（C_p、GTE、模量）	固化反应动力学 Curing reaction, kinetics	成分 Composition	Evaluation/experimental conditions 计算/实验条件	DSC / ADSC / IsoStep / TOPEM™	TGA / TGA-EGA	TMA / DLTMA	DMA	
玻璃化转变温度的 DSC 测量 Measurement of the glass transition temperature by DSC	•				•	•				70
玻璃化转变的计算方法 Evaluation possibilities for the glass transition	•				•	•				72
样品预处理对玻璃化转变的影响 Influence of sample pretreatment on the glass transition	•				•	•				78
玻璃化转变的 ADSC 测量 Measurement of the glass transition by ADSC	•					•				81
比热容测定 Determination of the specific heat capacity		•			•	•				83
动态固化：第一次和第二次升温 Dynamic curing: first and second heating runs			•		•	•				86
等温固化 Isothermal curing			•		•	•				89
后固化 Postcuring	•		•		•	•				91
玻璃化转变与转化率的关系 Glass transition as a function of the conversion	•		•		•	•				94
固化速率和动力学，等温测量 Rate of cure and kinetics, isothermal measurements			•		•	•				97
固化速率，动态测量 Curing rate, dynamic measurements			•		•	•				101
动力学计算和预测 Kinetic evaluations and predictions			•		•	•				102
玻璃化转变和后固化的分离（TOPEM™法） Separation of the glass transition and postcuring (TOPEM™)	•		•		•	•				105
紫外光固化的 DSC 测量 UV curing measured by DSC			•		•	•				107
加热热固性树脂时的质量变化 Mass changes on heating a thermoset				•			•			111
含量测定：水分、填料和树脂含量 Content determination: moisture, filler and resin content				•			•			113
苯酚—甲醛缩合反应的 TGA 分析 TGA analysis of a phenol-formaldehyde condensation reaction			•	•			•			115
线膨胀系数的测定 Determination of the linear expansion coefficient		•						•		117

续表

标题 Title	主题 Topics					方法 Methods				页码 Page
	玻璃化转变 Glass transition	Physical properties (C_p, GTE, modulus) 物理性能(C_p, GTE, 模量)	固化反应动力学 Curing reaction, kinetics	成分 Composition	Evaluation/experimental conditions 计算/实验条件	DSC / ADSC / IsoStep / TOPEM™	TGA / TGA-EGA	TMA / DLTMA	DMA	
用膨胀曲线测定玻璃化转变 Determination of the glass transition by means of the expansion curve	•							•		120
薄涂层软化温度的测定 Determination of the softening temperature of thin Coatings	•							•		122
由弯曲测试测定玻璃化转变 Determination of the glass transition from bending measurements	•							•		123
固化反应的弯曲测量研究 Investigation of the curing reaction using bending measurements			•					•		127
凝胶时间的 DLTMA 测定 Determination of the gelation time by DLTMA			•					•		130
玻璃化转变的 DMA 测定 Determination of the glass transition by DMA	•	•			•				•	132
玻璃化转变的频率依赖性 The frequency dependence of the glass transition	•	•							•	135
动态玻璃化转变 The dynamic glass transition	•	•			•				•	138
等温频率 Isothermal frequency		•							•	140
主曲线绘制和力学松驰频谱 Master curve construction and mechanical relaxation spectrum		•			•				•	142
固化的 DMA 测量 Curing measured by DMA			•						•	144
玻璃化转变 DSC、TMA 和 DMA 测量的比较 A comparison of the glass transition measured by DSC，TMA and DMA	•				•	•		•	•	146

应用一览表（第四至第九章）Application list (Chapter 4 to 9)

标题 Title	体系的开发 Development of systems	过程优化和控制 Process optimization and control	测定固化材料 Testing cured material	玻璃化转变,玻璃化 Glass transition, vitrification	样品制备 Sample preparation	物理性能(Cp, CTE, 模量) Physical properties (Cp, CTE, modulus)	固化, 后固化 Curing, postcuring	成分降解 Composition degradation	动力学计算 Kinetic evaluation	附件(EGA, UV) Accessories (EGA, UV)	DSC/ADSC/IsoStep/TOPEM™	TGA/SDTA	TGA/TGA-EGA	DMA	页码 Page
固化条件（温度、时间）的影响 Influence of curing conditions (temperature, time)				•	•	•					•				150
组分混合比的影响 Influence of the mixing ratio of the components		•	•	•		•					•				152
促进剂类型的影响 Influence of the type of accelerator	•										•				155
促进剂含量对固化反应的影响 Influence of accelerator content on the curing reaction		•					•				•				157
环氧树脂：转化率行为的预测和验证 EP: Prediction of conversion behavior and verification				•			•		•		•				160
环氧树脂固化的 DMA 测量 Curing of an EP resin measured by DMA		•		•	•	•	•							•	163
预浸料固化的 DMA 测量 Curing of a prepreg measured by DMA		•		•	•	•	•							•	166
粉末涂层的固化 Curing of a powder coating		•								•	•				168
重复后固化对玻璃化转变的影响 Effect of repeated postcuring on the glass transition				•	•						•				170
化学计量对固化和最终玻璃化转变温度的影响 The effect of stoichiometry on curing and the resulting glass transition temperature		•		•	•						•				173
活性稀释剂对最终玻璃化转变温度的影响 Influence of reactive diluents on the resulting glass transition temperature	•	•		•	•						•				176
玻璃化转变温度与转化率关系的测定 Determination of the dependence of the glass transition temperature on conversion		•		•	•	•	•				•				181
温度调制 DSC 测量等温固化反应过程中化学引发的玻璃化转变 Chemically induced glass transition in an isothermal curing reaction measured by temperature-modulated DSC				•			•				•				185

续表

标题 Title	体系的开发 Development of systems	过程优化和控制 Process optimization and control	测定固化材料 Testing cured material	玻璃化转变,玻璃化 Glass transition,vitrification	样品制备 Sample preparation	物理性能(Cp,CTE,模量) Physical properties (Cp,CTE,modulus)	固化,后固化 Curing, postcuring	成分降解 Composition degradation	动力学计算 Kinetic evaluation	附件(EGA,UV) Accessories (EGA,UV)	DSC/ADSC/IsoStep/TOPEM™	TGA/SDTA	TGA/TGA-EGA	DMA	页码 Page
非模型动力学和固化过程中的玻璃化 Model free kinetics and vitrification during curing		•		•			•		•		•				187
固化过程中玻璃化的测量 Measurement of vitrification during curing		•		•							•				191
TTT 图:从后固化实验测定 TTT diagram: Determination from postcuring experiments		•	•	•			•				•				193
TTT 图:温度调制 DSC 的应用 TTT diagram: Application of temperature-modulated DSC		•	•	•			•				•				195
玻璃化和非模型动力学 Vitrification and model free kinetics		•	•	•			•		•		•				197
固化反应中剪切模量的变化 Change of the shear modulus during the curing reaction		•												•	203
固化反应过程中剪切模量的频率依赖性 Frequency dependence of the shear modulus during a curing reaction		•				•								•	206
贮存后的后固化 Postcuring after storage			•		•		•				•				209
环氧树脂—碳纤维:贮存对预浸料的影响 EP-CF: Influence of storage on prepregs	•	•			•		•				•				211
玻璃化转变温度和"固化因子"按照 IPC-TM-650 的 DSC 测定 Glass transition temperature and "Cure Factor" by DSC according to ICP-TM-650				•							•				213
玻璃化转变温度和 z-轴热膨胀按照 IPC-TM-650 的 TMA 测定 Glass transition temperature and z-axis thermal expansion by TMA according to ICP-TM-650				•	•								•		214
印制电路板,纤维取向对膨胀行为的影响 Printed circuit boards, influence of fiber orientation on expansion behavior				•		•							•		216
碳纤维增强树脂的玻璃化转变的测定 Determination of the glass transition of CF-reinforced resins				•	•						•				218
复合材料纤维含量的热重分析测定 Determination of the fiber content of composites by thermogravimetric analysis	•		•					•					•		221

续表

标题 Title	体系的开发 Development of systems	过程优化和控制 Process optimization and control	测定固化材料 Testing cured material	玻璃化转变,玻璃化 Glass transition, vitrification	样品制备 Sample preparation	物理性能(C_p,CTE,模量) Physical properties (C_p, CTE, modulus)	固化,后固化 Curing, postcuring	成分降解 Composition degradation	动力学计算 Kinetic evaluation	附件(EGA,UV) Accessories (EGA,UV)	DSC/ADSC/IsoStep/TOPEM™	TGA/SDTA	TGA/TGA-EGA	DMA	页码 Page
预浸料中的碳纤维含量 Carbon fiber content in prepregs		•			•			•				•			224
印制电路板生产中的质量保证 Quality assurance in the production of printed circuit boards		•		•	•						•				226
碳纤维增强热固性树脂的玻璃化转变测定 Determination of the glass transition of carbon fiber reinforced thermosets				•	•		•				•				229
按照ASTM标准E1641和E1877求解的分解动力学和长期稳定性 Decomposition kinetics and long-term stability according to ASTM standards E1641 and E1877		•						•	•			•			233
印制电路板的老化 Aging of printed circuit boards		•		•	•	•					•				235
分解产物的TGA-MS分析 Analysis of decomposition products by TGA-MS		•						•		•		•	•		238
印制电路板分层的TMA-EGA测量 Delamination of printed circuit boards by TMA-EGA				•			•			•				•	240
印制电路板分层时间按照IPC-TM-650的TMA测定 Time to delamination of printed circuit board by TMA according to ICP-TM-650				•										•	242
质量保证,黏结层的失效分析 Quality assurance, failure analysis of adhesive bonds		•	•	•			•				•				244
油与增强环氧树脂管的相互作用 Interaction of oil with a reinforced EP resin pipe			•	•			•				•				246
进货控制:固化特性和玻璃化转变 Incoming goods control: curing characteristics and glass transition		•		•			•				•				249
不饱和聚酯:促进剂含量的影响 UP: Influence of the accelerator content	•	•		•			•				•				250
不饱和聚酯:硬化剂含量的影响 UP: Influence of the hardener content				•			•				•				252
抑制剂对等温固化的影响 Influence of the inhibitor on isothermal curing	•	•		•			•				•				254

续表

标题 Title	主题 Topics							方法 Methods					页码 Page
	体系的开发 Development of systems	过程优化和控制 Process optimization and control	测定固化材料 Testing cured material	玻璃化转变,玻璃化 Glass transition, vitrification	样品制备 Sample preparation	物理性能(Cp, CTE, 模量) Physical properties (Cp, CTE, modulus)	固化,后固化 Curing, postcuring	成分降解 Composition degradation	动力学计算 Kinetic evaluation	附件(EGA,UV) Accessories (EGA,UV)	DSC / ADSC / IsoStep / TOPEM™	TGA/SDTA	TGA / TGA-EGA	DMA	
不饱和聚酯:贮存后的固化行为 UP: Curing behavior after storage				•			•				•				255
乙烯基酯树脂:由催化剂引起的固化温度的移动 VE: Shift of curing temperature due to the accelerator	•	•		•							•				257
乙烯基酯—玻璃纤维:使用后管材的固化度 VE-GF: Degree of cure of a pipe after use			•	•							•				258
粉末涂料的紫外光固化 Curing of powder coatings using UV light		•					•			•	•				260
加工片状模塑料的模塑时间 Molding times for processing SMC		•					•				•				266
酚醛树脂:测试条件的影响 PF: Influence of measurement conditions		•					•				•				269
酚醛树脂:用TMA区别完全和部分固化的酚醛树脂 PF: Differentiation between completely and partially cured phenolic resins by TMA			•	•		•								•	270
酚醛树脂:树脂的软化行为 PF: Softening behavior of resins		•		•		•				•	•		•		272
两种不同的填充三聚氰胺甲醛/酚醛树脂模塑料 Two different filled MF/PF molding compounds				•							•				276
酚醛树脂:胶合板的纸预浸料 PF: Paper prepregs for plywood	•										•				278
酚醛树脂:缩聚反应的TGA/SDTA研究 PF: Condensation reaction investigated by TGA/SDTA	•										•	•			280
酚醛树脂:可溶性酚醛树脂的固化动力学 PF: Curing kinetics of resol resins							•		•		•				284
脲醛树脂模塑料:加工(模塑)的影响 UF molding compounds: Influence of processing (molding)				•							•				286
脲醛树脂:模塑料固化动力学 UF: Curing kinetics of molding compounds							•		•		•				288

续表

标题 Title	体系的开发 Development of systems	过程优化和控制 Process optimization and control	测定固化材料 Testing cured material	玻璃化转变，玻璃化 Glass transition, vitrification	样品制备 Sample preparation	物理性能(Cp, CTE,模量) Physical properties (Cp, CTE, modulus)	固化，后固化 Curing, postcuring	成分降解 Composition degradation	动力学计算 Kinetic evaluation	附件(EGA,UV) Accessories (EGA,UV)	DSC/ADSC/IsoStep/TOPEM™	TGA/SDTA	TGA/TGA-EGA	DMA	页码 Page
酚醛树脂：热导率的测定 PF：Determination of thermal conductivity			•			•				•	•				291
牙科复合材料的光固化 Light curing of a dental composite	•	•					•			•	•				296
聚氨酯：含溶剂的双组分体系 PUR：Two-component system with solvent	•	•		•			•				•				298
聚氨酯：在不同温度下加成聚合 PUR：Polyaddition at different temperatures				•		•	•				•				299
聚氨酯漆涂层的软化温度 Softening temperature of PUR lacquer coatings			•	•									•		301
聚氨酯模塑料：作为质量标准的玻璃化转变 PUR casting compounds：Glass transition as a quality criterion				•	•						•				303
双马来酰亚胺树脂—碳纤维：贮存温度对预浸料粘性的影响 BMI-CF：Influence of storage temperature on tackiness of prepregs				•		•					•				307
黏合剂的光固化 Light curing of adhesives	•	•	•	•			•		•	•	•				309

1 热分析概论 Introduction to Thermal Analysis

热分析是测量材料的物理和化学性能与温度关系的一类技术的名称[*]。在所有这些方法中,样品受控于升、降温或等温温度程序。

测量可在不同气氛中进行,通常使用惰性气氛(氮气、氩气、氦气)或氧化气氛(空气、氧气)。有时在测试期间气体从一种气氛切换到另一种气氛。另外有些时候有选择性变化的参数是气体压力。

Thermal analysis is the name given to a group of techniques used to measure the physical and chemical properties of materials as a function of temperature. In all these methods, the sample is subjected to a heating, cooling or isothermal temperature program.

The measurements can be performed in different atmospheres. Usually either an inert atmosphere (nitrogen, argon, helium) or an oxidative atmosphere (air, oxygen) is used. In some cases, the gases are switched from one atmosphere to another during the measurement. Another parameter sometimes selectively varied is the gas pressure.

1.1 差示扫描量热法(DSC) Differential Scanning Calorimetry

在 DSC 中,测量流入和流出样品的热流。DSC 可用于研究诸如物理转变(玻璃化转变、结晶、熔融和挥发化合物的蒸发)和化学反应这样的热效应,所获得的信息表征样品的热性能和组成。此外,也能测定诸如热容、玻璃化转变温度、熔融温度、反应热和反应程度等性能。

DSC 还可用于与能同步观察样品的仪器联用(DSC-显微镜法),或用不同波长的光照射(光量热法)。

In DSC, the heat flow to and from the sample is measured. DSC can be used to investigate thermal events such as physical transitions (the glass transition, crystallization, melting, and the vaporization of volatile compounds) and chemical reactions. The information obtained characterizes the sample with regard to its thermal behavior and composition. In addition, properties such as the heat capacity, glass transition temperature, melting temperature, heat and extent of reaction can also be determined. DSC can also be used in combination with instruments that allow the sample to be simultaneously observed (DSC microscopy) or exposed to light of different wavelengths (photocalorimetry).

1.1.1 常规 DSC Conventional DSC

常规 DSC 采用线性温度程序,样品和参比物(或只是空坩埚)以线性速率升、降温,或在某些情况下保持在恒定温度(即等温)。经常几个局部程序即所谓的程序段连接在一起形成一个完整的温度程序。典型的 DSC 曲线如图 1.1 所示。测试开始时曲线上的变化是由于初始的"启动偏移"(1)。在该瞬变区域,状态突然从等温模式变为线性升温模式。

Conventional DSC employs a linear temperature program. The sample and reference material (or just an empty crucible) are heated or cooled at a linear rate, or in some cases, held at a constant temperature (i.e. isothermally). Often several partial programs or so-called segments are joined together to form a complete temperature program. A typical DSC curve is shown schematically in Figure 1.1. The change in the curve at the beginning of the measurement is due to the initial "startup deflection" (1). In this transient region, the conditions suddenly change from an isothermal mode to a linear heating mode. The

---

[*] 国家标准 GB/T6425—2008《热分析术语》将热分析定义为:"在程序控温(和一定气氛)下,测量物质的某种物理性质与温度或时间关系的一类技术。"由此也可演绎出各种热分析方法的定义。详见国标的有关条款。——译注

启动偏移的大小取决于样品热容和升温速率。样品中如果存在挥发性物质如溶剂,会观察到由于蒸发产生的吸热峰(2),样品失重。通过称量测试前后的样品重量和通过使用不同种类的坩埚可获悉关于这种峰的进一步信息。不同于开口坩埚,完全密封的坩埚能防止样品的蒸发。在玻璃化转变区(3),样品的热容增加,因而可观察到一个吸热台阶,常伴有焓松弛峰。化学反应取决于所涉及反应的种类而产生放热或吸热效应(4)。最后,在较高的温度,开始分解(5)。实验中使用的吹扫气体的种类经常对发生的反应有影响,尤其在高温下。

magnitude of the startup deflection depends on the heat capacity of the sample and the heating rate. If volatile substances such as solvents are present in the sample, an endothermic peak (2) is observed due to the vaporization; the sample loses mass. Further information on such peaks can be obtained by weighing the sample before and after the measurement and by using different types of crucibles. In contrast to open crucibles, hermetically sealed crucibles prevent vaporization of the sample. At a glass transition (3), the heat capacity of the sample increases and therefore an endothermic step is observed. This is often accompanied by an enthalpy relaxation peak. Chemical reactions produce exothermic or endothermic effects (4) depending on the type of reaction involved. Finally, at higher temperatures, decomposition begins (5). The type of purge gas used in the experiment often has an influence on the reactions that occur, especially at high temperatures.

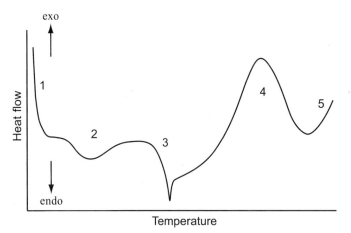

图 1.1 DSC 曲线示意图:1 初始启动偏移;2 水汽蒸发;3 带松弛峰的玻璃化转变;4 反应(例如固化);5 分解开始

Fig. 1.1 Schematic DSC curve: 1 initial startup deflection; 2 evaporation of moisture; 3 glass transition with relaxation peak; 4 reaction (e. g. curing); 5 beginning of decomposition.

可通过冷却样品和再次测量来区分转变和反应——化学反应是不可逆的,而熔化了的结晶材料在冷却或二次加热时会重新结晶。玻璃化转变也是可逆的,但经常在玻璃化转变的第一次升温测量中观察到的焓松弛是不可逆的。

Transitions and reactions can be differentiated by cooling the sample and measuring it again — chemical reactions are irreversible whereas crystalline materials melt then crystallize again on cooling or on heating a second time. Glass transitions are also reversible but not the enthalpy relaxation often observed in the first heating measurement of a glass transition.

1.1.2 温度调制 DSC(TMDSC) Temperature-modulated DSC

1.1.2.1 ADSC*

调制 DSC(ADSC)是一个专门类型的温度调制 DSC(TMDSC)。与常规 DSC 不同,线性温度程序被叠加上一个小的周期性温度变化。温度程序的特征为基础升温速率、温度振幅和温度周期性变化的持续时间(图 1.2)。采用准等温测试,基础升温速率也可为零。

Alternating DSC (ADSC) is a particular type of temperature-modulated DSC (TMDSC). In contrast to conventional DSC, the linear temperature program is overlaid with a small periodic temperature change. The temperature program is characterized by the underlying heating rate, the temperature amplitude and the duration of the periodically changing temperature (Fig. 1.2). With quasi-isothermal measurements, the underlying heating rate can also be zero.

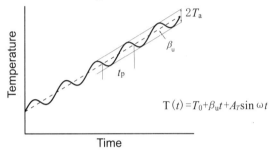

图 1.2 典型的 ADSC 温度程序:β_u 为基础加热速率;A_T 为温度振幅;t_p 为周期;$2\pi/P$ 为角频率 ω;P 为正弦波的周期。

Fig. 1.2 Typical ADSC temperature program:β_u is the underlying heating rate; A_T the temperature amplitude; t_p period;The angular frequency ω is defined as $2\pi/P$;where P denotes the period of the sine wave.

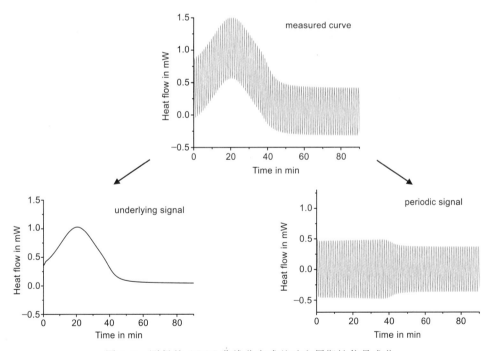

图 1.3 测得的 ADSC 曲线分离成基础和周期性信号成分。

Fig. 1.3 Separation of the measured ADSC curve into the underlying and the periodic signal components.

---

* 调制 DSC 的简称除 ADSC 外,尚有 MDSC、DDSC 和 ODSC 等。按国家标准 GB/T 6425-2008《热分析术语》的规定,应简称为 MTDSC,即温度调制式差示扫描量热法。——译注

由于温度调制,所测得的热流是周期性变化的。该热流能分成两部分,如图 1.3 所示。信号平均生成基础信号(总热流),它相当于常规 DSC 曲线。作为附加信息,还得到周期性信号成分。可逆热流为能够直接跟上加热速率变化的热流成分,可从同相热容计算得到。从总热流与可逆热流的差值可得到不可逆热流。该技术的优势之一是能将同时发生的过程分开。例如,可直接测量化学反应过程中的热容变化。

ADSC 曲线的计算基于傅立叶分析。复合热容 c_p^* 的复数模用下面的等式计算:

As a result of temperature modulation, the measured heat flow changes periodically. This can be separated into two parts as shown in Figure 1.3 Signal averaging yields the underlying signal (total heat flow), which corresponds to the conventional DSC curve. As additional information, one also obtains the periodic signal component. The reversing heat flow corresponds to the heat flow component that is able to follow the heating rate change directly and is computed from the in-phase heat capacity. The difference between the total heat flow and the reversing heat flow yields the non-reversing heat flow. One advantage of this technique is that it allows processes that occur simultaneously to be separated. For example, the change in heat capacity during a chemical reaction can be measured directly.

The evaluation of the ADSC curves is based on Fourier analysis. The modulus of the complex heat capacity c_p^* is calculated using the equation

$$|c_p^*| = \frac{A_\Phi}{A_\beta} \cdot \frac{1}{m}$$

式中 A_Φ 和 A_β 为调制热流和加热速率的振幅,m 为样品质量。ADSC 热流信号与加热速率之间的相角用于计算同相 c_p。

where A_Φ and A_β denote the amplitudes of the modulated heat flow and heating rate, and m the sample mass. The phase angle between the ADSC heat flow signal and the heating rate is used to calculate the in-phase c_p.

1.1.2.2 IsoStep

步进扫描 DSC IsoStep 是一种特殊类型的温度调制 DSC。在该方法中,温度程序由很多开始和结束为等温段的动态程序段组成(图1.4)。

IsoStep is a special type of temperature-modulated DSC. In this method, the temperature program consists of a number of dynamic segments that begin and end with an isothermal segment (Fig. 1.4).

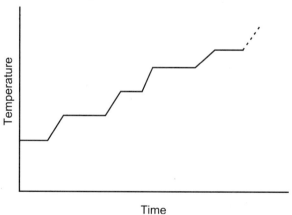

图 1.4 IsoStep 温度程序由不同的恒温和动态段组成。

Fig. 1.4 IsoStep temperature program consisting of different isothermal and dynamic segments

等温段能使动态段的等温漂移得到修正,这样能获得更准确的热容。等温步阶也可能包含动力学信息,例如化学反应。热容可用蓝宝石参比样进行测定,而动力学效应能从热容变化中分开。

The isothermal segments allow the isothermal drift of the dynamic segments to be corrected. This results in better heat capacity accuracy. The isothermal step may also contain kinetic information, for example of a chemical reaction. Heat capacity determinations can be made using a sapphire reference sample, and kinetic effects can be separated from changes in heat capacity.

1.1.2.3 TOPEM™

TOPEM™是根据对 DSC(仪器和样品)对随机调制基础温度程序响应的全面数学分析而设计的先进的温度调制 DSC 技术(图 1.5)。由于温度脉冲是随机分布的,因而是对系统在宽频范围施予温度振荡,而不只是某单一频率(如 ADSC)。振荡式输入信号(升温速率)和响应信号(热流)的相关分析能得到比常规温度调制 DSC 多得多的信息,不仅能将可逆与不可逆效应分开,而且还能测量样品的准稳态热容和测定频率依赖的热容值。这能用来在一次单独的测试中就区分开频率依赖的松弛效应(例如玻璃化转变)和非频率依赖的效应(例如化学反应)。

TOPEM™ is an advanced temperature-modulated DSC technique that is based on the full mathematical analysis of the response of a DSC (both the apparatus and the sample) to a stochastically modulated underlying temperature program (Fig. 1.5). Due to the randomly distributed temperature pulses, the system is subjected to temperature oscillations over a wide frequency range and not just at one single frequency (ADSC). An analysis of the correlation of the oscillating input signal (heating rate) and the response signal (heat flow) provides much more information than can be obtained using conventional temperature-modulated DSC. Not only can reversing and non-reversing effects be separated, but the quasi-static heat capacity of the sample is also measured and frequency-dependent heat capacity values are determined. This can be used to distinguish between frequency-dependent relaxation effects (e.g. glass transitions) and frequency-independent effects (e.g. chemical reactions) in one single measurement.

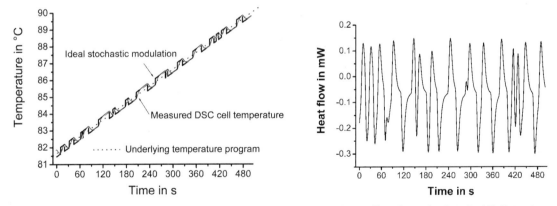

图 1.5 TOPEM™方法中炉体设定值的温度曲线(黑线),炉体温度(红线)产生在平均值上下变动的加热速率。如右图所示,流向样品的热流也不规则变动。

Fig. 1.5 Temperature curve of the furnace set value (black line) in a TOPEM™ method in which the furnace temperature (red curve) generates a heating rate that fluctuates around a mean value. The heat flow to the sample also fluctuates irregularly as shown in the diagram on the right.

1.2 热重分析(TGA) Thermogravimetric Analysis

当将样品升温时,经常开始失重。失重可能产生于样品的蒸发,或形成气体产物并逸出的化学反应。如果吹扫气氛不是惰性的,样品还能与气体反应。在某些情况下,样品质量也可能增加,例如在氧化反应中,如果形成的产物是固体。

热重分析(TGA)测量随温度或时间而变化的样品质量。

TGA 提供关于样品的性质及其成分的信息。如果样品分解产生于化学反应,则样品质量通常呈台阶状变化。台阶出现时的温度可表征该样品材料在所用气氛中的稳定性。

图1.6所示为典型的 TGA 曲线。通过分析单独质量台阶的温度和高度能确定材料的组成。

水、残留溶剂或添加油这样的挥发性化合物在相对低的温度逸出。这种化合物的排除取决于气体压力,在低压(真空)下,相应的失重台阶移到更低温度,就是说,加速了蒸发。分析在惰性气氛中的热解反应能确定含量(从台阶高度),甚至可确定材料的种类。

样品的碳黑或碳纤维含量可从切换到氧气气氛后的燃烧台阶的高度确定。由残留物测定残留填料、玻璃纤维或灰分。测试曲线因浮力效应和气流速率而产生的微小变化,可通过扣除空白曲线得到修正。

TGA 测量常用 TGA 曲线的一阶微分(称为 DTG 曲线)来表示。于是,TGA 曲线质量损失的台阶在 DTG 曲线上呈峰形。DTG 曲线相当于样品质量变化的速率。

When a sample is heated, it often begins to lose mass. This loss of mass can result from vaporization or from a chemical reaction in which gaseous products are formed and evolved from the sample. If the purge gas atmosphere is not inert, the sample can also react with the gas. In some cases, the sample mass may also increase, e. g. in an oxidation reaction if the product formed is a solid.

In thermogravimetric analysis (TGA), the change in mass of a sample is measured as a function of temperature or time.

TGA provides information on the properties of the sample and its composition. If the sample decomposes as a result of a chemical reaction, the mass of the sample often changes in a stepwise fashion. The temperature at which the step occurs characterizes the stability of the sample material in the atmosphere used.

Figure 1.6 shows a typical TGA curve. The composition of a material can be determined by analyzing the temperatures and the heights of the individual mass steps.

Volatile compounds such as water, residual solvents or added oils are evolved at relatively low temperatures. The elimination of such components depends on the gas pressure. At low pressures (vacuum), the corresponding mass loss step is shifted to lower temperatures, that is, vaporization is accelerated. The analysis of pyrolysis reactions in an inert atmosphere allows the content (from the step height) and possibly even the type of material to be determined.

The carbon black or carbon fiber content of a sample can be determined from the height of the combustion step after switching to an oxidative atmosphere. The residual filler, glass fiber or ash is determined from the residue. Small changes in the measurement curve due to buoyancy effects and gas flow rate can be corrected by subtracting a blank curve.

TGA measurements are often displayed as the first derivative of the TGA curve, the so-called DTG curve. Steps due to loss of mass in the TGA curve then appear as peaks in the DTG curve. The DTG curve corresponds to the rate of change of sample mass.

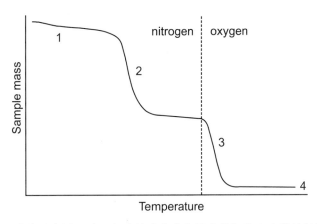

图 1.6 TGA 曲线示意图：1 由于挥发性成分蒸发导致的失重；2 在惰性气氛中的热解；
3 当切换到氧化气氛后碳的燃烧；4 残留物。

Fig. 1.6 Schematic TGA curve: 1 loss of mass due to the vaporization of volatile components; 2 pyrolysis in an inert atmosphere; 3 combustion of carbon on switching from an inert to an oxidative atmosphere; 4 residue.

气体产物扩散出样品的容易程度，在一定程度上将影响分解台阶的温度范围。当使用反应性气体时，样品表面气体的交换效率至关重要。可以使用合适的坩埚（例如浅皿的 30μL 氧化铝坩埚）和合适的样品几何形状（几个小颗粒或粉末）来降低测试时的扩散效应。

TGA 实验十分精确地测量样品质量的变化。然而令人遗憾的是，该技术无法提供关于逸出的气体分解产物的性质的任何信息。不过，可将 TGA 与合适的气体分析仪联用（逸出气体分析 EGA）来分析这些产物。

The temperature range of the decomposition steps is influenced to a certain extent by the ease with which the gaseous products are able to diffuse out of the sample. When reactive atmospheres are used, the efficiency of gas exchange at the surface of the sample is crucial. The effects of diffusion on the measurement can be reduced by using suitable crucibles (e. g. crucibles with low wall-heights such as the 30-μL alumina crucible) and by suitable sample geometry (several small pieces or powder).

In TGA, the change in mass of the sample is measured very accurately. Unfortunately, however, the technique does not provide any information about the nature of the gaseous decomposition products evolved. The products can however be analyzed by coupling the TGA to a suitable gas analyzer (evolved gas analysis, EGA).

1.3 热机械分析(TMA)　Thermomechanical Analysis

热机械分析测量样品升温时的尺寸变化。该项技术连续测量带一定力放置于样品表面的探头的位置或位移与温度或时间的关系。图 1.7 所示为典型的 TMA 曲线。探头施加的压力和样品的硬度决定了 TMA 实验事实上是膨胀测量还是针入测量。

在热膨胀测量中，探头在样品表面仅施加低应力。在所关心的温度范

Thermomechanical analysis measures the dimensional changes of a sample as it is heated. In this technique, the position or displacement of a probe resting on the surface of the sample with a certain force is continuously measured as a function of temperature or time. Figure 1.7 shows a typical TMA curve. The pressure exerted by the probe and the hardness of the sample determine whether the TMA experiment is in fact an expansion or a penetration measurement.

In the thermal expansion measurement, the probe exerts only a low pressure on the surface of the sample. The sample is heated

围线性升温。线性热膨胀系数(CTE)直接从测试曲线计算。

针入实验中,探头施加大得多的应力。可直接测量样品升温时的软化温度。材料在玻璃化转变温度处或熔融时软化。

如果对样品施加周期性变化的力,样品尺寸也呈周期性变化。该测试模式称为动态负载 TMA(DLTMA)。从振幅和样品厚度能估算出样品的弹性模量(杨氏模量)。DLTMA 模式也可编程,以使探头上下运动而交替接触或离开加热时的样品表面。对所得曲线形状的分析可确定凝胶点。

linearly over the temperature range of interest. The linear coefficient of thermal expansion (CTE) is calculated directly from the measurement curve.

In a penetration experiment, the probe exerts a much greater pressure. The softening temperature can be directly measured when the sample is heated. Materials soften at the glass transition temperature or on melting.

If a periodically changing force is applied to the sample, the sample dimensions also change periodically. This measurement mode is called dynamic load TMA, DLTMA. The elastic modulus (Young's modulus) of the sample can be estimated from the amplitude and the sample thickness. The DLTMA mode can also be programmed so that the probe moves up or down and so alternately touches or recedes from the surface of the sample as it is heated. Analysis of the resulting curve shape allows the gel point to be determined.

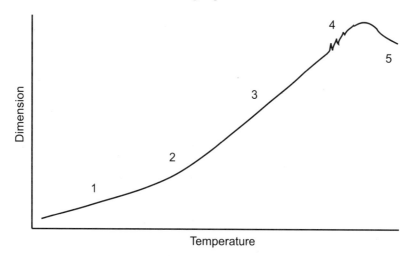

图 1.7 在低压缩应力下聚合物的 TMA 曲线示意图:
1 低于玻璃化转变的膨胀;2 玻璃化转变(斜率变化);3 玻璃化转变以上的膨胀;4 分层;
5 由于黏性流动或分解导致的塑性形变或穿透。

Fig. 1.7 Schematic TMA curve of a polymer under low compressive stress: 1 expansion below the glass transition; 2 glass transition (change of slope); 3 expansion above the glass transition; 4 plastic deformation 5 plastic deformation or penetration due to viscous flow or decomposition.

使用弯曲附件,用此技术还能测量硬样品的弯曲行为。

可用专门的样品支架测量纤维和薄膜的尺寸变化。一个独特的应用是测量材料在溶剂中的溶胀行为。为此,样品放入一个小容器内,加入相关的溶剂,于是,在等温下连续测量溶剂吸收导致的样品厚度随时间的

It is also possible to measure the bending behavior of hard samples with this technique using a bending accessory.

Special sample holders are available that allow the dimensional changes of fibers and films to be measured. A particular application is the measurement of the swelling behavior of materials in solvents. To do this, a sample is placed in a small container and the solvent of interest is added. The change in thickness of the sample due to solvent absorption is then

变化。

continuously measured isothermally as a function of time.

1.4 动态热机械分析(DMA) Dynamic Mechanical Analysis

动态热机械分析测定力学模量与温度、频率和振幅的关系。

施加于样品的周期性(通常为正弦)变化的力在样品中产生周期性的应力。样品对该应力作出反应,仪器测量相应的形变行为。由应力和形变测定力学模量 M。取决于所加应力的类型,测量剪切模量 G(施加剪切应力)或杨氏模量 E(拉伸或弯曲)。

样品不总是立即对周期性变化的应力作出反应——依赖于样品的黏弹性而有一定的时间滞后。这是产生施加应力和形变之间相位移的原因。将相位移考虑进去,动态测得的模量用实数部分 M' 和虚数部分 M'' 来描述。实数部分(贮存模量)描述与周期性应力同相的样品响应,它是样品(可逆的)弹性的量度。虚数部分(损耗模量)描述相位移为 90° 的响应部分,是转化为热(不可逆损失)的力学能量的量度。相位移的正切 $\tan\delta$ 也称作损耗因子,是材料阻尼性能的量度。模量和 $\tan\delta$ 与温度和测量频率有关。在室温下,橡胶材料的典型贮存模量值是在 0.1MPa 至 10MPa 之间。

图 1.8 的曲线表示树脂在固化过程之前、期间和之后的贮存和损耗模量的典型行为。

低温时树脂处于玻璃态(1),模量较高(约 2GPa)。

在玻璃化转变区(2),贮存模量在较窄温度范围内有若干数量级的变化。这是主松弛区,样品从玻璃态变化到黏液态;贮存模量下降,损耗模量呈最大值。主松弛发生的温度范围是与频率有关的,在较高测量频率下,松弛区移向较高温度,移动

In dynamic mechanical analysis, a mechanical modulus is determined as a function of temperature, frequency and amplitude.

A periodically changing force (usually sinusoidal) applied to the sample creates a periodic stress in the sample. The sample reacts to this stress and the instrument measures the corresponding deformation behavior. The mechanical modulus, M, is determined from the stress and deformation. Depending on the type of stress applied, either the shear modulus, G (with shear stress) or the Young's modulus, E (with stretching or bending) is measured.

The sample does not always immediately react to the periodically changing stress-a certain time delay occurs that depends on the viscoelastic properties of the sample. This is the cause of the phase shift between the applied stress and the deformation. To take this phase shift into account, the dynamically measured modulus is described by a real part M' and an imaginary part M''. The real part (storage modulus) describes the response of the sample in phase with the periodic stress. It is a measure of the (reversible) elasticity of the sample. The imaginary part (loss modulus) describes the component of the response that is phase-shifted by 90°. This is a measure of mechanical energy converted to heat (and therefore irreversibly lost). The tangent of the phase shift, $\tan\delta$, is also known as the loss factor and is a measure of the damping behavior of the material. The modulus and $\tan\delta$ depend on the temperature and the measuring frequency. At room temperature, rubbery materials show typical storage modulus values between 0.1MPa and 10MPa.

The curves in Figure 1.8 show the typical behavior of the storage and loss modulus of a resin before, during and after the curing process.

At low temperatures, the resin is in a glassy state (1) and the modulus is relatively high (about 2 GPa).

In the region of the glass transition (2), the storage modulus changes by several orders of magnitude over a relatively narrow temperature range. This is the main relaxation region, where sample changes from a glassy to a viscous liquid state; the storage modulus decreases and the loss modulus exhibits a maximum. The temperature range in which this occurs is frequency dependent. At higher measuring frequencies, the

的原因是由分子重排动力学所决定的松弛行为。对于未交联的树脂,样品是液态的,贮存模量很小,损耗模量下降超过5个数量级(3)。

relaxation region shifts to higher temperatures. The reason for the shift is that relaxation behavior is determined by the dynamics of molecular rearrangements. With uncrosslinked resins, the sample is liquid, the storage modulus is very small, the loss modulus decreases by more than 5 decades (3).

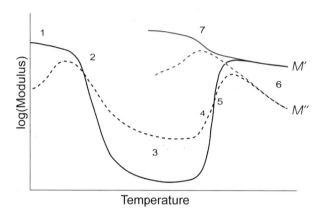

图1.8 活性树脂贮存模量 M'(实线)和损耗模量 M''(虚线)与温度关系的典型曲线:1 树脂的玻璃态(黑色曲线);2 主松弛区(未固化树脂的玻璃化转变);3 黏性流动;4 交联;5 凝胶点(贮存模量与损耗模量的交点);6 固化树脂的橡胶态;7 固化树脂的玻璃化转变(红色曲线)。

Fig. 1.8 Typical curves of the storage modulus, M', (continuous line) and the loss modulus, M'' (dashed line) as a function of temperature for a reactive resin: 1 glassy state of the resin (black curves); 2 main relaxation region (glass transition of the uncured resin); 3 viscous liquid, 4 crosslinking; 5 gel point (point of intersection of the storage modulus and the loss modulus); 6 rubbery state of the cured resin; 7 glass transition of the cured resin (red curves).

除了松弛行为,DMA 技术也能测量由于交联导致的材料变硬。这个反应期间(4)的行为与松弛转变中的不同。相应过程的温度范围与频率无关。M' 与 M'' 的交点通常定义为凝胶点(5)。倘如反应温度高于交联树脂的玻璃化转变,材料则进一步反应。最后,达到类橡胶态(6)。

Besides relaxation behavior, the DMA technique can also measure the stiffening of materials due to crosslinking. The behavior during this reaction (4) is different to that in a relaxation transition. The temperature ranges of the corresponding processes are not frequency dependent. The point of intersection of M' and M' is often defined as the gel point (5). The material reacts further provided the reaction temperature is higher than the glass transition of the crosslinked resin. Finally, a rubbery-like state is reached (6).

交联材料的 DMA 曲线标识为红色。由于交联,玻璃化转变与未交联树脂此时处于更高温度(7)。对未填充材料,贮存模量在玻璃化转变处变化约 3 个数量级。在约 1MPa 的模量值,达到所谓的橡胶平台。

The DMA curves of the crosslinked materials are marked red. Due to crosslinking, the glass transition now lies at higher temperatures compared to the uncrosslinked resin (7). With unfilled materials, the storage modulus changes by about 3 decades at the glass transition. At modulus values of about 1 MPa, the so-called rubbery plateau is reached.

除了所施加的应力(例如剪切或弯曲),在 DMA 测量中还可变化的是温度、频率和振幅。下表总结了用不同测量模式所能得到的信息。

Besides the applied stress (e.g. shear or bending), the temperature, frequency and amplitude can also be varied in DMA measurements. The following table summarizes the information that can be obtained using the different measurements modes.

温度变量 Variation of the temperature	频率变量 Variation of the frequency	振幅变量 Variation of the amplitude
• 松弛转变温度 Temperature of relaxation transitions • 玻璃化转变 Glass transition • 固化反应 Curing reaction • 组分的相容性 Compatibility of components • 阻尼行为 Damping Behavior	• 松弛行为 Relaxation behavior • 玻璃化转变 Glass transition • 分子的相互作用 Molecular interaction • 阻尼行为 Damping behavior	• 非线性力学行为 Non-linear mechanical behavior • 填料效应 Effect of fillers

1.5 与 TGA 的同步测量 Simultaneous Measurements with TGA

如 1.2 节所述,由于蒸发或在反应中产生的质量损失可用 TGA 十分灵敏地测量。然而,测量结果的解释常常需要附加的信息。这可以通过两个或两个以上技术组合成一个仪器系统来获得。常用技术为同步 DTA 或同步 DSC(总是包含在 TGA/DSC 系统中)和各种在线气体分析方法(例如质谱法和红外光谱法)。

As already mentioned in Section 1.2, the loss of mass due to vaporization or in reactions can be detected with great sensitivity using TGA. The interpretation of the measurement results, however, frequently requires additional information. This can be obtained by combining two or more suitable techniques into one instrument system. Techniques often used are simultaneous DTA or simultaneous DSC (this is always included in a TGA/DSC system) and various methods for on-line gas analysis (e.g. mass spectrometry and infrared spectroscopy).

1.5.1 同步 DSC 和差热分析(DTA、SDTA)
Simultaneous DSC and Differential Thermal Analysis

同步 DSC 的测量原理和功能与 1.1.1 节所述的常规 DSC 是一样的。
在差热分析(DTA)中,将样品和参比物在炉体中升温,用热电偶测量样品和参比物之间的温差。如果样品发生热效应(例如相转变或化学反应),额外的能量吸收或释放会改变样品的升温速率,导致样品和参比两端的温差。例如,在放热反应中,样品和参比的温差要比反应前后大。热效应表现为台阶和峰,正如 DSC 测试曲线一样。

在 SDTA(单式 DTA)中,未用参比

The measuring principle and functions of simultaneous DSC are same as those of a conventional DSC mentioned in Section 1.1.1. In differential thermal analysis (DTA), a sample and a reference material are heated in a furnace. The temperature difference between the sample and the reference material is measured using thermocouples. If a thermal event occurs in the sample (such as a phase transition or chemical reaction), the additional uptake or release of energy changes the heating rate of the sample. This results in a temperature difference between the sample and reference sides. For example, during an exothermic reaction, the temperature difference between the sample and reference is larger than before or after the reaction. Thermal effects are indicated by the presence of steps and peaks, just as in a DSC measurement curve.

In SDTA (single DTA), no reference sample is used. The

样。参比温度相当于程序温度,测量的是样品温度。

同步 DSC 技术可在 TGA 实验中同步测量 DSC 曲线,同步 DTA 或 SDTA 技术可在 TGA、TMA 和 DMA 实验中同步测量差热信号。这种信号常常有助于解释,因为它可测出并不伴随着质量或尺寸变化的热效应。例如,在 TMA 测量中,同步 SDTA 能够区分放热和吸热转变,测知化学反应。

reference temperature corresponds to the program temperature and the sample temperature is measured.

The simultaneous DSC technique enables the DSC curve to be simultaneously measured in TGA experiments, and the DTA or SDTA technique enables the differential thermal signals to be simultaneously measured in TGA, TMA and DMA experiments. This often aids interpretation because it detects thermal events that are not accompanied by a change in mass or dimensions. For example, in TMA measurements, simultaneous SDTA can distinguish between exothermic and endothermic transitions, and detect chemical reactions.

1.5.2 逸出气体分析(EGA) Evolved Gas Analysis

为了可靠解释 TGA 曲线,常常要更多地了解 TGA 中样品逸出气的性质。这些信息可通过将 TGA 仪器通过加热输送管连接到气体分析仪来获得。这能几乎同步地分析气体。两种最常用的在线联用是 TGA-MS(一台 TGA 耦联到一台质谱仪 MS)和 TGA-FTIR(一台 TGA 耦联到一台傅立叶变换红外仪 FTIR)。

这些技术的使用细节和应用见《热分析应用手册》的《热重-逸出气体分析》分册。

For a reliable interpretation of TGA curves, one would often like to know more about the nature of the gases evolved from the sample in the TGA. This information can be obtained by connecting the TGA instrument to a gas analyzer by means of a heated transfer line. This allows the gases to be analyzed almost simultaneously. The two most frequently used on-line combinations are TGA-MS (a TGA coupled to a mass spectrometer, MS), or TGA-FTIR (a TGA coupled to a Fourier transform infrared spectrometer, FTIR).

Practical details and applications of these techniques are given in "TGA-EGA" of "Application Handbooks Thermal Analysis".

1.5.2.1 TGA－MS

质谱仪(MS)由离子源、分析器和检测器组成。来自 TGA 的气体混合物(挥发性化合物/分解产物和吹扫气体)在离子源中离子化形成分子、离子和大量的碎片离子。离子按照质荷比(m/z)在分析器中被分开,然后在检测器系统中被记录。测得的离子谱显示出分子、离子和大量从不同分子的碎裂形成的碎片离子。图谱或碎裂方式是所测试的特定化合物的特征。此外,有些元素具有非常特征的同位素波谱,例如氯。

一般来说,逸出气体可通过测得的

A mass spectrometer (MS) consists of an ion source, an analyzer and a detector. The gas mixture (volatile compounds/decomposition products and purge gas) arriving from the TGA is ionized in the ion source with the formation of molecular ions and numerous fragment ions. The ions are separated according to their mass-to-charge ratio (m/z) in the analyzer and then recorded by the detector system. The resulting mass spectrum displays the molecular ions and a large number of fragment ions formed from the fragmentation of the different molecules. The spectrum or fragmentation pattern is characteristic of the particular compound measured. Furthermore, some elements have very characteristic isotope patterns, e.g. chlorine.

In general, evolved gases can be identified or characterized by the

碎片的质谱图来鉴定和表征。测得的图谱还可与图谱数据库中的图谱集对照。

在与 TGA 联用时，勿需在整个 TGA 测试范围内划分短的间隔进行反复测量（即扫描）。可单独选择特定物质特有的质荷比（m/z）的碎片离子，记录其强度与时间或温度的关系。连续测试有限量特定碎片的这个技术称为多离子检测（MID）或选择离子监测（SIM），可获得灵敏度的大大提高。

比较 MS 离子曲线与 TGA（或 DTG）曲线，可对热效应中质量损失所对应的物质加以鉴定或确认。

fragmentation pattern of the mass spectra measured. The spectra obtained can also be compared with collections of spectra in spectral databases.

In combination with a TGA it is not always necessary or usual to measure (i. e. scan) the entire mass range repeatedly at short intervals throughout the TGA measurement. Fragment ions of mass-to-charge (m/z) ratio characteristic for particular substances can be individually selected and their intensities recorded as a function of time or temperature. This technique of continuously measuring a limited number of specific fragments is known as multiple ion detection (MID) or selected ion monitoring (SIM) and results in a large increase in sensitivity.

A comparison of the MS ion curves with the TGA (or DTG) curve allows the presence of a substance responsible for the mass loss in a thermal effect to be identified or confirmed.

1.5.2.2 TGA－FTIR

也可用红外光谱法来鉴定 TGA 实验中形成的气体产物。TGA 仪器通过加热输送管与傅立叶变换红外光谱仪（FTIR）直接耦联，测量自 TGA 仪器到达 FTIR 气体池的气体混合物（挥发性化合物/分解产物和吹扫气体）在 4000 至 400cm^{-1}（波数）范围内的红外图谱。与分子中不同键的特定振动频率对应的波长处的红外能量被吸收，这能鉴定官能团和得到关于有关物质的性质的信息。

现代 FTIR 仪器能在几秒内测试高质量的图谱，在整个 TGA 实验范围内连续记录图谱，这意味着实际上对 TGA 实验的任何温度或时间点都可获得单独谱图，可将这些谱图可与数据库参比谱图对照，进而做出判断。

在实际中，为了简化数据处理，对利用了谱图中包含的全部或只是部分信息的以时间（或温度）为因变量的曲线进行记录，这些曲线称为 Gram-Schmidt 曲线或化学谱。

简单说来，Gram-Schmidt 曲线是通

Infrared spectroscopy can also be used to identify gaseous products formed in a TGA analysis. The TGA instrument is coupled directly to a Fourier transformation infrared spectrometer (FTIR) by a heated transfer line. Infrared spectra of the gas mixture (volatile compounds / decomposition products and purge gas) arriving in the FTIR gas cell from the TGA instrument are measured in the range 4000 to 400cm^{-1} (wavenumbers). Infrared energy is absorbed at wavelengths that correspond to the specific vibrational frequencies of the different bonds in the molecules. This allows functional groups to be identified and information on the nature of the substances involved to be gained.

A modern FTIR spectrometer is capable of measuring good quality spectra within a few seconds. The spectra are recorded continuously throughout the TGA analysis. This means that individual spectra are available for practically any temperature or point in time of the TGA analysis. These spectra can be compared with database reference spectra and evaluated.

In practice, to simplify the evaluation, curves are recorded as a function of time (or temperature) that make use of all or just part of the information included in the spectrum. These curves are known as Gram-Schmidt curves and chemigrams.

In simple terms, the Gram-Schmidt curve is obtained by

过对每个单独图谱的吸收波带强度的积分得到的,该信息表示为吸收强度与时间的关系曲线。曲线是任何瞬间到达 FTIR 光谱仪气体池内的物质量的量度。该曲线一般与 DTG 曲线相似,除了由于气体在连接两个仪器的传输管内输送产生的 DTG 和 Gram-Schmidt 峰之间明显存在的短滞后(通常只有几秒)。因为单独物质的红外吸收强度很不同,所以强度不是直接可比的。

化学谱表示特定波数区域的红外吸收与时间(或温度)的关系,能观察气体产物的形成与时间的关系(通过官能团)。随后,追溯 DTG 出峰期间测量的相应全部范围的 FTIR 图谱,并予以解释。

integrating the absorption band intensities of each individual spectrum. The information is presented as a curve of absorption intensity plotted as a function of time. The curve is a measure of the quantity of substance arriving in the gas cell of the FTIR spectrometer at any instant. The curves often resemble DTG curves except that there is obviously a short delay (normally just a few seconds) between the DTG and the Gram-Schmidt peaks due to the gas transport in the transfer line connecting the two instruments. The intensities are not directly comparable because individual substances have very different infrared absorption intensities.

Chemigrams display the infrared absorption of particular wavenumber regions as a function of time (or temperature and allow the formation of gaseous products (via functional groups) to be observed as a function of time. Afterward, the corresponding full range FTIR spectra measured during the DTG peak are recalled and interpreted.

2 热固性树脂的结构、性能和应用
Structure, Properties and Applications of Thermosets

2.1 概述 Introduction

聚合物分类有多种方法。一种分类法是依据其性能和应用，划分为热塑性塑料、弹性体或热固性树脂。生物高分子常常被当作单独一类，然而其热性能通常用类似于其他种类的方法论述。

下表总结了这些不同种类聚合物的一些独有特性。

There are many possible ways to classify polymers. One classification is based on their properties and application, and subdivides them into thermoplastics, elastomers or thermosets. Biopolymers are often treated as a separate group. The thermal properties can however usually be dealt with in a similar way to the other classes.

The following table summarizes some of the distinctive features of these different polymer classes.

表 2.1 聚合物的分类
Table 2.1 Classes of polymer.

种类 Type	特征 Features	举例 Examples
热塑性塑料 Thermoplastics	在负载下具有高强度、低应变的未交联聚合物。加热时它们软化或熔融，即使在反复的加热和冷却循环后也不失去加热时再成型的能力。 Uncrosslinked polymers with high strength and low strain under load. They soften or melt on heating, and do not lose their ability to be remolded on warming, even after repeated heating and cooling cycles.	聚乙烯 聚对苯二甲酸乙二醇酯 Polyethylene Polyethylene terephthalate
弹性体 Elastomers	轻度交联的聚合物。在低应力下它们显示高可逆性应变。按常识它们不能熔融，因为交联而不能流动。硫化或固化后，不可能再成型。 Lightly crosslinked polymers. Under low stress they exhibit highly reversible strain. They cannot melt in the conventional sense and cannot flow because of the crosslinks. Remolding after vulcanization or curing is no longer possible.	硫化天然橡胶 Vulcanized natural rubber
热固性树脂 Thermosets	高度交联的聚合物。交联（固化、硬化）后它们显示高力学强度和低应变。加工成最终产品涉及不可逆的化学反应。固化后，再成型是不可能的。 Heavily crosslinked polymers. After crosslinking (curing, hardening) they exhibit high mechanical strength and low strain. Processing to a finished product involves an irreversible chemical reaction. Remolding after curing is not possible.	固化的环氧树脂 酚醛树脂 聚酯 聚氨酯 Cured epoxy resins Phenolic resins Polyesters Polyurethanes

本应用手册主要阐述热固性树脂的化学与分析。为了对照，可在《热分析应用手册》的《热塑性聚合物》分

This application handbook deals mainly with the chemical and analytical aspects of thermosets. For comparison, many applications of thermal analysis to thermoplastics can be found in

册中找到许多热塑性聚合物的热分析应用,弹性体领域中热分析的使用在《热分析应用手册》的《弹性体》分册中有详细描述。同时,应注意到当今的聚合物材料常将不同类型的聚合物混配而兼具各自的性能。例如重要的热塑性弹性体,是通过具有弹性、热塑性或热固性性能的材料的共混或共聚获得的。

Application Handbook Thermal Analysis entitled THERMOPLASTICS. The use of thermal analysis in the field of elastomers is described in detail in Application Handbook Thermal Analysis entitled ELASTOMERS. At the same time, it is important to note that polymeric materials nowadays often have properties associated with different types of polymers, for example thermoplastic elastomers. This is achieved through blending or the copolymerization of materials with elastic, thermoplastic or thermosetting properties.

2.2 热固性树脂的化学结构 Chemical Structure of Thermosets

2.2.1 大分子 Macromolecules

就化学结构而言,热固性树脂是在致密三维网状结构内高度交联的网状大分子。就这方面来看,它们是由较小的大分子之间交联形成的复杂大分子体系。大分子是由许许多多(高达数千)通过化学键结合在一起的重复单元即单体单元组成的大的链状分子。许多工业聚合物的大分子并没有相同的结构,其分子大小且往往在组成上是不同的。

As far as their chemical structure is concerned, thermosets are network polymers that are heavily crosslinked in a dense three-dimensional network structure. In this respect, they are complex macromolecular systems formed by crosslinking between smaller macromolecules. Macromolecules are large chain-like molecules that consist of very many (up to several thousand) repeating units, the monomer units, held together by chemical bonds. The macromolecules in most technical polymers do not have a uniform structure. They differ in their size and often in composition.

关于聚合物链结构,通常分为线型、支化或交联即网状聚合物,不同的类型示意于下图。

With regard to polymer chain structure, polymers are usually classified as linear, branched, or crosslinked or network polymers. The different types are shown schematically in the following figure.

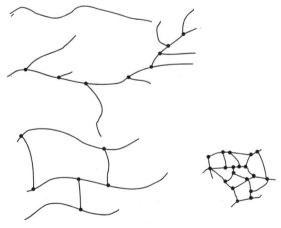

a) 轻度交联 lightly crosslinked   b) 高度交联 heavily crosslinke

图 2.1 高分子链结构

Fig. 2.1 Polymer chain structures.

线型聚合物仅由大分子主链构成,不存在侧链。可用平均摩尔质量和摩尔质量分布表征聚合物。预聚物常常具有较短的链结构,是为生产热固性树脂而贮存的中间体。

支化聚合物的大分子有侧链即不同长度的支链。为表征这些材料,必需了解侧链的长度和分布以及侧链及其分布的平均间隔。在聚合物网络形成过程中,起始分子(单体、预聚物)首先形成"树状结构",然后在反应进程中连接,变得越来越紧密。

交联聚合物是三维的大分子——具有不同尺寸网目的分子网络。在轻度交联的聚合物(弹性体)中,网目很大,仍可认为是只在为数不多分离的交联点化学连接在一起的线型或支化大分子。当材料被拉伸时,交联阻止了分子相互流动。高度交联的聚合物(热固性树脂),则交联点靠近得多。在极端情况下,在连接点之间只有一个单体单元。

In a linear polymer, a macromolecule consists of one main chain. Side chains do not occur. The polymer is characterized by the average molar mass and the molar mass distribution. Prepolymers often have such relatively short chain structures and are used as intermediates that can be stored for the manufacture of thermosets.

The macromolecules of branched polymers have side chains or branches of different lengths. To characterize these materials, it is necessary to know the length and distribution of the side chains as well as the average separation of the side chains and their distribution. During the formation of a polymer network, the starting molecules (monomers, prepolymers) first of all form "tree-like structures" that then become more and more closely linked during the course of the reaction.

Crosslinked polymers are three-dimensional macromolecules-a molecular network with meshes of different sizes. In lightly crosslinked polymers (elastomers), the meshes are so large that they can still be thought of as linear or branched macromolecules bound together chemically at just a few well-separated crosslinking points. The crosslinks prevent the molecules from flowing past each other when the material is stretched. In heavily crosslinked polymers (thermosets), the crosslinking points are much closer together. In the extreme case, there is just one monomer unit between the links.

2.2.2 热固性树脂概述 A general overview of thermosets

热固性树脂是指一类高度交联的三维网状结构聚合物,是热固性聚合物(预聚物)经历固化或交联反应(即固定化)而形成的不熔和通常不溶的坚硬的、刚性的材料。

与热塑性塑料不同,热固性树脂在固化后不能再次加热成型,只可机械加工。因此热固性树脂必须在固化反应期间即特定的模塑或涂层过程中转变成最终形状。起始材料(预聚物)或者是液态的,或者能通过升高温度成型然后固化。制成的热固性树脂一般是无定形的,即它们不熔而只在较高温度下(在玻璃化转变时)软化。这通常是它们应用的极限最高温度。热固性树脂由

The term thermoset refers to a group of highly crosslinked three-dimensional network polymers that are produced when thermosetting polymers (prepolymers) undergo a curing or crosslinking reaction (i.e. set) to form hard, rigid materials that cannot melt and are mostly insoluble.

Unlike thermoplastics, thermosets can no longer be thermally molded after curing, only mechanically machined. Thermosets therefore have to be converted to their final form during the curing reaction, that is, in a special molding or coating process. The starting materials (prepolymers) are either liquid or can be molded by increasing the temperature and then cured. In general, the thermosets produced are amorphous, that is, they do not melt but merely soften at higher temperatures (at the glass transition). This is usually the upper temperature limit for their application. Due to their high degree of crosslinking, thermosets are highly resistant to chemical attack and thermal

于高度交联,因而具有优良的抗化学腐蚀和热形变性能,并呈高力学强度和低蠕变倾向。

与加热时变为液体的热塑性塑料相对照,术语"热固性聚合物"或"热固性树脂"用于当加热时变为固体的液态树脂。当今"热固性树脂"一般用于变为不可逆坚硬的体系,不管这是通过热的作用、暴露于紫外光,还是添加活性组分达到的。所有这些情况下,均会发生不可逆的化学交联反应。

热固性树脂可根据用来生产它们的化合物的性质来分类。这包括很宽范围的树脂材料,诸如丙烯酸树脂、醇酸树脂、氨基树脂、双马来酰亚胺、环氧树脂、呋喃树脂、酚醛树脂、聚酰亚胺、不饱和聚酯、聚氨酯和乙烯酯类。新材料在不断开发中,诸如氰酸酯、硅树脂和芳香族 Arylzene 树脂,以及混合树脂。在随后的章节中,热固性树脂按照固化过程中涉及的官能团来分类。为有助于理解后面的章节,简述一下最重要的树脂体系和官能团及其应用。

为了改善树脂的性能,常将填料、增强纤维、催化剂、抑制剂和稳定剂等成分添加到树脂中。热固性材料用于各种各样的制造过程,例如大制件、薄涂层或黏合剂。

在工业上最重要的热固性聚合物材料酚醛树脂的大规模生产是在20世纪初。通过固化制造出一个绝缘体,1907年 Leo Baekeland 发明了第一个加热后保持其形状的塑料(酚醛树脂,一个苯酚甲醛热固性树脂)。商业化获得巨大成功,酚醛树脂在长时期内是塑料的同义词,因为这种材料使许多新的模塑工艺和应用成为可能并大大促进了更多树脂的开发。这个材料的成功有着特别重要的意义,不仅涉及化学领域,而且是实际的加工技术—苯酚甲醛

deformation and exhibit high mechanical strength and a low tendency to creep.

Originally, the term "thermosetting polymer" or "thermosetting resin" was used for liquid resins that become solid on heating-in comparison with thermoplastics, which become liquid on heating. Today "thermoset" is used generally for systems that become irreversibly hard, irrespective of whether this is achieved through the action of heat, exposure to UV light, or the addition of reactive components. In all these cases, an irreversible chemical crosslinking reaction takes place.

Thermosets can be classified according to the nature of the chemical compounds used to produce them. These include a wide range of resin materials such as acrylics, alkyds, amino resins, bismaleimides, epoxies, furanes, phenolics, polyimides, unsaturated polyesters, polyurethanes and vinyl esters. New materials are continually being developed, such as cyanate esters, silicones and Arylzene resins, as well as hybrid resins. In the sections that follow, thermosets are classified according to the functional groups involved in the curing process. The most important resin systems and functional groups and their applications are briefly reviewed as an aid to understanding the following chapters.

Components such as fillers, reinforcing fibers, catalysts, inhibitors and stabilizers are usually added to the resins in order to optimize the thermosetting properties. Thermosetting materials are used in the most diverse manufacturing processes, for example for large components, as thin coatings, or as adhesives.

The first technically important large-scale manufacture of a thermosetting polymeric material Bakelite was at the beginning of the 20th century. Setting out to make an insulator, Leo Baekeland in 1907 invented the first plastic (Bakelite, a phenol-formaldehyde thermoset) that held its shape after being heated. The commercialization was a great success and Bakelite was long a synonym for plastic because the material made possible many new molding processes and applications and greatly encouraged the development of further resins. Especially important for the success of this material was not only the chemistry involved, but also the actual processing techniques-the crosslinking reaction of the phenol-formaldehyde resins was performed under pressure and the action of heat. An important innovation was to prevent

树脂的交联反应是在热压作用下进行的。一个重要的革新是用压力防止形成水蒸汽气泡。商业成功来自使用经升高温度可永久成形的模塑料。随之迅速开发了各类树脂,满足特殊应用的材料性能得以改进。

the formation of bubbles of water vapor by using pressure. The commercial success resulted from the use of molding material that could be permanently shaped by temperature increase. The introduction of further types of resin and the optimization of material properties for specific applications quickly followed.

2.2.3 树脂 Resins

按照IUPAC(国际理论化学和应用化学联合会:聚合物科学的基本术语表,1996年推荐,Pure Appl. Chem.,68,2287-2311,1996),术语"树脂"是:"一个用于通常含有带反应性基团预聚物的软固体或高黏度物质的术语,历史上由天然树脂类推得名。这个术语在历史上曾用于宽泛的含义指任何是塑料基础材料的聚合物都为"树脂"。然而,近来它更多地用于狭窄的含义,指热固性聚合物和一些特别是珠状的软性网状聚合物。它有时候不仅用于热固性树脂的预聚物(热固性聚合物),而且用于固化了的网状结构聚合物(热固性树脂)(例如:环氧树脂、酚醛树脂)。很不赞成术语"树脂"对热固性树脂的这种二义性用法。

使用术语"树脂"的第二个例子是描述广泛应用于固相合成和聚合物载体、催化剂、反应物和净化剂应用领域的球状悬浮聚合的珠。"

树脂是生产热固性树脂的起始物质,通常由带两个或多个官能团的能够交联的线型低聚物构成。固化过程通常需要其他的反应组分,诸如硬化剂(固化剂)和促进剂。必要时这些可通过升温或UV光照来激活。对双组分体系,树脂和硬化剂分开贮存,然后混合进行固化反应。取决于官能团,可通过加聚、聚合或缩聚反应产生交联。后者,单体相互反应,伴有低分子量化合物的排除,例如酚醛树脂的情况。与

According to IUPAC, (International Union of Pure and Applied Chemistry, Glossary of Basic Terms in Polymer Science, Recommendations 1996, Pure Appl. Chem., 68, 2287-2311, 1996), the term "resin" is:
"A term used for soft solid or highly viscous substances, usually containing prepolymers with reactive groups, that are named historically by analogy with natural resins. This term was once used historically in a broad sense to designate any polymer that is a basic material for plastics. However, it became used more recently in a narrow sense to refer to thermosetting polymers and some soft network polymers especially in bead form. It is sometimes used not only for prepolymers of thermosets (thermosetting polymers) but also for cured network polymers (thermosets). (e.g., epoxy resin, phenolic resin). The ambiguous use of the term "resin" for thermosets is strongly discouraged.

The second example of the use of the term "resin" is to describe the spherical suspension-polymerized beads widely applied in solid-phase synthesis and polymer support, catalyst, reagent, and scavenger application fields."

Resins are the starting materials for the production of thermosets and consist mostly of linear, oligomeric substances with two or more functional groups that are capable of undergoing crosslinking. The curing process usually requires other reaction components such as hardeners (curing agents) and accelerators. These can if necessary be activated through temperature increase or exposure to UV light. With two-component systems, the resin and hardener are stored separately and then mixed to perform the curing reaction. Depending on the functional groups, crosslinking occurs either through polyaddition, polymerization or polycondensation reactions. In the latter case, the monomers react with each other with the elimination of low

缩聚反应不同,在加聚反应中没有低分子量化合物排除。这个类型的例子是二异氰酸酯与二元醇反应生成聚氨酯,或环氧树脂,其中一个氢原子从一个单体单元转移到另一个。在聚合反应中,并无低分子化合物排除,单体单元的组成保持不变,如不饱和聚酯、乙烯基酯树脂或甲基丙烯酸酯树脂。

molecular mass compounds, for example with phenolic resins. In polyaddition reactions, in contrast to condensation reactions, no low molecular mass compounds are eliminated. Examples of this type are the reaction of diisocyanates with diols to form polyurethanes, or epoxies, where a hydrogen atom is transferred from one monomer unit to the other. In polymerization, nothing is eliminated, the composition of the monomer units remains the same, for example with unsaturated polyesters, vinylesters or methacrylates.

2.2.3.1 环氧树脂 Epoxy Resins, EP

环氧树脂,可使用极为广泛的树脂和硬化剂。这为在很宽范围内的应用并进行性能优化提供了可能。因此将只讨论若干典型的代表性例子。

An extremely wide variety of resins and hardeners is available for use with epoxy resins. This allows properties to be optimized for a very broad range of applications. For this reason, only a few typical representatives will be discussed.

$$H_2C\overset{O}{\underset{\diagdown\diagup}{-}}CH-$$

环氧即氧化乙烷环
Epoxy or oxirane ring

环氧基团与伯胺反应形成仲胺(或进一步形成叔胺):

The epoxy group reacts with primary amines to form a secondary (or further, tertiary) amine:

$$R-NH_2 + Epoxy \longrightarrow R-NH-CH_2-CHOH-$$

与酸反应生成酯桥:

The reaction with acids produces ester bridges:

$$R-COOH + Epoxy \longrightarrow R-CO-O-CH_2-CHOH-$$

醇与环氧反应形成醚桥(例如羟基与过量环氧进一步反应):

Alcohols react with epoxies to form ether bridges (e.g. further reaction of the hydroxyl group with excess epoxy):

$$ROH + Epoxy \longrightarrow R-O-CH_2-CHOH-$$

在与酸酐的固化反应中,末端的环氧基团与位于中间的羟基受到攻击而形成醚键和酯桥。

要获得一个能够反应并同时按应用裁制的体系,需要适当的配方。例如这可以是单组分体系,或为了更好的贮存而采用树脂和硬化剂分开的双组分体系。一个典型的配方包含诸如表2.2所示的化合物。

In curing reactions with anhydrides, terminal epoxy groups and centrally located hydroxyl groups are attacked with the formation of ether and ester bridges.

To obtain a system that is capable of reacting and at the same time tailored to the application requires an appropriate formulation. This could for example be a one-component system, or for better storage, as a two-component system with separate resin and hardener. A typical formulation contains compounds such as those shown in Table 2.2.

在单罐体系中,两部分合在一起贮存在耐贮存条件下备用。

In one-can systems, both sides are combined together and are stored under shelf-stable conditions ready for use.

表 2.2 典型的环氧树脂配方
Table 2.2 Typical epoxy resin formulation.

树脂部分 Resin side	硬化剂部分 Hardener side
环氧树脂 Epoxy resin	固化剂 Curing agent
含环氧的活性稀释液 Epoxy-containing reactive diluents	催化剂和促进剂 Catalysts and accelerators

在单组分或双组分中: On either or both sides:
非活性稀释液和树脂改性剂 Nonreactive diluents and resinous modifiers
填料(增强的和/或非增强的) Fillers (reinforcing and/or non-reinforcing)
着色剂(颜料和染料) Colorants (pigments and dyes)
流变添加剂(触变胶和黏性抑制剂) Rheological additives (thixotropes, viscosity suppressants)
性能促进剂(润湿剂、附着力促进剂、阻燃添加剂) Property promoters (wetting agents, adhesion promoters, flame retardant additives)
加工助剂(去泡剂、脱模剂) Processing aids (deaerating agents, mold release agents)

树脂 Resins

DGEBA(双酚 A 二缩水甘油醚)是一种广泛使用的环氧树脂,由双酚 A 和环氧氯丙烷合成。

DGEBA (the diglycidyl ether of bisphenol A) is a widely used epoxy resin. It is synthesized from bisphenol A and epichlorohydrin.

氧化乙烷环
α-环氧官能团
Oxirane ring
α-epoxy function

双酚 A
Bisphenol A

高熔点固态环氧化物的环氧基团之间能有多至 20 个重复单元($n=20$):

In high melting point solid epoxies, there can be up to 20 repeating units ($n=20$) between the epoxy groups:

此外,有许多树脂类型在低聚物分子中含双官能和多官能环氧基团,例如环氧酚醛树脂(详见酚醛树脂部分):

Furthermore, there are many resin types with di-functional and multi-functional epoxy groups in the oligomer molecules, such as the epoxidized novolacs (see PF resins):

$$\text{(epoxy novolac structural formula)}$$

继续开发着眼于获得更好的热稳定性和最佳力学性能。由聚乙二醇、腰果油和脂肪酸制备的环氧树脂系列,糅合起来改善力学柔性。此外,也用端羧基丁腈弹性体(CTBN)来改性树脂。

在 DGEBA 基础上氟化或溴化的化合物用于改进耐候性和抗菌性或作为阻燃剂。

Continued development is focused on achieving ever better temperature stability and optimum mechanical properties. To incorporate mechanical flexibility, one series of epoxy resins is based on polyglycols, cashew nut oils and fatty acids. Furthermore resins are also modified with carboxyl-terminated polybutadiene acrylonitrile elastomers (CTBN).

Fluorinated or brominated compounds on DGEBA basis are used to improve weathering stability and bacterial resistance or as flame retardants.

硬化剂 Hardeners

液态反应,即热塑性环氧树脂反应形成热固性树脂是用硬化剂作为交联固化剂完成的。通常用多官能团的胺或酸酐来与双官能团的环氧生成网状结构。这些加聚反应不排除任何低分子量组分。硬化剂与环氧树脂按化学计量反应。在与胺的反应中,可以形成能再与环氧反应的羟基基团。与酸酐反应形成酯键。胺固化通常在室温或稍高于室温进行,而酸酐固化则需要在高温下进行。

硬化剂通常也含有使反应时间和温度适应所期望的工艺条件的催化剂和促进剂(咪唑或叔胺)。

The reaction of the liquid or thermoplastic epoxy resin to form thermosets is carried out using hardeners as crosslinking or curing agents. Multifunctional amines or acid anhydrides are mostly used to build up a network with the bifunctional epoxies. These polyaddition reactions do not eliminate any low molecular mass components. The hardener reacts stoichiometrically with the epoxy. In the reaction with amines, hydroxyl groups are formed that can again react with epoxies. In the reaction with acid anhydrides, ester bonds are formed. Amine curing usually takes place at room temperature or slightly above, whereas high temperatures are needed for acid anhydride curing.

The hardener usually also contains catalysts and accelerators (imidazole or tertiary amines) that adapt the reaction times and temperatures to the desired processing conditions.

胺 Amines

脂肪胺是最大一类环氧树脂固化剂。其特性是适用期短且反应放热高。欲在室温下固化,且当允许玻璃化转变温度低于或等于约 100℃

Aliphatic amines constitute the largest group of epoxy curing agents. They are characterized by short pot lives and high exothermic reactions. They are used with bisphenol A resins when room-temperature curing is desired and when glass

时,则将它们用于双酚 A 树脂。普通采用的多胺隶属于下列同系物:二乙烯基三胺(DETA)、三乙烯基四胺(TETA)和四乙烯基五胺(TEPA)。对于大多数体系,伯胺和仲胺有相似的反应性。

transition temperatures below or at about 100℃ can be tolerated. The most common polyamines used are those that belong to the following homologous series: diethylenetriamine (DETA), triethylenetetramine (TETA), and tetraethylenepentamine (TEPA). In the majority of systems, the primary and secondary amines have similar reactivities.

$$H_2N-CH_2-CH_2-NH-CH_2-CH_2-NH_2$$
DETA

二乙烯基三胺(DETA)有两个伯胺和一个仲胺基团,按化学计量关系能够与五个环氧乙烷环反应形成多连接的网状结构。

Diethylenetriamine (DETA) has two primary and one secondary amine groups and can react stoichiometrically with five oxirane rings to form a multiple-linked network.

芳香胺通常赋予胺固化环氧树脂最好的性能,但必须升高温度进行固化,例如 4,4-二氨基二苯甲烷(DDM)。

Aromatic amines generally provide the best properties for amine cured epoxies but curing has to performed at elevated temperatures, e.g. 4,4-diaminodiphenylmethane (DDM).

氨基化合物和聚酰胺也用作固化剂。

Amides and polyamides are also employed as curing agents.

双氰胺(DICY)经常用于单组分体系,因为反应是通过精细分布的 DICY 晶体的熔化进行的,这确保长期贮存。

Dicyandiamide (DICY) is often used in one-component systems because the reaction proceeds through the melting and dissolution of the finely distributed DICY crystals, which ensures a long shelf life.

酸酐 Acid Anhydrides

酸酐的选择对最终产品的性能有很大的影响,也影响价格和加工工艺。例如酐硬化的双酚 A 体系适用期长、放热反应焓低、黏合和介电性能优异,玻璃化转变温度在 125℃ 和 170℃ 之间,但要求升高温度固化。

The choice of the acid anhydride again has a large influence on the properties of the finished product and also influences the price and processing. As an example: anhydride-hardened bisphenol A systems offer long pot life, low exothermic reaction enthalpies, excellent adhesion and dielectrical properties, and glass transition temperatures between 125℃ and 170℃, but require curing at elevated temperatures.

除酐外,也用苯酚和有机酸作固化剂。不过常作为反应促进剂与其他硬化剂一起使用。

Besides the anhydrides, phenols and organic acids are also used. These are however usually employed as reactive accelerators together with other hardeners.

2.2.3.2 酚醛树脂 Phenolic Resins, PF

苯酚甲醛树脂由三官能团(两个邻位和一个对位)的苯酚或甲酚与双官能团的甲醛(以 40% 的甲醛水溶液)通过缩聚反应生成。第一步形

Phenol-formaldehyde resins are produced through the polycondensation reaction of trifunctional phenol (2 ortho-and one para-positions) or cresol with bifunctional formaldehyde (as a 40% formalin solution). In a first step, methylol compounds

成羟甲基化合物,然后第二步,通过碱性催化剂反应生成碱催化酚醛预聚物 Resol 树脂或通过酸性催化剂生成酸催化酚醛预聚物 Novolacs,依照通用方程式进行:

are formed which, then, in a second step, react through alkaline catalysis to Resol resins or through acid catalysis to Novolacs, according to the general equation:

$$苯酚 + 甲醛 \rightarrow 聚合物 + H_2O$$
$$Phenol + Formaldehyde \rightarrow Polymer + H_2O$$

Novolacs 是与亚甲基桥交联的可溶、未改性时可熔的多环苯酚。其分子量在 600 至 1500 范围内。用固化剂(通常是环己胺伴有氨的排除)将其转化为不溶的热固性树脂而不单是通过热的作用。

Novolacs are soluble, unmodified meltable polycyclic phenols crosslinked with methylene bridges. Their molecular masses are in the range 600 to 1500. They are converted to insoluble thermosets by curing agents (usually hexamethylene amine with the elimination of ammonia), but not through the action of heat alone.

Resols 是在约 100℃ 下生成的液态中间体,易溶解,是能够只通过热的作用固化的单环或多环羟甲基苯酚。在 150 至 160℃ 的固化过程中,先在中间步骤中从合适的苯酚生成部分缩合的半熔酚醛树脂 Resitols,在 160 至 200℃ 最终完全缩合形成不溶不熔酚醛树脂 Resites。

Resols are liquid intermediates that are produced at about 100℃. They are readily soluble, mono-or polycyclic hydroxymethyl phenols that can be cured through the action of heat alone. In the curing process at 150 to 160℃, partially condensed Resitols are first produced from suitable phenols in an intermediate step. These finally condense completely at 160 to 200℃ to form Resites.

通过苯酚和甲醛的反应伴有水的排除形成酚醛树脂的过程:
A) Novolacs 通过酸催化剂与过量甲醛(甲醛对苯酚比例 0.75:1)反应
B) Resols 通过碱催化剂与过量甲醛反应。
Novolacs 和 Resols 间的若干差异汇总于表 2.3。

Production of PF resins through the reaction of phenol and formaldehyde with the elimination of water:
A) Novolacs through acid catalysis with excess formaldehyde (formaldehyde to phenol ratio 0.75:1)
B) Resols through base catalysis with excess formaldehyde.

Several differences between novolacs and resols are summarized in Table 2.3.

表 2.3 Novolacs 和 resols 之间的不同
Table 2.3 Differences between novolacs and resols

	碱催化酚醛预聚物 Novolacs	酸催化酚醛预聚物 Resols
室温下状态 State at room temperature	固态 Solid	液态 Liquid
类型 Type	二阶段树脂 two-stage resin	一阶段树脂 one-stage resin
固化 Curing	从预聚物,例如与环己胺 from prepolymers, for example with hexamethyleneamine	经过 B 阶段 via the B stage
催化剂 Catalyst	酸 Acids	碱 Bases
固化时排除物 Elimination products on curing	NH_3	H_2O
保存期限 Shelf life	很长 very long	<1 年 <1 year
应用 Application	模塑品 Molding	浇铸件、黏合剂 casting, adhesive

Baekeland 采用的缩聚反应由三个主要阶段组成:

1. A 阶段即碱催化酚醛预聚物 Resol：初始的缩合产物主要是醇。树脂是热塑性的,可溶于水。生产的干燥预缩合物作为半成品在市场上有售,例如,100℃下凝胶化时间为 10 分钟。

2. B 阶段即半熔酚醛树脂 Resitol：有较高的缩合度,带有很少量交联、不完全固化。预缩合物在模塑期间进一步交联。例如,在高压下于 150℃将 Resol 醇溶液注入纸张或布片或带,压制成板。如果 Resol 在被加热的轧机上与填料混合,进一步缩合生成模塑料。

3. C 阶段即不溶不熔酚醛树脂 Reist：几乎完全固化的树脂,即不熔也不溶。加工这些树脂至关重要的是,起始组分的反应和特别是从 B 阶段到 C 阶段的转变是分步独立进行的。在可固化模塑料的生产中,树脂与填

The polycondensation reaction introduced by Baekeland consists of three main steps:

1. Stage A or Resol: the initial condensation products are mainly alcohols. The resin is thermoplastic and soluble in water. The resulting dried precondensate is commercially available as a semifinished product, for example with a gelation time of 10min at 100℃.

2. Stage B or Resitol: there is a higher degree of condensation with few crosslinks, not fully cured. The precondensates are further crosslinked during molding. For example, plates are pressed from paper or fabric sheets or strips impregnated with an alcoholic Resol solution (prepregs) at 150℃ under high pressure. If Resols are mixed with fillers on heated rolling mills, molding compounds are produced on further condensation.

3. Stage C or Resit: the resin, almost completely cured, is infusible and insoluble.

It is crucial for the processing of these resins that the reaction of the starting components and in particular the transition from the B to the C state is performed in separate steps. In the production of curable molding compounds, the resin is precondensed with

料和增强化合物一起预缩合,按配方由此生成预制品。然后模塑料能贮存起来,直到经由温度升高导入模具(B阶段),然后在更高的温度下进一步固化。

如同所有的树脂体系,酚醛树脂也是按配方使用的。一个典型的例子如表2.4所示。

filler and reinforcing compounds and thereby formulated ready-made. The molding compound can then be stored until it is introduced into the mold through an increase in temperature (B state) and then further cured at a still higher temperature.

As with all resin systems, PF resins are also used as formulations. Table 2.4 shows a typical example.

表 2.4 复合木制品(胶合板)用黏合剂的典型配方(加工温度100至150℃、压力5至40MPa)。
Table 2.4 Typical formulation for adhesives used in composite wood products (plywood) for a process temperature of 100 to 150℃ and a pressure of 5 to 40 MPa.

成分 Constituent	每百份树脂份数 Parts per hundred resin
酚醛树脂 PF resin	100 100
水 Water	多达20 up to 20
填料 Filler	5 至 10 5 to 10
增补剂 Extender	多达10 up to 10
硬化剂(甲醛) Hardener (formaldehyde)	0 至 10 0 to 10
增溶剂(NaOH) Solubilizer (NaOH)	3 至 10 3 to 10

2.2.3.3 氨基树脂 Amino Resins,UF,MF

起始物质是脲和甲酰胺(UF)或甲胺和甲酰胺(MF)。如同酚醛树脂,在微碱性条件下形成网状结构。先经加成反应成羟甲基,而后经缩聚反应排除水。

The starting materials are urea and formamide (UF) or methylamine and formamide (MF). As with phenolic resins, network formation takes place under slightly alkaline conditions, first via an addition reaction to methylols and followed by the elimination of water in the polycondensation reaction:

$$三聚氰胺或脲 + 甲醛 \rightarrow 聚合物 + H_2O$$
$$\text{Melamine or Urea} + \text{Formaldehyde} \rightarrow \text{Polymer} + H_2O$$

① 加成反应形成羟甲基: ① Addition reaction to methylol:

$$三聚氰胺或脲 + 甲醛 \rightarrow 聚合物 + H_2O$$
$$R-C-NH_2 + H_2C=O \rightarrow R-C-NH-CH_2OH$$

② 缩合形成交联聚合物(带有支化点 R′)：

② Condensation to the crosslinked polymer (with branching points R′):

$$R-C-NH-CH_2OH + HNR'-CH_2OH \rightarrow R-C-NH-CH_2-NR'-CH_2OH + H_2O$$

三聚氰胺
Melamine

脲
Urea

除形成的水外,甲醛也可在反应中蒸发。表 2.5 是 MF 模塑料的一个典型配方。由于甲醛对健康有害,所以现已用糠醛(或类似物)取代。

Besides the water formed, formaldehyde can also vaporize in the reaction. A typical formulation of an MF molding compound is given in Table 2.5. Since formaldehyde is injurious to health, it has nowadays been replaced by furfural (or analogs).

表 2.5　MF 模塑料的一种配方
Table 2.5　Formulation of an MF molding compound.

30%	脱色纤维素 bleached cellulose
50%	MF 树脂 MF resin
7%	水分 Moisture
1%	脱模剂（硬脂酸盐）mold release agent (stearate)
5%	着色剂 Colorant
7%	增塑剂 Plasticizer

2.2.3.4　醇酸树脂,不饱和聚酯树脂 Alkyds, Unsaturated polyester resins, UP

例如,分子量为几千的不饱和聚酯树脂(UP)是通过马来酸酐与乙二醇(通常是不饱和二羧基酸与二醇)的缩合生产的。这些线型预聚物也称为醇酸树脂。

Unsaturated polyester resins (UP) with a molecular mass of several thousand are for example produced through the condensation of maleic anhydride with ethylene glycol (in general: unsaturated dicarbonic acids with diols). These linear prepolymers are also known as alkyd resins.

预聚物交联成热固性树脂是通过与苯乙烯的自由基聚合发生的,苯乙烯同时用作溶剂并降低黏度。在交联过程中不生成排除产物。

The crosslinking of the prepolymer to the thermoset occurs through radical copolymerization with styrene, which at the same time serves as solvent and reduces the viscosity. No elimination products are formed in this crosslinking process.

通过改变酸、醇和乙烯基单体,热固性树脂的性能和树脂的反应性能够适应特殊的应用。例如,也采用饱和二羧基酸诸如脂肪酸、对苯二酸或邻苯二甲酸酐,苯乙烯和醋酸乙烯生成较短的交联桥,与更加正电性的单体如甲基丙烯酸甲酯相比,这

By varying the acids, alcohols and vinyl monomers, the properties of the thermoset and the reactivity of the resin can be adapted to the particular application. For example, saturated dicarboxylic acids such as adipic acid, terephthalic acid, or phthalic anhydride are also used. Styrene and vinyl acetate result in shorter crosslinking bridges and hence to polymerizates that are harder than more electropositive monomers such as methyl

种聚合物更坚硬，甲基丙烯酸甲酯在线型聚酯链之间形成较长的桥。

methacrylate, which form longer bridges between the linear polyester chains.

由丙二醇、马来酸酐和邻苯二甲酸酐的酯化制备不饱和 UP 预聚物

Preparation of an unsaturated UP prepolymer by esterification of propylene glycol, maleic anhydride and phthalic anhydride.

过氧化物可用作不饱和聚酯树脂的硬化剂。过氧化物分解为自由基，由此通过促进剂（含钴催化剂的酮过氧化物，在室温）的加成、通过加热（例如过氧化苯甲酰，>100℃）或通过光（UV，在室温）引发交联反应。抑制剂（苯醌）确保提高贮存稳定性和延长开始时的处理时间。例如，常将 TBC（叔丁基苯磷二酚）作为聚合的稳定剂和抑制剂添加到丁二烯、苯乙烯和其他活性单体液流中。在 60℃下，它的聚合抑制作用比对苯二酚高 25 倍。一个典型的树脂合成物是由 60% 的邻苯二甲酸酐、丙二醇和马来酸酐以摩尔比 1.5∶2.7∶1.0 生成的预缩合物和 40% 的苯乙烯（含 0.01% 作为稳定剂的对苯二酚）组成的。

Peroxides are used as hardeners for UP resins. The peroxides decompose to radicals and thereby initiate the crosslinking reaction through the addition of accelerators (ketone peroxides with cobalt catalysts, room temperature), through warming (e. g. benzoyl peroxide, >100℃) or through light (UV, room temperature). Inhibitors (quinones) ensure that storage stability is increased and that the processing time at the beginning is increased. For example, TBC (4-tert-butylcatechol) is usually added as a stabilizer and an inhibitor of polymerization to butadiene, styrene and other reactive monomer streams. Its polymerization inhibitory effect is 25 times higher than that of hydroquinone at 60℃. A typical resin composition consists of 60% of a precondensate made of phthalic anhydride, propylene glycol and maleic anhydride in the molar ratio 1.5∶2.7∶1.0 and 40% styrene together with 0.01% hydroquinone as stabilizer.

线型聚酯与苯乙烯桥形成热固性树脂的交联（*网状结构延续）

Crosslinking of the linear polyester with styrene bridges to thermosets (* network continuation).

2.2.3.5 乙烯基酯树脂 Vinyl ester resins，VE

乙烯基酯树脂是环氧树脂和不饱和聚酯树脂的中间族。它们通常是带甲基丙烯酸酯活性基团的双酚 A 或酸催化酚醛预聚物 Novolac 类。

VE resins are an intermediate group between EP and UP resins. They are usually of bisphenol A or Novolac type with the methacrylate reactive groups：

双酚 A 类乙烯基酯树脂
Vinylester resin of the bisphenol A type.

交联成热固性树脂是通过乙烯基的均聚或通过与苯乙烯的共聚发生的,用与不饱和聚酯类似的方法。
也以乙烯基酯树脂为基础制备紫外光固化黏合剂和密封剂,特别是不饱和丙烯酸树脂。如同其他配方,偶氮化合物或苯甲酮用作光引发剂。

Crosslinking to thermosets takes place through homopolymerization of the vinyl groups or through copolymerization with styrene，in a similar way to unsaturated polyesters.
UV curing adhesives and sealants are also based on VE resins，especially the unsaturated acrylic resins. As with the other formulations，azo-compounds or benzophenone are used as photoinitiators.

2.2.3.6 烯丙基、DAP 模塑料 Allylics，DAP molding compounds

20 世纪 50 年代应航空航天工业要求,需要能耐特殊条件的材料。这激励了聚合物化学工作者去创造一族基于二烯丙基邻苯二甲酸酯(DPA)的塑料。

The need for materials to withstand the extraordinary conditions required by the aerospace industry in the 1950s encouraged polymer chemists to create a family of plastics based on diallyl phthalate (DAP).

二烯丙基邻苯二甲酸酯(DAP)形成预聚物的反应
Reaction of diallyl phthalate (DAP) to the prepolymer.

分子量 6000 至 12000 的预聚物的聚合和部分交联是通过过氧化物引发的烯丙基双键的加成反应(例如含 3% 过氧化二枯基)而发生的。相对高黏度 DAP 树脂的完全交联

The polymerization and partial crosslinking of prepolymers with molecular mass 6,000 to 12,000 takes place through a peroxide-induced addition reaction (e. g. with 3% dicumyl peroxide) of the allylic double bond. Complete crosslinking of the relatively high viscosity DAP resin occurs at temperatures around 150℃.

发生在大约150℃温度下。通常以短和长玻璃纤维和石粉作填料。

Mostly short and long glass fibers and powdered stone are used as fillers.

2.2.3.7 聚丙烯酸酯 Polyacrylate, PAK

聚丙烯酸酯(PAA)在150至300℃之间的温度经交联形成聚丙烯酸酐：

Polyacrylic acid (PAA) undergoes crosslinking to form polyacrylic anhydride at temperatures between 150 and 300℃：

α-氰基丙稀酸甲酯通过弱碱催化剂（水汽）以阴离子机理聚合（离子链聚合）。这些快速反应体系的典型应用是单组分快速黏合剂。就它们的结构而言，固化后的黏合剂相当于热塑性塑料。

α-cyanacrylic acid methyl ester polymerizes through weak base catalysis (moisture) in an anionic mechanism (ion chain polymerization). Typical applications of these very fast reacting systems are the one-component instant adhesives. As far as their structure is concerned, the cured adhesives correspond to thermoplasts.

热塑性丙烯酸酯（例如 PMMA）能像乙烯基酯树脂那样交联。耐溶剂性和抗银纹性得以改善。用作涂料的配方含有丙烯酸树脂、低聚体和单体交联剂、催化剂和添加剂，通过高能光的作用（紫外光→自由基）进行交联，这能让更环境友好的活性体系取代环境污染的含溶剂涂料体系。例如，用甲基丙烯酸酯基团（Chemlink）两头封端的聚苯乙烯与1,6-己二醇二丙烯酸酯或四氢化糠基丙烯酸酯和光引发剂交联（配方见表2.6）。

Thermoplastic acrylates (e.g. PMMA) can be crosslinked like VE resins. This achieves improved solvent and craze resistance. Formulations used for coatings contain the acrylic resins, oligomers and monomer crosslinkers, catalysts, and additives and are crosslinked through the action of high-energy light (UV →radicals). This allows solvent-containing coating systems that pollute the environment to be replaced with more environmentally-friendly reactive systems. For example, polystyrene capped at both ends with methacrylate groups (Chemlink) is crosslinked with 1,6-hexanediol diacrylate or tetrahydrofurfuryl acrylate and photoinitiators (for formulation see Table 2.6).

表2.6 一个紫外光固化的聚丙烯酸酯体系的配方
Table 2.6 Formulation of a UV-curing polyacrylate system.

甲基丙烯酸酯基团 Chemlink	50%
四氢化糠基丙烯酸酯 Tetrahydrofurfuryl acrylate	37%
三羟甲基丙烷三丙烯酸酯 Trimethylolpropane triacrylate	5%
润湿剂 Wetting agent	1%
光引发剂 Photoinitiator	2%
其他 Others	5%

2.2.3.8 聚氨酯体系 Polyurethane systems，PUR

聚氨酯是由异氰酸酯与多元醇、伯胺或仲胺或其他至少带两个活性氢原子的化合物的反应生成的。二异氰酸酯和二醇的聚合在低温下以加成反应(含一些引发剂)依次进行。甲苯二异氰酸酯(TDI)常作为二异氰酸酯使用,是有毒单体,现在更常用于预聚物中。除了芳香族二异氰酸酯,也用脂肪族二异氰酸酯(例如,正己烷二异氰酸酯 HDI)作为单体。

Polyurethanes are based on the reaction of isocyanates with polyols, primary or secondary amines, or other compounds with at least two active hydrogen atoms. The polymerization of a diisocyanate and a diol proceeds as an addition reaction at low temperatures and with certain initiators. Toluene diisocyanate (TDI) is often used as the diisocyanate. As a monomer, this is poisonous and is nowadays more often used in prepolymers. Besides aromatic diisocyanates, aliphatic diisocyanate are also used (e.g. HDI, hexane diisocyanate).

交联反应是用过量二异氰酸酯按照下列反应式进行的：

The crosslinking reaction occurs with an excess of the disocyanate according to the following reaction schemes：

A. 氨基甲酸酯与羟基反应：

A. Urethane Reaction with hydroxyl groups：

$$R-N=C=O + HO-R_3 \rightarrow R-NH-CO-O-R_3$$

B. 聚脲与伯胺反应：

B. Polyurea Reaction with primary amines：

$$R_1-N=C=O + H_2N-R \rightarrow R_1-NH-CO-NH-R$$

C. 线性脲与仲胺反应：

C. Linear urea Reaction with secondary amines：

$$R_1-N=C=O + HNR-R_2 \rightarrow R_1-NH-CO-NR-R_2$$

与胺形成脲基。不过,产物属于聚氨酯族。与伯胺的反应是很快的,没有催化剂便可发生。有过量二异氰酸酯,则交联发生在胺基的支化点上(上述反应 B 和 C)。二异氰酸酯与水反应形成相应的氨基甲酸(R_3 当作氢的反应 A),伴有 CO_2 排除进一步形成胺。生产塑性泡沫是在小心控制的条件下进行的。硬泡沫塑料用大量过量的二异氰酸酯制造,而软泡沫塑料从柔性聚合物和聚醚制造。

With amines, urea groups are formed. Nevertheless, the products belong to the urethane family. The reaction with primary amines is very fast and takes place without the presence of a catalyst. Crosslinking occurs at the branching point of the amine groups with excess diisocyanate (reactions B and C, above).

With water, the diisocyanate reacts to form the corresponding carbamic acid (reaction A with R_3 as hydrogen) and further to amine with elimination of CO_2. In plastic foam production this is performed under carefully controlled conditions. Rigid foams are made using a large excess of diisocyanate, and soft foams from flexible polymers and polyethers.

可制造弹性迥然不同的各种聚氨酯,从非常软的到非常硬的聚氨酯。通过适当的配方配置和反应组分选择(胺、多元醇、异氰酸酯、环氧化物)可使所需产品的这些性能得以优化。

A large variety of polyurethanes are manufactured with very different elastic properties, from very soft to very hard. These properties can be optimized for the desired product by appropriate formulation and selection of reaction partners (amines, polyoles, isocyanates, epoxides).

2.2.3.9 二氰酸酯树脂 Dicyanate resins

二氰酸酯单体一般通过加成三聚来固化,无挥发性副产物形成,先形成多氰尿酸酯预聚物,然后生成具有高玻璃化转变温度和低水分吸附的热固性树脂。反应温度在 150 至 270℃之间:

Dicyanate monomers generally cure via an addition trimerization without the formation of volatile byproducts to first form polycyanurate prepolymers, and then thermosets with high glass transition temperatures and low moisture absorption. The reaction temperatures lie between 150 and 270℃:

2.2.3.10 聚酰亚胺、双马来酰亚胺树脂 Polyimides (PI), Bis maleimide (BMI) resins

聚酰亚胺是从各种二酐和二胺单体制备的,可用大分子主链中亚胺重复结构单元来表征。该结构赋予其不同寻常的热氧化稳定性。交联聚酰亚胺是活性预聚物(短链低聚体)通过高温下均聚得到的,例如双马来酰亚胺(以过氧化二枯基作为自由基催化剂)制得的交联聚酰亚胺。其中存在双马来酰亚胺基本结构的 BMI 树脂是聚酰亚胺的亚族:

Polyimides are prepared from a variety of dianhydride and diamine monomers and are characterized by repeating imide structural units in the polymer backbone. This structure contributes to their exceptional thermal and oxidative stability. Crosslinked polyimides are obtained from reactive prepolymers (short chain oligomers) through homopolymerization at high temperatures, e.g. from bismaleimide (with dicumyl peroxide as free-radical catalyst).

BMI resins are a subgroup of the polyimides in which the basic structure of bismaleimide is present:

酰亚胺结构
Imide structure

聚酰亚胺 PI,特别是双马来酰亚胺 BMI 树脂,反应形成具有非常致密网状结构因而具有高玻璃化转变温度的热固性树脂。这种材料很脆。

The PI and especially the BMI resins react to form thermosets with very dense networks and hence high glass transition temperatures. The materials are very brittle. One therefore attempts to loosen the networks a little through copolymerization

因此,为提高其强度,人们尝试通过共聚(不降低玻璃化转变温度)或通过添加其他组分形成互穿网络来使网状结构略加松散。Sidney H. Goodman 的著作(302 至 467 页)详细阐述了不同类型的聚酰亚胺。

(without lowering the glass transition temperature) or by creating interpenetrating networks through the addition of other components in order to increase their strength. A detailed description of the different types of PI is given in the Sidney H. Goodman's book, pages 302 to 467.

2.2.3.11 硅树脂 Silicone resins, SI

硅树脂(聚硅氧烷)也是交联体系,因而有时也将其列入"热固性树脂"的话题下。不过,这些聚合物的性质通常如弹性体,故不在这里讨论。聚合物链结构从双官能团的硅烷醇通过排除水得到(—Si—O—)$_n$ 键生成。通过三官能团的硅烷醇进行交联。

在溶胶与凝胶状态相互转换的反应中,含 Si 结构发生交联。聚合物陶瓷材料也是基于硅氧烷的交联。无机网状结构的转换是可变的。但在此不讨论这些材料。

Since silicones (polysiloxanes) are also crosslinking systems, they are also sometimes included under the topic "thermosets". These polymers, however, usually behave as elastomers and will not be discussed here.
The polymer chain structure arises from difunctional silanols through the elimination of water to (—Si—O—)$_n$ bonds. Crosslinking takes place via trifunctional silanols.
In sol-gel reactions, the Si-containing structures are crosslinked. Polymer ceramic materials are also based on the crosslinking of siloxanes. The transition to inorganic networks is fluid. These materials will not however be discussed here.

2.3 固化反应 The curing reaction

交联反应是放热的。通常主要关注反应温度、反应速率和反应焓(反应热)。在技术上,了解起始物质、促进剂、抑制剂、增塑剂和填料对反应行为的影响是重要的。在反应过程中,因摩尔质量和交联程度的增加而使黏度和弹性模量增加,密度增大。

Crosslinking reactions are exothermic. In general, the temperature of the reaction, the reaction rate and the reaction enthalpy (heat of reaction) are the main points of interest. Technologically, it is important to know the influence of the starting substances, accelerators, inhibitors, plasticizers and fillers on reaction behavior. During the reaction, the viscosity and the modulus of elasticity increase due to an increase in the molar mass and crosslinking, and the density increases.

2.3.1 交联步骤 Crosslinking steps

交联过程是复杂的,因为涉及到不同的反应步骤。高分子结构的生长基本上可分成两种链形成机理:
1. 通过双官能团基元反应的步阶式生长。
2. 通过单体连接到交联高分子上的链生长。

The crosslinking process is complex because different reaction steps are involved. Basically, the growth of the polymer structure can be divided into two mechanism of chain formation:
1. Stepwise growth through elementary reactions of two functional groups.
2. Chain growth through the attachment of monomers to the crosslinking polymer.

在第一个过程中,通过两个共活性中心的反应产生一个共价键,从而将单体、低聚体和大分子连接在一起。高分子的结构主要由单体的官能度和共活性中心的摩尔比决定。在第二个过程中,只是单体连接到链上,聚合物的结构由单体的官能度、引发与增长之间的反应速率比和单体与引发剂的浓度决定。
热固性树脂的形成可以两种方式发生:
1.通过含有三个或更多官能度的单体。
2.通过已有的线型或支化大分子(预聚物)的化学交联。
在已完全反应的聚合物网状结构中,实际上所有的组成单元以共价键结合在巨大的三维结构中。在凝胶点,最初呈现明显的巨大网状结构。

In the first process, a covalent bond is created through the reaction of two co-reactive centers and thus joins monomers, oligomers and macromolecules together. The structure of the polymer is mainly determined by the functionality of the monomers and the molar ratios of the co-reactive centers. In the second process, only monomers are attached to the chain and the structure of the polymer is determined by the functionality of the monomer, by the ratio of the reaction rates between initiation and growth, and by the concentrations of monomer and initiator. The formation of a thermoset can take place in two ways:

1. Through monomers with a functionality of three or more.
2. Through chemical crosslinking of linear or branched macromolecules (prepolymers) that are already present.

In the fully reacted polymer network, practically all the constituent units are incorporated covalently in the infinite three-dimensional structure. At the gel point, the infinite network becomes apparent for the first time.

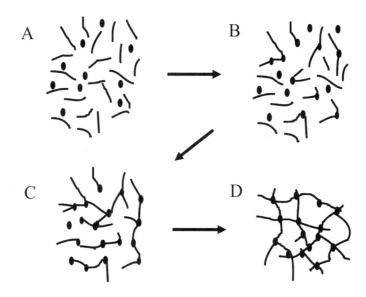

图 2.2 交联步骤示意图
A:含低分子量单体的起始树脂物质("A 阶段单体");
B:线型生长和支化(B-阶段物质即预聚物);
C:凝胶化的但仍不完全的网状结构;D:完全固化的聚合物(C-阶段热固性树脂).
Fig. 2.2 Schematic diagram illustrating the steps involved in crosslinking.
A: starting resin material with low molecular mass monomers ("A-stage monomers");
B: linear growth and branching (B-stage material or prepolymer);
C: gelated, but still incomplete network; D: fully cured polymer (C-stage thermoset).

如图 2.2 所示,反应随着更大分子的形成而开始,大分子也可能已支化。这个所谓的预聚合是提高聚合物的分子量到某个中间值的途径,因而仍有足够的剩余反应点在固化环境中进一步反应。预聚合的一些原因是为了提高黏度、减小毒性和降低活性以控制凝胶化时间。对优化固化条件来说,考虑化学计量是重要的。不过实际上,混合比可以由于聚到一起反应的反应物质的空间位阻而调整。副反应和链终止杂质可能降低计算得到的活性点数量。交联是通过三官能团和多官能团在链之间形成的,于是通过交联从黏性液体形成弹性凝胶体。交联在凝胶体中继续,直到形成几乎完全致密的网状结构,或反应由于玻璃化而停止。

当在交联反应中形成两个或多个聚合物网状结构时,则称之为形成互穿聚合物网络(IPN)。IPN 是两个或多个互相盘绕交联的聚合物的联合体。只观察到一个玻璃化转变是合乎理想的,然而,经常观察到两个玻璃化转变,表明有一定程度的相分离。

As shown in Figure 2.2, the reaction begins with the formation of larger molecules that may also already be branched. This so-called prepolymerization is a way to increase the molecular weight of a polymer to some intermediate value so that there are still enough residual reacting sites to further react in a curing environment. Some reasons for prepolymerization are to increase the viscosity, decrease the toxicity, and reduce reactivity for control of gelation time.

Stoichiometric considerations are important for optimum curing conditions. However, in reality, the mixing ratio may be adapted due to steric hindrance of reacting species to come together and react. Side reactions and chain-stopping contaminants may reduce the calculated number of reactive sites. Crosslinks are formed between the chains via trifunctional and multifunctional groups. An elastic gel is then formed from the viscous liquid through crosslinking. In the gel, the crosslinking continues until the network density is practically complete or the reaction comes to a stop due to vitrification.

When two or more polymer networks are formed in a crosslinking reaction, the term interpenetrating polymer network (IPN) is used. IPNs are combinations of two or more interwinding crosslinked polymers. Ideally, only one glass transition is observed. Often, however, two glass transitions are observed, which indicates a certain degree of phase separation.

2.3.2 TTT 图 TTT diagram

在树脂固化中可能出现的三个不同物理状态通常用 TTT(时间—温度转换)固化图(图2.3)来表示。这是温度对反应时间对数的曲线图,表示在反应温度 T_R 下,经一定反应时间后树脂的状态。低于 T_{g0},树脂处于玻璃态,实际上反应是受阻止的。在固化温度 T_{R2},在相对短时间后到达凝胶化线(虚线),材料凝胶化(凝胶点 G_2),转化为橡胶态,交联继续直到完全固化。因此,固化温度总是比可能的最大玻璃化转变温度 $T_{g\infty}$ 要高。对于更长的反应时间,可能开始分解。

如果选择反应温度 T_{R1},到达凝胶

The three different physical states that can occur in curing are usually displayed in a TTT (time-temperature transformation) cure diagram (Fig. 2.3). This is a plot of temperature versus the logarithm of reaction time. It shows the state of the resin after a certain reaction time at a reaction temperature T_R. Below T_{g0}, the resin is in the glassy state and the reaction is practically blocked.

At a curing temperature T_{R2}, the gelation line (dashed line) is reached after a relatively short time, the material gels (gel point G_2) and is transformed to the rubbery state, and crosslinking continues until complete curing. The curing temperature is thus always higher than the maximum possible glass transition temperature $T_{g\infty}$. With longer reaction times, decomposition can begin.

If a reaction temperature of T_{R1} is chosen, it takes somewhat

点 G_1 前需时长一些。交联在橡胶态弹性材料中继续，直到在 V_1 点发生玻璃化。玻璃化转变温度因持续的交联而提高，直到到达反应温度。转变为玻璃态时，反应停止。材料是硬的，看起来好像固化业已完全。

如果在使用中温度超过了用该法固化的热固性树脂的温度 T_{R1}，则材料会变软而产生缺陷。在玻璃态，反应并不是完全受阻的，而是仍然能以很低的速率继续进行。材料性能显然是不稳定的。

因此，连续的 S 形线表示在等温固化温度下树脂变为玻璃态物体的时间。图 2.3 是对活性树脂体系反应时间和反应温度的表征。

longer until the gel point at G_1 is reached. Crosslinking continues in the rubbery elastic material until vitrification occurs at V_1: The glass transition temperature increases due to continued crosslinking until it reaches the reaction temperature. At the transition to the glassy state, the reaction stops. The material is hard and it seems as if curing is complete.

If the temperature T_{R1} of a thermoset cured in this way is exceeded during use, softening occurs, which can lead to a material defect. In the glassy state, the reaction is not completely hindered but can still proceed at a very low rate. The properties of the material are clearly not stable.

The continuous S-shaped line therefore shows when the resin vitrifies at an isothermal curing temperature. Fig 2.3 thus characterizes a reactive resin system with regard to reaction temperature and reaction time.

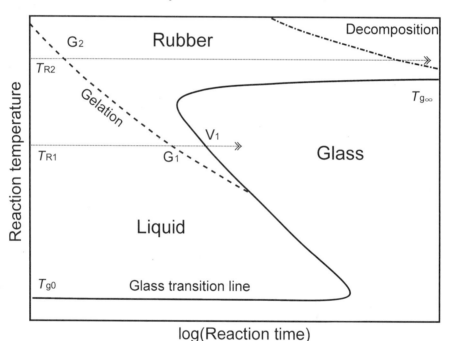

图 2.3 固化过程 TTT 图。它显示固化时物质的三种状态：液体、橡胶弹性材料、玻璃态。连续的 S 形曲线是反应在等温温度 T_R 时树脂变为玻璃态的时间。

Fig. 2.3 The curing process illustrated by the TTT diagram. It shows the three states of the material on curing: liquid, rubbery elastic material, glass. The continuous S-shaped curve shows the time it takes for the resin to vitrify when the reaction is performed at an isothermal temperature T_R.

从技术的观点来看，特定温度下的凝胶化时间、适用期和贮存时间是十分重要的，不管交联是用催化、通过光照还是温度升高启动的。

在凝胶点，黏度显著升高（理论上是

From the technical point of view, the gelation time, the pot life and the storage time at a particular temperature are of great importance, irrespective of whether crosslinking is started catalytically, through exposure to light or temperature increase. At the gel point, the viscosity increases markedly (theoretically

无限的),因而也就不可能以浇注、涂层、注入等方式使用这种树脂(按 DIN 16916 规定的 B-时间)。

适用期(加工时间)也与之相关。它表征在正常条件下,在反应开始后和在混合物变为难以加工或用其他方法难以处理前,可用于进行反应热固性树脂配方操作的时间(例如在模塑机中进行无故障模塑并得到无缺陷制件的停留时间)。

保存期可指热固性树脂体系实际贮存的任意时间,不管是单组分体系还是组分混合后的体系。保存期也用来描述未混合组分的贮存稳定性。例如,某些固化剂由于吸收大气中的水汽会失去活性。

业已指出,热固性树脂的运送、加工和性能强烈依赖于凝胶化和玻璃化。已预反应变为玻璃态物体的树脂能容易地贮存,简单加热就开始固化。这简化了树脂的运送和加工,因为树脂和硬化剂无须先以合适的比例混合。事实上,Baekeland 在温度控制的生产中引入了步骤概念。温度对固化的影响提供了另一个实用的生产控制机制,分段概念:在 A 阶段,热固性树脂准备反应(即混合后,但交联还未开始),然后随着时间和交联的进行经历 B 阶段。经常通过降低到足以获得无活性时段(长至一年)的温度来停止该过程,例如对活性涂料粉末或黏性胶带,以方便应用。C 阶段指最终成形的完全交联阶段。

infinitely) and thereby ends the possibility of using the resin (casting, coating, pumping) (the B-Time according to DIN 16916).

The pot life (processing time) is also connected with this. It characterizes the time available to process a reacting thermosetting resin formulation under normal conditions after the start of the reaction before the mixture becomes intractable or otherwise difficult to process (e. g. residence time in a molding machine for trouble-free molding and defect-free parts).

Shelf life can mean an arbitrary time for practical storage of a thermoset system, either a one-component system or a system after mixing the components. Shelf life is also used to describe the storage stability of unmixed components. For example, some curing agents will lose reactivity due to the uptake of atmospheric moisture.

As already indicated, the handling, processing and properties of a thermoset depend strongly on gelation and vitrification. A resin in the prereacted, vitrified state can be easily stored and simply heated to start curing. This simplifies the handling and processing of resins because resin and hardener do not first have to be mixed in the right proportions. In fact, Baekeland introduced the step concept in temperature-controlled production. The influence of temperature on curing provides another practical production control mechanism, the concept of staging: in the A-stage, the thermoset is ready to react (i. e. after mixing, but crosslinking has not begun), then it goes through the B-stage as time and crosslinking progresses. Often, the process is stopped by lowering the temperature sufficiently to achieve an inactive period (up to 1 year), for example for reactive coating powders or tacky adhesive tapes for easy application. The C-stage represents the fully crosslinked part in its final configuration.

2.3.3 固化动力学 Curing Kinetics

为了控制固化反应的温度和时间,关于树脂的反应动力学的信息是重要的。另一个重要因素是反应焓,特别当成型大制件时。树脂和热固性树脂是不良热导体,这就是为什么固化时间在某种程度上依赖于起

Information about the reaction kinetics of a resin is important in order to control the temperature and duration of a curing reaction. Another important factor is the reaction enthalpy, especially when large parts are molded. Resins and thermosets are poor conductors of heat, which is why the curing time is to some extent dependent on the initial reaction temperature and the

始反应温度和树脂量的原因(即半绝热行为)。由于放热反应产生的自加热可导致局部过热和成品中很大的机械应力乃至热分解。

反应的动力学分析通常有两个目的:
- 首先是描述反应速率(目的1)。
- 其次是研究反应机理(化学反应路径)(目的2)。

固化反应涉及许多复杂的过程(从液态到凝胶态到固态的转变、引发、链增长和形成交联、反应终止、扩散)。因此,第二个目的是非常费时的,需要采用许多不同的分析方法诸如热分析、光谱学、色谱法、化学分析等。但除了热分析,这些方法均不在本手册讨论之列。

对于许多用途,就反应进行现象学描述(目的1)通常就已足够,例如,设计和优化工业固化过程,或研究和区分树脂/硬化剂类型、催化剂、填料和添加剂的影响。在此,人们在寻找能用尽可能少的测试手段来充分描述固化反应的方法。就实用而言,这指这样一种方法,对于因反应时间太长或太短而无法直接测试的不同条件下的反应行为,能对转化率进程与时间和温度的关系作出预测。

依赖于必要条件仅采用一个升温速率的简单DSC测试,实际上是一个可能的方法。计算基于非常简化的假定,例如,反应是按一步进行的,可用简单的一级模型(n级动力学)描述。因此预测是非常有限的,应该十分小心地对待。该种分析对于区分不同的添加剂仍然可能还是足够的。

因为固化反应的复杂性,所以所谓的等转化率法更适于提供可靠的预测。不过,要花费稍多一些的测试时间,因为必须进行至少三个(更多更好)不同升温速率的测试。无需考虑关于特定化学反应机理的信息或假定。因此这个方法称为"非模型

amount of resin used (i.e. semi-adiabatic behavior). Self-heating due to an exothermic reaction can result in local overheating and lead to large mechanical stresses in the product or even to thermal decomposition.

The kinetic analysis of a reaction usually has two goals:
- First, to describe the reaction rate (Goal 1).
- Second, to investigate the mechanism (chemical reaction pathway) (Goal 2).

The curing reaction involves many complex processes (transition from liquid to gel to solid; initiation, chain growth and formation of crosslinks, termination of the reaction, diffusion). Therefore, the second goal is very time consuming and requires the use of many different analytical techniques such as thermal analysis, spectroscopy, chromatography, chemical analysis and so on. These methods except for thermal analysis will not however be discussed in this handbook.

For many purposes, the phenomenological description of the reaction (Goal 1) is usually adequate, e.g. to design and optimize a technical curing process or to investigate and differentiate the influence of resin/hardener types, catalysts, fillers and additives. Here one is looking for methods that can describe the curing reaction sufficiently well with as few measurements as possible. In practical terms, this means a method that can predict the reaction behavior or the course of the conversion as a function of temperature and time for different conditions, that is, for conditions that cannot be measured directly because reaction times would be too long or too short.

Depending on the requirements, this is in fact possible with a simple DSC measurement at one heating rate. The evaluation is based on very simplified assumptions, for example, that the reaction is one-step and can be described by a simple first order model (n^{th} order kinetics). The predictions are therefore very limited and should be treated with great caution. All the same, this type of analysis could be sufficient to differentiate different additives.

Because of the complexity of the curing reaction, the so-called iso-conversion methods are much better suited to provide reliable predictions. The investment in measurement time is, however, somewhat greater because at least three (or better more) measurements at different heating rates have to be performed. Information or assumptions about a particular chemical reaction mechanism are not necessary. The method is therefore called

动力学"(MFK)。更先进的方法"高级非模型动力学"(AMFK)使用动态和等温两种测试。

动力学计算和预测的基础是要获知反应速率 r。例如这可由转化率(α)与时间(t)的关系来测定：

$$r = \frac{d\alpha}{dt}$$

反应速率是转化率 $f(\alpha)$ 的函数并受温度的影响：

$$\frac{d\alpha}{dt} = k(T)f(\alpha)$$

反应速率与温度的关系用 Arrhenius 方程描述：

$$k(T) = k_0 e^{-E_A/RT}$$

式中 E_A 是活化能，R 是气体常数，k_0 是速率常数。

在下面的应用实例中所描述的动力学计算，使用了两种计算方法：

- n 级动力学，由此

- 非模型动力学，其中活化能与转化率有关。

非模型动力学基于 Sergey Vyazovkin 的研究和出版物(见下面的参考文献)*。该法假定对于一定转化率的反应速率只与温度有关。这称为等转化率原理。

固化速率易于用量热来测量，因为分子间每个单独化学键的形成产生一定量的热。因此，在只有一种类型反应发生的简化假设下，测得的热流量直接与反应速率即每单位时间的成键数成正比：

$$d\alpha/dt = 热流量(t)/总反应热$$

转化率(固化度或固化程度)与从反应开始起测得的反应焓成正比，因此等于至时间 t 的热流积分归一化到总反应焓：

"Model Free Kinetics"(MFK). The more advanced method, "Advanced Model Free Kinetics"(AMFK), makes use of both dynamic and isothermal measurements.

The basis for kinetic evaluations and predictions is a knowledge of the reaction rate, r. This can for example be determined from measurements of conversion (α) as a function of time (t):

$$r = \frac{d\alpha}{dt}$$

The reaction rate is a function of the conversion, $f(\alpha)$, and is influenced by temperature.

$$\frac{d\alpha}{dt} = k(T)f(\alpha)$$

The temperature dependence of the reaction rate is described by the Arrhenius equation:

$$k(T) = k_0 e^{-E_A/RT}$$

where E_A is the activation energy, R the gas constant, and (k_0) the rate constant.

In the kinetic evaluations described in the following application examples, two evaluation methods are used:

- n^{th} order kinetics, whereby

$$\frac{d\alpha}{dt} = k(T)(1-\alpha)^n$$

- model free kinetics, in which the activation energy depends on the conversion.

Model free kinetics is based on the research and publications of Sergey Vyazovkin (see the literature references below). In this method, the assumption is made that the reaction rate for a certain conversion is only a function of the temperature. This is known as the iso-conversion principle.

The curing rate can be easily measured calorimetrically because the formation of each individual chemical link between molecules produces a certain amount of heat. For this reason, the measured heat flow is directly proportional to the reaction rate or the number of links per unit of time under the simplified assumption that only one type of reaction occurs.

$$\frac{d\alpha}{dt} = \frac{\text{heat flow}(t)}{\text{total heat of reaction}}$$

The conversion, i.e. the degree of cure or extent of cure, is proportional to the reaction enthalpy measured from the beginning of the reaction, and therefore corresponds to the integral of the heat flow up until time t, normalized to the total reaction enthalpy.

---

* 类似方法最早是由 Qzawa(1965)和 Flyun wall(1965)提出的。——译注

$$\alpha(t) = \frac{partial\ heat\ of\ reaction(t)}{total\ heat\ of\ reaction}$$

This method, however, only measures the relative conversion; it yields only the fraction of reacted groups out of reactive groups that could react within the topological and time constraints of the experiment. The absolute extent of reaction is defined as the fraction of reactive groups that have actually reacted. In reality, there may be more groups that remain unreacted but are not able to react because of topological limitations. These topological limitations are caused by the crosslinked structure of the polymer. A number of reactive groups are constrained by the network from moving too far, thus preventing them from reacting with other constrained groups. It is therefore impossible to achieve true 100% chemical conversion in a thermosetting polymer formed by the polymerization of multifunctional monomers. Since the sterically hindered groups cannot react, the measured reaction enthalpy shows only the reaction enthalpy possible under the chosen conditions. It does not therefore correspond to the theoretical value expected from the stoichiometric composition and can vary depending on the measurement conditions. All the same, kinetic determination is useful because it shows the course of the conversion, that is, how many of the reactive groups that can realistically react have already reacted.

The main application of the kinetic description of curing reactions is the possibility of making predictions regarding conversion behavior under "extreme" conditions. With just a few measurements, reaction behavior can studied under conditions that are impractical from the measurement point of view, such as at high temperatures or long reaction times. This, for example, allows the behavior of a material in very short processing cycles (just a few seconds) or the long-term stability (over months) of a prepreg to be investigated.

References:

S. Vyazovkin, Int. J. Chem. Kinet. 28, 95 (1996);

S. Vyazovkin, J. Therm. Anal. 49, 1493 (1997);

S. Vyazovkin, J. Comput. Chem. 18, 393 (1997).

2.4 热固性树脂的应用 Applications of Thermosets

热固性树脂的应用是极其广泛的。主要用途之一是作为复合材料的基体。现在复合材料在许多领域如汽车、航空航天和航海业中是公认的高性能工程材料。它们是用从热压罐模塑到树脂传递模塑的各种技术制造的。

黏合剂是热固性树脂的另一个主要应用领域,发展迅速。例如,几年前黏合剂受限于相对低性能的应用,当今的配方范围能够装配飞机和汽车部件。建筑是主要市场。

大量的热固性树脂用于涂料工业。这近年来经历了快速变化,由于生产商致力于减少使用溶剂和引进紫外光固化配方以降低挥发性有机化合物的排放。重复利用是生产商正在解决的另一个环境问题。热固性树脂材料是按长期使用设计的,因而难以毁坏。不过确有重新使用和循环利用的方法。材料的革新包括在复合材料中使用天然纤维增强和应用生物聚合物作为基体材料。

当今没有热固性树脂的日常生活是难以想象的。使用量巨大,这是无可质疑的。据可靠估计,总量为20,000,000 吨/年。

The applications of thermosets are extremely varied. One of the primary uses is as the matrix in composite materials. Composites are now accepted as high performance engineering materials in many fields such as the automotive, aerospace and marine industries. They are made using a range of techniques from autoclave molding to resin transfer molding.

Adhesives are another major application area for thermosets and progress has been rapid. For example, a few years ago adhesives were restricted to relatively low performance applications, while today the range of formulations makes it possible to assemble aircraft and automotive components. Construction is a major market.

A large volume of thermosets is used in the coatings industry. This has undergone rapid change in recent years as manufacturers strive to reduce VOC emissions, by reducing the use of solvents and bringing in UV curing formulations. Recycling is another environmental issue that is being tackled by manufacturers. Thermoset materials are designed for a long lifetime and can therefore be difficult to break down. However, methods of re-use and recycling are available. Innovations in materials include the use of natural fiber reinforcement in composites and the application of biopolymers as the matrix materials.

Everyday life without thermosets is hard to imagine today. This is clear from the large quantities used. Reliable estimates give a total amount of 20,000,000 t/year.

2.4.1 热固性树脂的性能 Properties of Thermosets

用于特定应用的反应树脂依赖于非常多的因素。在评价指标中力学性能往往是首先要考虑的,其他包括物理的、化学的、经济的和生态的标准(表2.7)。

本手册主要讨论与温度有关的性能。同时,兼顾其他性能。下述章节略述一些最重要的应用,但难于讨论任何工艺细节。

The reaction resin used for a particular application depends on very many factors. Mechanical behavior is often high up in the rating list, which includes physical, chemical, economical and ecological criteria (Table 2.7).

This handbook deals mostly with temperature-dependent properties. At the same time, other properties must also be taken into account. The following sections touch on some of the most important applications, but it is hardly possible to discuss any processing details.

表 2.7 评价热固性树脂的典型性能

Table 2.7 Typical properties for assessing thermosets.

力学性能 Mechanical	强度、刚度、硬度、韧性 strength, stiffness, hardness, toughness 尺寸准确性 dimensional accuracy 脆性 brittleness
热学性能 Thermal	膨胀行为 expansion behavior 导热性 thermal conductivity 耐热性、软化 resistance to heat, softening 防火性、防发热 fire resistance, resistance to glow heat
电学性能 Electrical	电阻、介电强度、漏电电流 resistance, dielectric strength, leakage current 介电性能 dielectric properties
化学性能 Chemical	抗酸碱、抗氧和其他化合物 resistance toward acids and bases, oxygen and other chemicals 抗溶剂、防吸水 resistance to solvents, water absorption
加工工艺性能 Technical processing	贮存、模塑、固化过程、黏结强度、颜色、染色 storage, molding, curing processes, adhesive strength, color, dyeing 表面、添加剂迁移、气味、生物相容性 surface, migration of additives, smell, biological compatibility

2.4.2 加工 Processing

热固性树脂的加工和反应树脂成热固性最终产品的加工工艺是各式各样的。它依赖于树脂类型和应用，只能简要提及。一个要点当然是交联反应是怎样引发的，即是由混合、升温还是由光引发完成的，树脂是怎样转移到铸模然后固化的。表 2.8 总结了许多加工技术。用相应的时间、温度和压力来控制反应始终是讨论的中心问题。

The processing of thermosets and fabrication of reaction resins to the thermosetting end products is very varied. It depends on the type of resin and the application and can only be briefly mentioned. An important point is certainly how the crosslinking reaction is initiated, that is, whether it is done by mixing, temperature increase, or photoinitiation, and how the resin is transferred to the mold and then cured. Table 2.8 summarizes a number of processing techniques. The control of the reaction with the corresponding times, temperatures and pressures is always at the forefront of any discussion.

表 2.8 用于热固性树脂的加工技术
Table 2.8 Processing techniques used for thermosets.

模塑树脂的加压成型 Pressing of molding resins	浇注： Casting：
拉挤工艺（树脂浴中纤维牵伸） Pultrusion (pulling of fibers through a bath of resin)	热压罐成型 Autoclave molding
连续层压 Continuous lamination	接触成型、手糊成型 Contact molding, hand lay-up molding
纤维缠绕 Filament winding	冷压和真空袋成型 Cold-press and vacuum bag molding
黏接 Adhesion	热压成型 Hot-press molding
喷涂沉积 Spray deposition	树脂注射或传递成型 Resin injection or transfer molding
粉末涂层 Powder coating	树脂渗透成型 Resin infusion molding
发泡 Foaming	

2.4.3 各种树脂的应用领域和性能
Areas of application and properties of individual resins

2.4.3.1 环氧树脂 Epoxy Resins，EP

环氧树脂有很广的应用范围，可用作浇注树脂、预浸料坯、玻璃纤维增强模塑料、层压料、黏合剂和粉末涂料，实例包括管件和金属板涂层、印制线路板、封装电子元件和牙科应用的涂层等等。不过最大用量是建筑业的涂料和结构件（53%）、电子工业的印制线路板（13%）和黏合剂（9%）。预浸料含有高达约 60% 的纤维玻璃织物，在高达 160℃ 温度和高达 7bar 压力下模塑和固化。预浸料只能在有限的时间内贮存。环氧树脂的性能变化很大，与所用的树脂和硬化剂有关。环氧树脂以低收缩、作为黏合剂具有良好耐湿性（极性基团）、高强度、良好的热和电绝缘性以及某种程度的高温稳定性等性能而著称。不过，耐酸性差、对紫外光敏感

EP resins have a very wide range of applications and are used as cast resins, prepregs, glass-fiber reinforced moldings, laminates, adhesives and powder coatings. Examples include coatings of tubing and metal sheets, for printed circuit boards, for encapsulating electronic components and for dental applications. The largest quantities are however used for coatings and constructional elements in the building industry (53%), for printed circuit boards in the electronics industry (13%), and for adhesives (9%). Prepregs contain up to about 60% fiberglass fabric and are molded and cured at temperatures up to 160℃ and pressures up to 7 bar. Prepregs can only be stored for a limited time.

The properties vary significantly depending on the resin and hardener used. In general, EP resins are well known for properties such as their low shrinkage, good wettability as an adhesive (polar groups), high strength, good thermal and electrical insulation, and, to some extent, high temperature

和吸水量相对大。用于环氧体系的填料有许多不同的功能,其中一些汇总于表 2.9。

stability. However, they exhibit low resistance to acids, are UV sensitive and absorb a relatively large amount of moisture. The fillers used in EP systems have a number of different functions. Some of these are summarized in Table 2.9.

表 2.9　用于环氧树脂改性的填料
Table 2.9　Fillers used in EP resins for modifying properties.

填料/增强材料 Filler/Reinforcing material	对性能的影响 Effect on the property
玻璃或碳纤维 Glass or carbon fibers	机械增强 Mechanical reinforcement
银和金粉、石墨 Silver and gold powder, graphite	提高导电性 increases electrical conductivity
氧化铝 Aluminum oxide	改善导热性 improves thermal conductivity
石墨 Graphite	改善润滑性 improves lubricity
滑石粉 Talc	提高机械加工性 increases machinability
沙、硅土 Sand, silica	降低成本 lowers costs
硅土 Silica	降低热膨胀 lowers thermal expansivity

2.4.3.2　酚醛树脂 Phenol-formaldehyde resins, PF

生产的酚醛树脂总量的三分之一用作木料的黏结剂和结合剂(例如胶合板、硬纸版、碎料板)。其他的重要产品是油漆与清漆、装饰层压板和电气绝缘纸的浸渍树脂。苯酚甲醛树脂是复合材料(例如车辆中的水泵)和耐磨损耐摩擦衬套(例如圆盘刹车中的制动瓦)的经济的基体树脂。在航空与航天工业则是应用于高温。

纯酚醛树脂的热固性树脂是很脆的。因此它们总是用木材或石粉或玻璃纤维增强。酚醛模塑料具有高抗蠕变、优异的刚性和尺寸稳定性、良好的高温电性能和低吸水性。

A third of the volume of PF resins produced are used as binders and bonding agents for wood (e.g. for plywood, chipboard, particle board). Other important products are paints and varnishes, decorative laminates and impregnating resins for electrical insulating paper. Phenol-formaldehyde resins are economical matrix resins for composites (e.g. water pumps in vehicles), for abrasives and friction linings (e.g. brake shoes in disk brakes). They are used for high temperature applications in the aerospace industry.

Thermosets from pure PF resin are very brittle. For this reason they are always reinforced with wood or powdered stone, or glass fibers. PF molding materials exhibit high creep resistance, excellent rigidity and dimensional stability, good electrical properties at high temperatures and low water absorption.

表 2.10　一些用于酚醛树脂改性的填料
Table 2.10　Some fillers used in PF resins for modifying properties.

填料/增强材料 Filler/Reinforcing material	对性能的影响 Effect on the property
纤维素纤维 Cellulose fibers	良好的冲击强度 good impact strength
矿物填料(云母、黏土) Mineral fillers (mica, clay)	改善电气绝缘性 improves electrical insulation
石墨纤维 Graphite fibers	改善强度和摩擦性能 improves strength and frictional properties

续表

玻璃纤维 Glass fibers	改善冲击强度、热和电气绝缘性 improves impact strength, heat and electrical insulation
纸 Paper	电气绝缘性 electrical insulation
滑石粉 Talc	增强刚度、抗蠕变 enhanced stiffness, creep resistance
硅土 Silica	磨损、电气和热绝缘性能 for abrasives, electrical and heat insulation properties
硅土 Silica	降低热膨胀 lowers thermal expansivity

专门的酚醛树脂用作适合于进一步机械加工的半成品（珠宝）、硬泡沫（做插花）、纸层压板（低成本印制线路板）、制作木材硬质纤维板产品和炼煤（燃烧后的碳化酚醛物块）。由酚醛模塑料制造的产品实例有电闸盖、显像管支架、插座、控制旋钮、熨斗柄和车辆中的排气管。

除了树脂的化学结构外，填料对性能和固化过程也有重要影响（表2.10）。填料还降低了成本和固化反应期间的温度升高。

Technical phenolic resins are used as semi-finished products for further mechanical processing (jewelry), as rigid foams (for arranging flowers), as paper laminates (low-cost printed circuit boards), for wood hardboard products and for artificial coal (burned carbon PF mass). Examples of products from PF molding compounds are switch covers, tube holders and sockets, control knobs, handles of irons and air ducts in vehicles.

Fillers also have important influence on the properties and curing processes (Table 2.10), besides the chemical structure of the resin. Fillers also lower costs and the temperature rise during the curing reaction.

2.4.3.3 氨基树脂 Amino resins, UF/MF

生产的 UF/MF 树脂超过 70% 用作复合木材的黏结剂。应用于油漆、清漆和涂料以及装饰层压板（家用）也是重要的。尤其是，产品显示出高硬度、浅色、良好的染色特性和电气绝缘强度（电气装置材料）。作为木材黏合剂的氨基树脂是在相对低的温度（70℃）和压力（1.4MPa）下用所谓的冷压工艺在 24 小时内加工的。不过，如果没有催化剂，MF 树脂只能在高温高压下（热压）加工。这些条件能确保木材颗粒是部分浸渍的，使其结构更强和更防水。

More than 70% of the UF/MF resins manufactured are used in composite wood materials as binders. The applications in paints, varnishes and coatings and in decorative laminate sheets (household) are also important. In particular, the products exhibit hardness, light color, good dyeing characteristics and electrical dielectric strength (electrical installation material). Amino resins as adhesives for wood are processed within 24 hours at relatively low temperatures (70℃) and pressures (1.4 MPa) in so-called cold pressing processes. MF resins can however only be processed at high temperatures and pressures (hot pressing) without catalysts. These conditions ensure that the particles of wood are partially impregnated, which makes the structure stronger and more moisture resistant.

2.4.3.4 聚酯树脂 Polyester resins，UP

不饱和聚酯树脂(以及醇酸树脂)约70％用作玻璃纤维增强复合材料(见表2.11)。产品经由浇注、喷涂、手工层压、拉挤、纤维缠绕、注塑成型(BMC 团状模压料)或热压(SMC 片状模压料)制成最终形状。管子、管道、船体和汽车部件用此方法制造。高填充树脂(例如用$CaCO_3$)用作聚合物混凝土和厨房、浴室用人造石(盆、便池)。其他无机填料用作阻燃剂($Al(OH)_3$)、用于 SMC 片状模压料和胶衣(用玻璃微球，例如用玻璃纤维增强塑料 GFRP 作涂料)。

更多的应用包括透明屋顶、混凝土涂层、机器外壳、点火线圈和变压器线圈、转换器和其他模塑电气零件。对于户外应用必须使用紫外光稳定剂。

UP resins (and alkyd resins) are used to about 70％ as glass-fiber reinforced composites (see Table 2.11). The products are given their final shape by casting, spraying, manual lamination, pultrusion, filament winding, injection molding (bulk molding compounds, BMC) or hot pressing (sheet molding compounds, SMC). Tubes, pipes, hulls of boats, and automobile parts are manufactured in this way. Highly filled resins (e. g. with $CaCO_3$) are used as polymer concrete and artificial stone for kitchens and baths (basins, WC bowls). Other inorganic fillers are used as fire retardants ($Al(OH)_3$), and for sheet molding compounds, SMC, and gelcoats (with glass microspheres, e. g. as a coating with glass-fiber reinforced plastics, GFRP).

Further applications include: transparent roofing, concrete coatings, machine housings, ignition and transformer coils, converters and other molded electrical components. UV stabilizers have to be used for external applications.

表 2.11 不饱和聚酯树脂的重要应用领域
Table 2.11 Important application areas for UP resins

车辆制造 Vehicle construction	16％
容器、管材/管道 Containers，tubing/pipes	14％
卫生设备、建筑物 Sanitary, building	13％
波纹板 Corrugated sheet	8％
造船 Boat building	8％
聚合物混凝土 Polymer concrete	8％
电子/电气器件 Electronics/electrical	7％

可用各种不同的工艺来固化不饱和聚酯树脂。工艺影响产品的性能和制造过程的生产率。在不饱和聚酯树脂的固化过程中，需注意的重要一点是模塑料可能收缩高达8％，在约10％反应转化率时已到达凝胶点。固化类型的更多影响见诸表2.12。如同对于一切树脂，必须在玻璃化转变温度以上完成后固化。

Various different processes are used to cure UP resins. The processes influence the properties of the product and the productivity of the manufacturing process. In the curing process of UP resins, an important point to note is that the molding compounds can contract by up to 8％ and that the gel point is already reached at about 10％ reaction conversion. Further influences of the type of curing are given in Table 2.12. As with all resins, postcuring must be carried out above the glass transition temperature.

表 2.12 不同不饱和聚酯固化工艺对性能的影响
Table 2.12 Influence of different UP curing processes on properties

固化 Curing	优势 Advantages	劣势 Disadvantages
用钴促进剂冷固化 Cold curing with cobalt accelerator	色浅 light color 缓慢固化 slow curing 相对消除模塑制件的应力 molded parts relatively free of stress	铸模移去时间长 long mold removal times
用胺促进剂冷固化 Cold curing with amine accelerator	即使在室温下固化相对较快 relatively fast curing, even at room temperature	变黄 becomes yellow
热固化 Hot curing	非常快速固化 very fast curing 适用期长 long pot life	反应期间收缩较大和冷却后内应力较大的风险 risk of greater contraction during reaction and internal stresses after cooling
紫外光固化 UV curing	非常快速固化 very fast curing 适用期极长 xtremely long pot life	变黄 becomes yellow 仅适合于薄涂层 only for thin coatings

2.4.3.5 乙烯基酯树脂 Vinyl ester resins,VE

从价格来看,乙烯基酯树脂介于不饱和聚酯和环氧树脂之间。它呈现较高的强度和较好的化学稳定性,这就是为什么将其用于耐冲击应力的汽车部件、仪器、管子和管道、储罐的原因。末端乙烯基化合物使其具有很大的活性。通过甲基丙烯酸酯终止,它们比不饱和聚酯树脂更耐水解。它还具有较好的对玻璃纤维的润湿和黏结性能。

From the point of view of price, VE resins lie between UP and EP resins. They exhibit greater strength and better chemical resistance, which is why they are used for automobile parts subject to impact stress, apparatus, tubing and pipes, and storage tanks. The terminal vinyl compounds make them very reactive. Through methacrylate termination they are more resistant to hydrolysis than the UP resins. They also have better wetting and adhesive properties with regard to glass fibers.

2.4.3.6 苯二酸二烯丙酯模塑料 DAP molding compounds

烯丙基类树脂具有尺寸稳定性程度高、耐气候和光以及良好的热绝缘性能等特征。应用主要在航空航天工业和汽车技术,在这些领域中必需具备高力学强度和热稳定性,例如接线条、电子附件。

Allylics are characterized by their high degree of dimensional stability, resistance to weathering and light, as well as good thermal insulating properties. The applications are mainly in the aerospace industry and automobile technology, where high mechanical strength and thermal stability is necessary, e. g. terminal strips, electronic accessories.

2.4.3.7 丙烯酸酯树脂 Acrylate

聚丙烯酸（PAA）交联成聚丙烯酸酐作整形植入物。

反应非常快的氰基丙烯酸酯化合物的典型应用是单组分快速黏合剂。结构完全固化的黏合剂是热塑性的。它们是理想的装配助剂，但是不耐气候，受各种溶剂、碱和酸的腐蚀。

能紫外固化的丙烯酸酯通常作为黏合剂和涂料用于薄层中。它们取代了含溶剂的黏合剂，循环时间短，这是生产率的重要标准。主要应用领域是微电子零件（表面贴装器件 SMD）、电信通讯和光导装置。

Polyacrylic acid (PAA) is crosslinked as polyacrylic anhydride for orthopedic implants.

Typical applications of the very fast reacting cyanacrylates compounds are the one-component instant adhesives. Structurally, the fully cured adhesives correspond to thermoplasts. They are ideal assembly aids but are not resistant to weathering and are attacked by different solvents, alkalis and acids.

UV-curable acrylates are mostly used in thin layers as adhesives and coatings. They replace solvent containing adhesives and allow short cycle times, an important criterion for productivity. The main application areas are in microelectronic components (surface mount devices, SMD), telecommunications and optical light guides.

2.4.3.8 聚氨酯 Polyurethane，PUR

聚氨酯非常耐碱和耐酸水解。这一点，以及异氰酸酯基团可反应的多样性，产生了许多具有非常不同应用的聚合物，例如耐候涂层、软鞋底、软和硬泡沫、挡风玻璃黏合剂、航空航天工业上的三明治/蜂窝状结构。比如说聚氨酯浇注树脂用于变压器涂层和用作地板涂层以保护受磨损（耐磨损性）。个别的产品能耐高达 220℃ 的连续温度，或特别耐酸雨、鸟粪或光照变色。硬质整皮泡沫的形成是通过 RIM（反应注射成型）控制的，从而产生具有多孔芯和无孔边的模塑制件，例如用于负载扬声器外壳、椅座和商店设备。由于交联非常快，所以组分能在喷嘴中直接混合，作喷涂应用。聚氨酯浇注树脂的交联反应焓低，因此大的模塑制件能快速固化，不会因过热而损坏。

如同所有的树脂配方，聚氨酯树脂除了通常的填料和增强料外还使用许多添加剂。添加剂包括交联剂、

Polyurethanes are very resistant toward alkaline and acid hydrolysis. This, and the multitude of possible reactions of the isocyanate group lead to a range of polymers with very different applications, for example, for weather-resistant coatings, soft shoe soles, soft and rigid foams, adhesives for windshields, and sandwich/honeycomb structures for the aerospace industry. PUR casting resins are used for example for coating transformers and as floor coatings to protect against wear (scuff-resistance). Individual products can withstand continuous temperatures of up to 220℃ or are especially resistant toward acid rain, bird droppings or discoloration through light. The formation of hard integral foam is controlled through RIM (reaction injection molding) so that molded parts with cellular core and cell-free edges are produced, e.g. for load speaker housings, chair seats and shop equipment.

Since crosslinking is very fast, the components can be mixed directly in a spray nozzle and applied as a coating. PUR casting resins have low crosslinking reaction enthalpy. Large molded parts can therefore be rapidly cured without the possibility of overheating and hence damage.

As with all resin formulations, PUR resins use many additives besides the usual filler and reinforcing materials. The additives include crosslinkers, surfactants, emulsifiers, cell size

表面活性剂、乳化剂、孔尺寸控制剂和用于生产硬泡沫的低沸点液体、其他阻燃剂、防老剂和紫外稳定剂，以及抗静电剂(硫酸盐、碳黑)和生物杀灭剂(抗生物降解的磷酸三丁酯锡)。反应物是分开贮存的，必须保护多元醇和异氰酸酯不受潮。

controllers and low boiling liquids for the production of rigid foams, other fire retardants, antiaging agents and UV-stabilizers, as well as antistatic agents (sulfate, carbon black) and biocides (tributyl tin against biological degradation). The reactants are stored separately and the polyols and isocyanates have to be protected against moisture.

2.4.3.9 聚酰亚胺 PI 和 BMI　Polyimide, PI and BMI

聚酰亚胺是能在300℃以上使用的特别耐热的聚合物材料。例如，聚酰亚胺薄膜用作高性能印制线路板的底板。涂铝的薄膜因为其良好的低温性能而用于航空和航天工业(保护层、宇宙飞行服)。聚酰亚胺模塑制件用于喷气发动机的不同压缩机段的密封，用作无润滑剂的切断装置和滑行轨。

由于其特殊的性能(例如用加压成型进行高效制造，减轻重量)，双马来酰亚胺模塑制件用来代替金属。应用实例包括叶轮泵、打印机导向装置、集成电路母板和火箭头锥。

与环氧树脂(<2.5%吸水率)相比，氰酸酯几乎不吸水。它们也比许多环氧树脂更加热稳定、机械强度更高。氰酸酯固化时几乎不收缩，因此在黏合接合处或复合材料中产生较小应力。

Polyimides are special heat-resistant polymer materials that can be used at temperatures well above 300℃. For example, PI films are used as supports for high performance printed circuits. Aluminum-coated films are used in the aerospace industry (protective layers, space suits) because of their good low-temperature properties. PI molded parts are used for seals in different compressor stages of jet engines and as lubricant-free shut-off devices and slide rails.

BMI molded parts are used instead of metals because of their special properties (for example efficient manufacture by pressing, weight saving). Application examples include vane pumps, printer guides, mother boards for integrated circuits, and rocket nose-cones.

Cyanate esters absorb very little water compared with epoxy resins (< 2.5%). They are also thermally more stable and mechanically stronger than many epoxy systems. Cyanate esters shrink very little on curing and therefore generate less stresses in adhesive joints or composites.

2.4.3.10　硅树脂 Silicone resines

硅树脂的特征是高温稳定性。它们用来浸渍电动机和变压器中的电绝缘体及用作耐热漆。作为原材料，它们用来改性不饱和聚酯、环氧和丙烯酸树脂漆，这些漆因此变得有弹性和耐风化。

溶胶与凝胶状态相互转换的反应从硅或醇盐金属单体前驱体产生各种无机网状结构。在低温下形成整体无机凝胶，并转化为不经高温熔融

Silicone resins are characterized by high temperature stability. They are used to impregnate electrical insulators in motors and transformers and as heat-resistant paints. As raw materials, they are employed to modify UP, EP and acrylic resin paints, which through this become elastic and resistant to weathering.

The sol-gel reaction produces a variety of inorganic networks from silicon or metal alkoxide monomer precursors. Monolithic inorganic gels are formed at low temperatures and converted to glasses without a high temperature melting process. Through

过程的玻璃态。通过这个过程，能在室温下生成具有所期望的硬度、透光、耐化学、单一空隙率和耐热性能的均质无机氧化物材料。

this process, homogeneous inorganic oxide materials with desirable properties of hardness, optical transparency, chemical durability, tailored porosity, and thermal resistance can be produced at room temperatures.

2.4.3.11 使用范围和应用概述 Overview of the areas of use and application

各种树脂可能使用的范围汇总于表 2.13。具有代表性应用的各类产品和行业列于表2.14。尽管列表并非包罗万象，但仍可明了能用热分析进行表征之处。当今，现代材料是带着最终用途的意向进行特定设计的，其不仅具有热固性树脂的性能，而且是类橡胶的，或热塑性的。含高填料含量的材料甚至能表现陶瓷行为(假牙、聚合物混凝土)。领域之间的边界经常是不固定的，明智的是参考有关的应用书籍。

The large range of possible uses of individual resins is summarized in Table 2.13. The individual product groups and industries with typical applications are listed in Table 2.14. Although the list is not complete, it nevertheless indicates where thermal analysis can be used for characterization purposes. Nowadays, modern materials are designed specifically with their final use in mind and can have not only thermoset properties but also be rubbery-like or thermoplastic. Materials with high filler content can even exhibit ceramic behavior (artificial teeth, polymer concrete). The borders between the fields are often fluid and it is wise to consult relevant application books.

表 2.13 不同树脂最重要用途一览表
Table 2.13 Overview of the most important uses of the different resins.

	EP	UP, VE	PF	MF, UF	PAK	PUR	PI, BMI	SI
含增强纤维的复合材料 Composites with reinforcing fibers	×	×	×	×	×	×	×	×
黏合剂 Adhesives	×	×	×		×	×		
涂料、油漆和清漆 Coatings, paints and varnishes	×	×	×	×	×	×		
地板、层压板 flooring, laminates	×	×		×				
电绝缘材料、印制线路板(PCB)、浇注树脂、封装材料 Electrical insulation, printed circuit boards (PCB), casting resins, encapsulation	×	×	×	×		×	×	×
木材黏接剂 Binders for wood			×	×	×			
混凝土黏结剂 Binders for concretex			×					

续表

	EP	UP, VE	PF	MF, UF	PAK	PUR	PI, BMI	SI
模塑化合物材料 Materials, molding compounds	×	×	×	×				
热绝缘材料（硬泡沫） Thermal insulation (rigid foam)						×		

表 2.14 不同行业和应用实例

Table 2.14　Examples of different industries and applications.

医学界 Medicinal	假牙、填补物、植入物 artificial teeth, fillings, implants
汽车工业 Automobile	窗结合剂和密封剂、水泵、车身部件、车架 window bonding and sealants, water pumps, car body parts, tram carriages 颜色涂层、刹车片 color coatings, brake shoes
建筑业 Building and construction	聚合物混凝土、地板涂层、密封和绝缘用硬泡沫 polymer concrete, floor coatings, rigid foam for sealing and insulation 矿物棉黏结剂 binders for mineral wool
电子行业 Electronics	印制线路板、表面贴装黏合剂（SMD 黏合剂）、芯片包 printed circuit boards, surface mount adhesive (SMD adhesive), chip packages 热固性高导电性银墨 thermosetting highly conductive silver ink
电气业 Electrotechnical	接线条、开关罩、变压器浇注树脂 terminal strips, switch covers, casting resins for transformers
家用 Household	陶器、椅子、电视柜、冰箱绝缘材料、黏合剂 crockery, chairs, television cabinets, refrigerator insulation, adhesives 业余爱好者用浇注树脂 casting resins for hobby workers
运动装备 Sports equipment	网球拍、滑雪板、跳高杆 Tennis rackets, skis, high jump bars
木材行业 Wood industry	纤维板、胶合板、层压板 fiber boards, plywood, laminates
航空和航天 Aerospace	直升机轮叶、航空器直尾翼和尾部、防护性结构、黏合剂 helicopter rotor blades, aircraft fins and tails, shields, adhesives
能源生产 Energy production	风力发电站的风轮叶片 wind rotor blades of wind power stations

工厂建筑 Plant construction	管道和管子、箱体涂层 pipes and tubing, tank coatings
光学器件 Optics	镜头黏合剂、光纤涂料 adhesives for lenses, coating of optical fibers
研磨料 Abrasives	切割轮和砂纸用黏结剂 Binder for cutting wheels and sand paper
机器和工具行业 Machine and tool industry	手柄、外壳部件、开关旋钮、构造材料 handles, parts of housings, switch knobs, constructional materials 3-维模型 3-D prototyping
铸造和耐火产品 Foundry and refractory products	热固性树脂作为沙芯和铸模用黏结剂用于金属铸造中 thermosetting resins as binders for sand cores and molds used in metal casting
印刷墨水及相关应用 Printing inks and associated applications	基于热固性树脂的丝网油墨,能紫外固化 Screen inks based on thermosets, UV-curables 汽车、器具和家具应用的装饰末道漆 decorative finish for automotive, appliance, and furniture applications

2.5 热固性树脂的表征方法 Characterization methods for thermosets

2.5.1 所需信息的概述 Overview of information required

多种不同测试和分析方法用于表征原材料、中间产物和固化过程。例如,对未固化的树脂的测试用来表征可能的过程行为,并包括黏度和黏性等的测定。对固化产物本身进行另外的许多测试,这些测试的目的是表征其热学、力学或化学性能。填料、聚合物和添加剂含量的定量测定也是重要的。

对大多数产品,热固性树脂的生命周期可划分为四个截然不同的时期:树脂的生产、热固性树脂的加工、产品的使用和最后的处理。取决于所关心的领域,表征的问题有:

- 树脂类型(环氧树脂、不饱和聚酯树脂、酚醛树脂……)
- 树脂生产(填料、促进剂……)
- 树脂加工(反应时间、适用期、手

A large number of different tests and analytical methods are used to characterize the raw materials, intermediate products and curing processes. Tests on the uncured resin, for example, serve to characterize possible processing behavior and include the determination of viscosity, stickiness, etc. A large number of additional measurements are performed on the cured product itself. The purpose of these tests is to characterize its thermal, mechanical or chemical properties. The quantitative determination of the content of fillers, polymers and additives is also important.

For most products, the lifecycle of a thermoset can be divided into four distinct periods: the production of the resin, the processing to a thermoset, its use as a product, and finally its disposal. Depending on the field of interest, there is the question of the characterization of:

- The resin type (EP, UP, PF, ...)
- Resin manufacture (fillers, accelerator, ...)
- Resin processing (reaction time, pot life, hand lay-up molding,

糊成型……) ...)
- 模塑料(力学强度、弹性模量、玻璃化转变温度)
- 后处理可能性(燃烧、再生……)

- The molding compound (mechanical strength, elastic modulus, glass transition temperature) and
- Disposal possibilities (burning, recycling, ...)

航空和航天、汽车、建筑工业和其他行业已经就其特定需要开发了最优化的测试方法。有许多测定专门性能的标准方法和对各类产品的测试方法。涉及热分析的一些标准方法汇总在2.5.4节。

可用热分析技术测试的热固性树脂的效应和过程汇总于表2.15。

The aerospace, automobile, building industry and other industries have developed test methods optimized for their particular needs. There are a large number of standard methods for the determination of specific properties and tests for individual product groups. Some of the standard methods that involve thermal analysis are summarized in Section 2.5.4.

Effects and processes of thermosets that can be measured by thermal analysis techniques are summarized in Table 2.15.

表 2.15 与树脂和热固性树脂有关的性能
Table 2.15　Properties of interest of resins and thermosets.

起始物质、树脂配方 Starting substances, resin formulation	组成、保存期、黏合性能、树脂的活性、黏度 Composition, shelf life, adhesive properties, reactivity of resins, viscosity
加工、固化 Processing, curing	动力学、固化速率 Kinetics, curing rate 冷、热、光引发 cold, hot, photoinitiation 催化剂和其他添加剂的影响 influence of catalysts and other additives 凝胶化时间、适用期 gelation time, pot life 交联度、玻璃化 degree of crosslinking, vitrification 收缩行为 shrinkage behavior
材料/树脂浇注料 (固化树脂) Material / resin casting compound (cured thermoset)	机械性能：刚度、模量、阻尼、疲劳行为、蠕变行为 X Mechanical properties: Stiffness, modulus, damping, fatigue behavior, creep behavior 热性能：玻璃化转变、膨胀系数、热容和导热性、在相应温度下的长、短时间稳定性、阻燃性 Thermal properties: Glass transition, expansion coefficient, heat capacity and conductivity, short and long time stability at the corresponding temperatures, flame retardation 化学性能：残留单体含量、耐风化、溶剂中的溶胀行为、吸水、组分、增强纤维含量 Chemical properties: Residual monomer content, weathering resistance, swelling behavior in solvents, water absorption, composition, content of reinforcing fibers
后处理 Disposal	热解、燃烧、残渣、分解气体 Pyrolysis, combustion, residues, decomposition gases

2.5.2 表征热固性树脂的热分析技术
TA techniques for the characterization of thermosets

这里描述的分析热固性树脂的热分析方法包括差示扫描量热法(DSC)、热重分析(TGA)、热机械分析(TMA)和动态热机械分析(DMA)。

这些方法的应用可能性是各种各样的,所以只能给予简短的概述。

DSC用于测量样品发生物理转变和化学反应的温度及其焓变。从中可获得关于材料及其浓度、化学活性和物理结构的信息。例如,玻璃化转变的分析提供了关于固化度和所使用的聚合物类型和填料的信息。此外,可研究可能的后固化反应或者热稳定性。

而且,这些方法能表征催化剂和添加剂对固化过程的作用。

TGA经常用来分析热固性树脂的组分。例如,可以较容易地测量水分含量和填料或纤维含量。常常能够获得所用聚合物性质的信息。联用的逸出气体分析可得到所用阻燃剂类型的详细信息。也能研究不同气氛中材料的稳定性。

TMA用于测量样品的尺寸性能。经常感兴趣的性能是树脂或者复合材料的膨胀系数及其在玻璃化转变处的变化。也能测定机械应力对尺寸稳定性的影响。通过测定软化、分解或分层行为能研究材料的可使用温度范围。软化区域也表示玻璃化转变的温度范围。此外,能估算杨氏模量。TMA还能表征高聚物在不同溶剂中的溶胀行为。

DMA:当样品受到机械应力时,所施加的机械应力的性质起着决定性作用。应力发生的时间或频率(例如突然冲击或经过一段长时间的作用)是重要的因素。用DMA进行动态力学研究能计算当施加这样的

The thermoanalytical methods for the analysis of thermosets described here include differential scanning calorimetry (DSC), thermogravimetric analysis (TGA), thermomechanical analysis (TMA) and dynamic mechanical analysis (DMA).

The application possibilities of these methods are so diverse that only a brief overview can be presented.

DSC is used to measure the temperature at which physical transitions and chemical reactions take place and to determine the enthalpy involved in such processes. From this, information can be obtained on materials, their concentration, chemical reactivity and physical structure. For example, the analysis of glass transitions provides information on the degree of cure and the polymer type and fillers used. In addition, possible postcuring reactions or thermal stability can be investigated.

Furthermore, these methods allow catalysts and additives to be characterized with regard to their effect on the curing process.

TGA is often used to analyze the composition of thermosets. For example, it is relatively easy to measure the moisture content and the filler or fiber content. Very often information can be obtained that indicates the nature of the polymer used. Combined evolved gas analysis yields detailed information on the type of fire retardant used. The stability of materials in different atmospheres can also be investigated.

TMA measures the dimensional properties of samples. A property often of interest is the expansion coefficient of either the resin or the composite and the change at the glass transition. The dimensional stability after mechanical stress can also be determined. The usable temperature range of the material can be investigated by determining its softening, decomposition or delamination behavior. The softening region also indicates the temperature range of the glass transition. Furthermore, Young's modulus can be estimated. TMA also allows the swelling behavior in different solvents to be characterized.

DMA: When the sample is subjected to mechanical stress, the nature of the applied stress plays a decisive role. The time or frequency over which or at which the stress occurs (e. g. as a sudden impact or over a long time) is an important factor. Dynamic mechanical investigations with DMA allow differences in behavior to be evaluated when such stresses are applied. The

应力时行为的不同。能在大的频率和温度范围内测定阻尼行为和模量。通过研究非线性行为也能得到有关填料作用的信息。因为力学模量依赖于分子状态,所以能获得关于所用的聚合物、关于交联和关于聚合物共混物和添加剂的相容性的信息。能获得关于材料模量和阻尼行为的大小以及关于这些变化的温度范围(玻璃化转变)的数据。下表汇总了可用热分析进行研究的热固性树脂的某些效应和性能。取决于方法的相对重要性,用Ⅰ或Ⅱ来标记。

damping behavior and modulus can be determined over a wide frequency and temperature range. Information on the effect of fillers can also be obtained by investigating non-linear behavior. Since the mechanical modulus depends on molecular conditions, information can be gained about the polymers used, the crosslinking and the compatibility of polymer blends and additives. Data on the size of the modulus and the damping behavior of materials is obtained as well as on the temperature range of these changes (glass transition).

An overview of some of the effects and properties of thermosets that can be investigated using thermal analysis is given in the table below. The methods are marked with Ⅰ or Ⅱ depending on the relative importance of the method.

表 2.16 用四个最重要的热分析方法能研究的效应(Ⅰ主要方法,Ⅱ选择性方法)
Table 2.16 Effects that can be investigated with the four most important thermoanalytical methods
(Ⅰ Main method, Ⅱ alternative method).

	DSC	TGA	TMA	DMA
玻璃化转变 Glass transition	Ⅰ		Ⅰ	Ⅰ
黏弹性行为,弹性模量 Viscoelastic behavior, elastic modulus			Ⅱ	Ⅰ
阻尼行为 Damping behavior				Ⅰ
软化温度 Softening temperature			Ⅰ	Ⅱ
膨胀,热膨胀系数,收缩 Expansion, CTE, contraction			Ⅰ	
固化行为 Curing behavior	Ⅰ		Ⅱ	Ⅱ
凝胶化 Gelation			Ⅰ	Ⅰ
反应焓和动力学 Reaction enthalpy and kinetics	Ⅰ			
催化剂活性 Catalyst activity	Ⅰ			Ⅱ
组分 Composition	Ⅱ	Ⅰ		Ⅱ

	DSC	TGA	TMA	DMA
填料含量,增强纤维含量 Filler content, content of reinforcing fibers	Ⅱ	Ⅰ		
热稳定性/分解 Thermal stability/decomposition	Ⅱ	Ⅰ	Ⅱ	
氧化稳定性 Oxidation stability	Ⅰ	Ⅱ		
溶剂中溶胀 Swelling in solvent			Ⅰ	
蒸发/解吸附/汽化 Evaporation/desorption/vaporization	Ⅱ	Ⅰ		
熔融和结晶 Melting and crystallization	Ⅰ			Ⅱ

2.5.3 玻璃化转变 Glass transition

2.5.3.1 玻璃化转变和松弛:热学和动态玻璃化转变 Glass transition and relaxation:thermal and dynamic glass transition

玻璃化转变是表征热固性树脂的重要热效应。因此,玻璃化转变温度(T_g)的测定是热分析中最常用的方法之一。在玻璃化转变处,物理性能如比热容、热膨胀系数和力学模量显著变化。这意味着可用DSC、TMA和DMA技术来测量玻璃化转变。

降温时,在玻璃化转变区,材料从过冷液态或类橡胶态变化到玻璃态固体。升温时则以反方向发生玻璃化转变。在液体状态,分子能够相互相对运动,发生所谓的协同重排。重排中涉及的区域的体积为几个nm^3。在玻璃态,协同重排被冻结。

协同重排是指以一定的速率发生的松弛过程,因而具有特征频率。在较低温度下,重排频率降低,即重排变得较慢。

可用Deborah数(D)来表征时间或频率依赖的效应。D是协同重排的

The glass transition is an important thermal effect for the characterization of thermosets. The determination of the glass transition Temperature (T_g) is therefore one of the most frequent applications in thermal analysis. At the glass transition, physical properties such as the specific heat capacity, the coefficient of thermal expansion and the mechanical modulus change noticeably. This means that the glass transition can be measured using DSC, TMA and DMA techniques.

At the glass transition, on cooling, a material passes from a supercooled liquid or rubbery-like state to a glassy solid state. The glass transition also occurs in the reverse direction on heating. In the liquid state, molecules are able to move relative to one another. So-called cooperative rearrangements occur. The volume of the region involved in the rearrangement is several nm^3. In the glassy state, the cooperative rearrangements are frozen-in.

Cooperative rearrangements, a relaxation process, take place at a certain rate and hence have a characteristic frequency. The frequency of the rearrangements decreases at lower temperatures, that is, the rearrangements become slower.

The Deborah number (D) can be used to characterize time-or frequency-dependent events. D is the ratio of the characteristic

特征时间 T_a 和观察时间 T_b 之比，所以：

$$D = T_a / T_b.$$

T_a 在高温下变小，而 T_b 依赖于测试参数(降温速率、频率)。

对于 $D<1$，协同重排的特征时间比观察时间短，材料表现为液态或类橡胶态。在玻璃态，$D>1$。这时，协同重排如此之慢以致于在测试期间不发生重排，似乎处于被冻结状态。由此可见玻璃化转变与测试即观察条件有关。

可观察到两种玻璃化转变：

1. 玻璃化(反玻璃化)：降温时冻结协同重排，观察到从液态到玻璃态的转变(热学玻璃化转变)。观察时间决定于降温速率：与慢速降温相比，则快速降温观察时间短。因此，快速降温测试的玻璃化温度比慢速降温测试的高。

2. 动态玻璃化转变：当测试频率大约等于协同重排的特征频率时，则可观察到重排。换言之，如果在高频下分析材料(例如施以高频振动的机械应力)，则材料表现为硬的，因为协同重排跟不上测试频率。如果应力施加得较慢(低频)，材料在同样的温度下看起来是软的。由于施加振动的机械应力，所以称为动态力学玻璃化转变。此温度可能与动态热玻璃化转变(例如用 ADSC 测量的结果)不同。

动态玻璃化转变温度经常比玻璃化温度高。

在调制技术如 DMA(及 DLTMA 和 TMDSC)中，两个重叠的应力同时施加到样品上。一个是周期性成分；另一个是由基础升温或降温程序给予的。动态玻璃化转变由频率依赖性成分测试；而(反)玻璃化是由基础温度程序(线性升温或降温)测定的。

time of the cooperative rearrangement, T_a, and the observation time, T_b, so that:

$$D = T_a / T_b.$$

T_a becomes smaller at higher temperatures, whereas T_b depends on the measurement parameters (cooling rate, frequency).

For $D < 1$, the characteristic time of the cooperative rearrangement is shorter than the observation time. The material appears to be liquid or rubbery-like. In the glassy state, $D>1$. The cooperative rearrangements are so slow that they do not occur during the measurement – they appear to be frozen-in. These considerations show that the glass transition depends on the measurement or observation conditions.

Two kinds of glass transitions can be observed:

1. Vitrification (resp. devitrification): On lowering the temperature, the cooperative rearrangements freeze and one observes the transition from the liquid to the glassy state (thermal glass transition). The observation time is determined by the rate of cooling: at high cooling rates the observation time is shorter than at low rates. The glass transition temperature measured at high cooling rates is therefore higher than that measured at low cooling rates.

2. Dynamic glass transition: This is observed when the measurement frequency corresponds approximately to the characteristic frequency of the cooperative rearrangements. In other words, if a material is analyzed at high frequency (e. g. mechanically stressed), it appears hard because the cooperative rearrangements cannot follow the measuring frequency. If the stress is applied more slowly (at lower frequency), the material appears soft at the same temperature. With mechanical stress, one refers to the dynamic mechanical glass transition. This temperature can differ from the dynamic thermal glass transition (e. g. ADSC).

The dynamic glass transition temperature is often higher than that of vitrification.

In modulated techniques such as DMA (and DLTMA and TMDSC), two overlapping stresses are simultaneously applied to the sample. One is the periodic component; the other is given by the underlying heating or cooling program. The dynamic glass transition is measured by the frequency-dependent part, whereas the (de-) vitrification is determined by the underlying temperature program (linear heating or cooling).

玻璃化转变时分子的运动性显著增强,导致弹性模量(例如杨氏模量)的极大变化。从玻璃态变化到橡胶态时,聚合物的模量从约 1000 MPa 降低到约 1MPa,即降了三个数量级。巨大变化意味着玻璃化转变能高灵敏地测试。DMA 甚至能测试次级松弛过程和 c_p 或 CTE 变化非常小的玻璃化转变。

弯曲测试尤其适合于复合材料和自支撑体系。薄样品如薄膜或纤维通常用拉伸测试。对于反玻璃化后黏度很低的样品,实际上只能用剪切样品夹具测试。这是 DMA 剪切测试的一个重要优势。

由于频率范围宽(6 个数量级)和灵敏性,所以用 DMA 来测定玻璃化转变而获得分子运动性信息,从而可研究分子结构对力学性能的影响(力学谱)。

说明协同重排的一个例子:

协同重排可以通过思考一辆挤满了乘客的每站必停的公共汽车的情形来理解。在行使中,每一个人紧贴站立在一起,不能移动,即运动是被冻结的。如果某个提着大包的人必须在下一车站下车,则许多乘客都不得不一起移动(协同)才能使这成为可能。此外需要一点额外的自由空间(自由体积)来提高运动性——也许靠近车门的某个乘客必须下车。所有这些都需要一点时间;它是个动力学过程。公共汽车的乘客经历了一系列协同重排。

参考文献:

J. E. K. Schawe,接近热玻璃化转变的聚苯乙烯的热松弛的描述,J. of Polymer Science, Part B: Polymer Physics, Vol. 36, (1998) pp 2165-2175

At the glass transition, the mobility of molecules increases greatly leading to a large change in the modulus of elasticity (e. g. Young's modulus). The modulus of a polymer decreases from about 1000MPa to about 1MPa, i. e. by three decades, on passing from the glassy to the rubbery state. The large change means that the glass transition can be measured with great sensitivity. DMA can even measure secondary relaxation processes and glass transitions with very low changes in c_p or CTE.

Bending measurements are especially suitable for composites and self-supporting systems. Thin samples such as films or fibers are usually measured in tension. Samples that have a very low viscosity after devitrification can in practice only be measured in the shear sample holder. This is an important advantage of shear measurements in the DMA.

Due to the wide frequency range (6 decades) and sensitivity, DMA is used to determine the glass transition and obtain information on molecular mobility. This allows the influence of molecular structure on mechanical properties to be investigated (mechanical spectroscopy).

An example to illustrate cooperative rearrangements:

Cooperative rearrangements can be understood by considering the situation in a local bus crammed full of passengers. During the journey, everyone is standing close together and cannot move, i. e. movement is frozen. If someone carrying a large bag has to get out of the bus at the next bus stop, a number of passengers all have to move together (cooperatively), to make this possible. A little additional free space (free volume) is also needed to increase mobility-perhaps someone near the door has to get out. This all needs a little time; it is a kinetic process. The bus passengers experience a series of cooperative rearrangements.

Reference:

J. E. K. Schawe, Description of Thermal Relaxation of Polystyrene Close to the Thermal Glass Transition, J. of Polymer Science, Part B: Polymer Physics, Vol. 36, (1998) pp 2165-2175

2.5.3.2 玻璃化转变温度的测定 Determination of the glass transition temperature

兹就如何测定玻璃化转变温度的应用描述如下。

各种热分析技术 DSC、TMA 和 DMA 对玻璃化转变时物理性能的变化，呈现不同的灵敏度。表2.17 比较了这三种技术。中间一栏指出玻璃化转变时物理性能的大致变化。

The following applications describe how the glass transition temperature is determined.

The individual thermal analysis techniques, DSC, TMA and DMA show different sensitivities with respect to changes in physical properties during the glass transition. Table 2.17 compares the three techniques. The middle column indicates the approximate change in the physical property at the glass transition.

表 2.17 各种 TA 技术测定玻璃化转变时的灵敏度
Table 2.17 Overview of the sensitivity of individual TA techniques used to determine the glass transition

物理性能 Physical property	玻璃化转变时的变化 Change at the glass transition	技术 Technique
比热容 Specific heat capacity	5 到 30% 5 to 30%	DSC（TMDSC） DSC (TMDSC)
膨胀（CTE） Expansivity (CTE)	50 到 300% 50 to 300%	TMA（膨胀测量法） TMA (Dilatometry)
力学模量 Mechanical modulus	1 到 3 个数量级 1 to 3 decades	DMA（DLTMA） DMA (DLTMA)

除了上表，也可参考表 2.18 来帮助决定在特定情况下最好用哪种技术来测定玻璃化转变温度。

为了比较来自不同测试的结果，应该考虑表 2.19 中列出的影响因素。也可参见实例(3.5节)。

Besides the above table, Table 2.18 can also be consulted to help decide which technique is best to use for the determination of the glass transition temperature in a particular situation.

To compare the results from different measurements, the influences given in Table 2.19 should be taken into account. See also the example (Section 3.5).

表 2.18 玻璃化转变热分析测定法的优缺点比较
Table 2.18 Advantages and disadvantages of the TA techniques used to determine the glass transition

技术 Technique	优点 Advantages	缺点 Disadvantages	要点 Important points
DSC	使用方便；标准方法；也适合于低黏度材料例如溶剂；不需要特殊的样品制备。 Ease of use; standard methods; also suitable for low viscosity materials e.g. solvents; no special sample preparation.	用低升/降温速率，样品热流曲线上只测试到一个小的台阶。 With filled samples and low heating/cooling rates, only a small step in the heat flow curve is measured.	传感器和样品之间需要良好的热传递，由于其他效应例如松弛可能发生重叠。 Good heat transfer required between sensor and sample, possible overlapping due to other effects, e.g. relaxation.

续表

技术 Technique	优势 Advantages	劣势 Disadvantages	重要点 Important points
TM DSC (ADSC, IsoStep, TOPEM®)	分离不可逆效应;能研究频率的影响;不需要特殊的样品制备。 Separation of non-reversing effects; Influence of frequency can be investigated; No special sample preparation is necessary.	由于所用的升温速率低,通常测试时间较长。 Measurements usually take longer due to the low heating rates used.	从传感器到样品需要非常好的热传递。 Very good heat transfer from sensor to sample is required.
TMA	样品操作容易;即使低升温速率给出良好的结果;TMA能测试非常薄的膜(软化/穿透);标准方法。 Easy sample handling; Even low heating rates give good results; TMA can measure very thin films (softening/penetration); Standard methods.	T_g 以上,黏度必须足够大;对于填充材料只测试到小的变化;由于松弛效应会发生重叠。 Above the T_g, the viscosity must be sufficiently large; Only small changes are measured with filled materials; Overlapping due to relaxation effects.	样品上测试探头的压力;测定CTE要求两个平行的表面。 Pressure of the measurement probe on the sample; Two parallel surfaces are required to determine the CTE.
DLTMA	杨氏模量测定;高填充材料(弯曲测试)。 Young's modulus determination; Highly filled materials (bending measurement).	对于刚硬材料模量值经常太低。 Modulus values are often too low with stiff materials.	样品制备;可能用弯曲模式。 Sample preparation; Possibly bending measurements.
TGA/DSC SDTA	高温;使用TGA和TMA技术时的附加信息。 High temperatures; Additional information using TGA and TMA techniques.	比DSC灵敏度低;需要较多样品。 Less sensitive than DSC; More sample is needed.	样品中温度梯度。 Temperature gradients in the sample.
DMA	对于玻璃化转变灵敏度高;准确的模量值;标准方法;频率范围宽;测试力学性能。 High sensitivity for glass transitions; Accurate modulus values; Standard methods; Wide frequency range; Measurement of mechanical properties.	相对长的测试时间,由于升温速率低。 Relatively long measurement time due to the low heating rate.	选择合适的样品几何形状和夹具;样品制备时间长。 Proper choice of sample geometry and holder; Sample preparation can be lengthy.

表 2.19 影响玻璃化转变温度（T_g）的测定因素
Table 2.19 Influences on the glass transition temperature（T_g）

随着下列量（例如频率）的增加，T_g 增大（↗）、保持相同（＝）或下降（↘）。
T_g increases（↗）, remains the same（＝）, or decreases（↘） with an increase of the following quantities（e. g. frequency）.

增加 Increase of	T_g	T_g 变化的经验规则 Rule of thumb for the change of T_g
升温速率（≥降温速率） Heating rate（≥ cooling rate）	↗	4K/每数量级 4 K/decade
样品热阻 Thermal resistance in the sample	台阶变宽 Step becomes broader ↗	样品制备依赖性强 Depends strongly on sample preparation
样品上 TMA 探头的压力 TMA probe pressure on the sample	＝	用起始点（onset）作为 T_g Use the onset for T_g
DLTMA 周期（频率降低） DLTMA period（frequency decreases）	↘	5K/每频率数量级 5 K/frequency decade
ADSC 周期（频率降低） ADSC period（frequency decreases）	↘	4K/每频率数量级 4 K/frequency decade
DMA 振幅 DMA amplitude	＝	
DMA 频率 DMA frequency	↗	5 到 10K/每频率数量级 5 to 10K/frequency decade

2.5.4 热固性树脂分析的标准方法 Standard methods for thermoset analysis

对于热固性树脂（和通常对于聚合物）的大量标准方法和测试步骤，本书无法全面总结。现只列出许多国际标准作为例子（有些是德语的）。

The large number of standard methods and test procedures used for thermosets（and polymers in general）makes it impossible to give a comprehensive overview in this booklet. Instead, a number of international standards are listed below as examples（some are in German）.

组织 Organizations：

ASTM	美国材料试验协会 American Society for Testing and Materials	www. astm. org
ISO	国际标准化组织 International Organization for Standardization	www. iso. ch
EN	CEN，欧洲标准化委员会 CEN, European Committee for Standardization	www. cenorm. be

续表

CSA	加拿大标准协会 Canadian Standards Association	www.csa.ca
DIN	德国标准协会 Deutsches Institut für Normung	www.din.de
JIS	日本工业标准 Japanese Industrial Standards	www.jsa.or.jp
CEI/IEC	CEI/IEC:国际电工委员会 CEI/IEC: The International Electrotechnical Commission	www.iec.ch
IPC	IPC-美国电子工业协会 IPC-Association Connecting Electronics Industries	www.ipc.org

标准 Standard	标题 Title	简述 Short description
ASTM D 696	塑料的线性热膨胀系数 Coefficient of Linear Thermal Expansion of Plastics	膨胀测试。TMA 使用较短的样品。 Dilatometric determination. TMA uses shorter specimen.
ASTM D 3386	电气绝缘材料的线性热膨胀系数的标准测试方法 Standard test method for coefficient of linear thermal expansion of electrical insulating materials	TMA,用高度2.5到7.5mm样品,热电偶尽可能靠近样品。 TMA using specimen height of 2.5 to 7.5 mm and thermocouple as close to the sample as possible.
ASTM D 3417	用热分析测定聚合物的熔融热和结晶热 Heat of fusion and crystallization of polymers by thermal analysis	在氮气中用10K/min的升温和降温速率的DSC测试。 DSC with heating and cooling rates of 10 K/min in nitrogen.
ASTM D 3418	用热分析测定聚合物的转变温度。 Transition temperature of polymers by thermal analysis	DSC测量:以10K/min在氮气中的熔融和结晶;以20K/min在空气或氮气中骤冷样品的玻璃化转变。 DSC measurements: melting and crystallization at 10 K/min in nitrogen; Glass transition of shock-cooled samples at 20 K/min in air or nitrogen.
ASTM D 3850	热重方法测定固体电子绝缘材料的快速降解。 Rapid degradation of solid electrical insulation materials by thermogravimetric method	测定失重为10、20、30、50和75%时的温度。 Temperatures at mass losses of 10, 20, 30, 50 and 75% are determined.
ASTM D 4092	塑料的动态热机械测量 Dynamic mechanical measurements on plastics	技术术语的定义和描述。 Definitions and descriptions of technical terms.
ASTM D 4473	塑料标准测试方法: 动态热机械性能: 固化行为 Standard test method for plastics: dynamic mechanical properties: cure behavior	热固性无支撑树脂和基板支撑树脂在各种振动形变(剪切、压缩)作用下:模量、tanδ、0.01至100Hz的阻尼因子。 Thermosetting unsupported resins and resins supported on substrates are subjected to various oscillatory deformations (shear, compression): modulus, tanδ, damping factor for 0.01 to 100 Hz.

续表

标准 Standard	标题 Title	简述 Short description
ASTM D 5023	用三点弯曲测试塑料的动态力学性能 Measuring the dynamic mechanical properties of plastics using three-point bending	热固性树脂和复合材料体系；弹性（贮存）、损耗（黏性）和复合模量；tanδ与频率、时间或温度的关系；0.01到100Hz；样品几何形状：64×13×3 mm。 Thermosetting resins and composite systems; elastic (storage), loss (viscous) and complex moduli; tanδ as a function of frequency, time, or temperature; 0.01 to 100 Hz; Sample geometry: 64×13×3 mm.
ASTM D 5024	压缩测试塑料的动态力学性能 Measuring the dynamic mechanical properties of plastics in compression	见D5023 样品几何形状：25×5 mm，圆柱状。 see D5023 Sample geometry: 25×5 mm cylindrical.
ASTM D 5026	拉伸测试塑料的动态力学性能 Measuring the dynamic mechanical properties of plastics in tension	见D5023 样品几何形状：76×13×1mm。 see D5023 Sample geometry: 76×13×1 mm.
ASTM D 5028	用热分析测试拉挤树脂的固化性能 Curing properties of pultrusion resins by thermal analysis	测定固化参数。 Determination of curing parameters.
ASTM D 6090	树脂软化点（梅特勒杯球法） Softening point resins (METTLER cup and ball method)	用FP83HT带8.7mm钢球的6.35mm孔样品杯 Sample cup with 6.35 mm hole with steel ball 8.7 mm using FP83HT.
ASTM E 698	热不稳定材料的Arrhenius动力学常数 Arrhenius kinetic constant for thermally unstable materials	基于不同升温速率的三次或三次以上DSC测试：峰温。 DSC, based on three or more measurements at different heating rates: peak temperatures.
ASTM E 831	热机械分析固体材料的线性热膨胀 Linear thermal expansion of solid materials by thermomechanical analysis	直径为2到5mm和厚度为2到10mm样品的TMA测试。 TMA measurements of samples with a diameter of 2 to 5 mm and a thickness of 2 to 10 mm.
ASTM E 1131	组分分析的热重分析法 Compositional analysis by thermogravimetry	组分测定：高挥发性的、中等挥发性的、分解产物、灰分。 Determination of components: highly volatile, medium volatile, decomposition products, ash.
ASTM E 1269	比热容的DSC测定法 Determining specific heat capacity by DSC	根据熟知的蓝宝石比热容。 Based on known specific heat capacity of sapphire.

续表

标准 Standard	标题 Title	简述 Short description
ASTM E 1356	玻璃化转变的 DSC 和 DTA 测量 Glass transition temperature by DSC or DTA	一条曲线上测试 6 个温度。 6 temperatures are determined from one curve.
ASTM E 1545	玻璃化转变温度的热机械分析（TMA）测量 Assignment of the glass transition temperature by thermomechanical analysis (TMA)	用直径 4 到 6mm 探头和 0 到 5mN 力的膨胀模式；用直径 2 到 4mm 探头和 20 到 50mN 力的穿透模式 Expansion mode with a probe of 4 to 6 mm diameter and a force of 0 to 5 mN; Penetration mode with a probe with 2 to 4 mm diameter and a force of 20 to 50 mN.
ASTM E 1640	玻璃化转变温度的动态热机械分析（DMA）测量 Assignment of the glass transition temperature by dynamic mechanical analysis (DMA)	$\tan\delta$、损耗模量的峰温或储存模量的起始点（对数或线性刻度） Peak temperature of $\tan\delta$, loss modulus or onset on storage modulus (log or linear scaled).
ASTM E 1641	热重法分解动力学 Decomposition kinetics by thermogravimetry	需要 4 次或 4 次以上的热重测试。测定 5、10、15、20％时的温度。对于模拟和预测，设定反应级数为 1。 4 or more thermogravimetric measurements are needed. Determination of the temperatures at 5, 10, 15 and 20％. For simulations and predictions, the reaction order is set to one.
ASTM E 1824	玻璃化转变温度的 TMA 拉伸测量 Assignment of a glass transition temperature using TMA under tension	薄膜和纤维。T_g 用切线交点确定。 Films and fibers. T_g is given by the tangents intercept.
ASTM E 1868	干燥失重的热重分析标准测试方法 Standard test method for loss-on-drying by thermogravimetry	测定在特定的温度和时间条件下从测试样品中逸出的任何种挥发性物质的量 The amount of volatile matter of any kind that is evolved from a test specimen under a specific set of temperature and time conditions is determined.
ASTM E 1877	从热重分解数据计算材料的耐热性 Calculating thermal endurance of materials from thermogravimetric decomposition data	为求出热稳定曲线和得到材料的相对热指数而对测试方法 E1641 所测定的 Arrhenius 活化能数据的附加处理。Additional treatment of the Arrhenius activation energy data determined by test method E1641 to develop a thermal endurance curve and derive a relative thermal index for materials.
ASTM E 2041	Borchard 和 Daniels 方法的 DSC 动力学参数 Kinetic parameters by DSC using Borchard and Daniels method	基于一次测试的 n 级动力学；方法 A 假定 $n=1$，方法 B 用多次线性回归。 n^{th} order kinetics based on one measurement; method A assumes $n=1$, method B uses multiple linear regression.

续表

标准 Standard	标题 Title	简述 Short description
ASTM E 2070	用等温方法的DSC动力学参数标准测试方法 Standard test method for kinetic parameters by DSC using isothermal methods	方法A在一个小的的温度范围(~10K)内从一系列等温实验测定活化能、指前因子和反应级数等动力学参数。 测试方法B还测定一组时间效应和等温数据的活化能。 Method A determines kinetic parameters for activation energy, pre-exponential factor and reaction order from a series of isothermal experiments over a small (~10K) temperature range. Test method B also determines the activation energy of a set of time-to-event and isothermal temperature data.
ASTM E 2160	热反应性材料反应热的DSC标准测试方法 Standard test method for heat of reaction of thermally reactive materials by DSC	反应性材料可包括热不稳定或热固性材料。测定放热反应的外推起始温度和热流峰温。 Reactive materials may include thermally unstable or thermosetting materials. The extrapolated onset temperature and peak heat flow temperature for the exothermic reaction are determined.
ASTM E 2347	压痕软化温度的TMA标准测试方法 Standard test method for indentation softening temperature by TMA	用压入(1mm^{2}探头)测试样品比模量6.65MPa(方法A)或33.3MPa(方法B)(等同于测试方法D 1525)时的温度。 Temperature at which the specific modulus of either 6.65 (Method A) or 33.3 MPa (Method B) (equivalent to test method D 1525) of a test specimen is realized by indentation (1 mm^2 probe).
CSA Z245.20	钢管的外部熔结环氧涂层 External fusion bond epoxy coating for steel pipe	T_g(起始点)和后固化,20K/min;工厂应用熔结环氧(FBE)涂层的规格 T_g (onset) and postcuring, 20K/min; qualification of plant-applied fusion bond epoxy (FBE) coating.
DIN 16916	反应树脂、酚醛树脂;术语、分类、测试方法 Reaction resins, phenolic resins; terms, classification, test procedures	例如凝胶时间、游离酚和非挥发性含量及水稀释剂的测定。 For example, determination of the gelation time, of free phenol and nonvolatile content, and water dilution.
DIN 51007	热分析;差热分析;原理 Thermal Analysis; Differential Thermal Analysis; Principles	用DSC测定玻璃化转变温度、比热容和一级转变. Determination of glass transition temperature, specific heat capacity and first order transitions by DSC.

续表

标准 Standard	标题 Title	简述 Short description
DIN 53752	塑料测试；线性热膨胀系数的测定 Plastics testing; determination of the coefficients of linear thermal expansion	用石英玻璃管膨胀计，样品长度20到35mm。 A quartz glass tube dilatometer is used, samples lengths of 20 to 35 mm.
DIN 53765	塑料和弹性体的测试；热分析，DSC方法 Testing of plastics and elastomers; thermal analysis, DSC method	计算：升温或等温下的比热、玻璃化转变、熔融、结晶、化学反应。 Evaluation: specific heat, glass transition, melting, crystallization, chemical reaction by heating or isothermally.
DIN 65467	航空和航天；含增强或不含增强物的有机聚合物材料的测试；DSC方法 Aerospace; testing of organic polymer materials with and without reinforcement; DSC methods	测定玻璃化转变、熔融或固化反应。 Determination of glass transition, melting or curing reactions.
DIN 65583	航空和航天；纤维增强塑料；动态负载下纤维复合材料玻璃化转变的测定 Aerospace; Fiber reinforced plastics; Determination of the glass transition of fiber composites under dynamic load	通过弯曲（DMA）用贮存模量测定 T_g。3K/min A1：单悬臂 $30\times10\times2$ mm A2：3点弯曲 $50\times10\times2$ mm A3：双悬臂 $50\times10\times2$ mm T_g=外推起始点或2%时的温度 用153.4℃（$\tan\delta_{max}$）的聚碳酸酯校准温度 T_g determination using storage modulus by bending (DMA). 3 K/min A1: single cantilever $30\times10\times2$ mm A2: 3-point-bending $50\times10\times2$ mm A3: dual cantilever $50\times10\times2$ mm T_g = extrapolated onset or temperature at 2% Temperature calibration with polycarbonate at 153.4℃ ($\tan\delta_{max}$).
IEC 1006	测定电气绝缘材料玻璃化转变温度的测试方法 Test methods for the determination of glass transition temperature of electrical insulating materials	方法C：动态热机械分析。 Method C: dynamic mechanical analysis.
IPC_TM_650 No 2.4.24	用TMA测定玻璃化转变温度和Z-轴热膨胀 Glass transition temperature and Z-axis thermal expansion by TMA	用于印制电路板中测定介电材料的 T_g 和 T_g 前后的CTE。 T_g and CTE is determined before and after T_g of dielectric materials used in printed circuit boards.
IPC_TM_650 No 2.4.24.1	（层压板和印制电路板的）分层时间（TMA方法） Time to delamination (TMA method) (for laminates and printed circuit boards)	以10K/min从35升温到260℃，保持10min或直到破坏（不可逆效应）。 Heating from 35 to 260℃ at 10 K/min and holding 10 min or to failure (irreversible event).

续表

标准 Standard	标题 Title	简述 Short description
IPC_TM_650 No 2.4.25	用DSC测定预浸料、层压板和印制电路板的玻璃化转变温度和固化因子 Glass transition temperature and cure factor by DSC for prepreg, laminate and printed boards	半台阶高度测定为T_g,加热速率20K/min。固化因子是第一次和第二次加热的T_g之差。 The T_g is determined as the half step height, heating at 20 K/min. The cure factor is the difference between the T_g's of the first and second heating runs.
ISO 1043	塑料—符号和缩略术语 Plastics-Symbols and abbreviated terms	提供已确定聚合物的缩略术语和表示组分和专门特性的符号。 Provides abbreviated terms for established polymers and symbols for components and special characteristics.
ISO 4899	玻璃纤维增强反应树脂模塑料;测试方法 Glass-fiber-reinforced reaction resin molding compounds; test procedures	如DIN 16946。 as DIN 16946.
ISO 6721	塑料—动态力学性能的测定 Plastics-Determination of dynamic mechanical properties	用DMA弯曲(第5部分)、剪切(第6部分)和拉伸模式(第4部分)。 Using DMA in flexural (part 5), shear (part 6) and tensile mode (part 4).
ISO 11357	塑料—差示扫描热法(DSC) Plastics-Differential scanning calorimetry (DSC)	玻璃化转变(第2部分)、比热容(第4部分)、聚合动力学(第5部分)。 Glass transition (part 2), specific heat capacity (part 4), polymerization kinetics (part 5).
ISO 11358	塑料—热重分析法 Plastics-Thermogravimetry	第2部分:TGA动力学。 Part 2: TGA kinetics.
ISO 11359	塑料—热机械分析(TMA) Plastics-Thermomechanical analysis (TMA)	第2部分:线膨胀系数和玻璃化转变温度的测定。 第3部分:穿透温度的测定。 Part 2: determination of the linear expansion coefficient and the glass transition temperature. Part 3: determination of the penetration temperature.
ISO 11409	塑料—酚醛树脂—用DSC测定反应热和温度 Plastics-Phenolic resins-Determination of heats and temperatures of reaction by DSC	钢坩埚,2MPa,反应热,峰温。 Steel pan, 2 MPa, heat of reaction, peak temperature.
JIS K 6911	热固性塑料的测试方法 Test method for thermosetting plastics	此日本工业标准详细说明了如硬度、强度、磨损、沸水、紫外光照射的一般方法。 This Japanese industrial standard specifies general methods such as hardness, strength, abrasion, boiling water, UV light exposure.

标准 Standard	标题 Title	简述 Short description
JIS K 7197	塑料线性热膨胀系数的 TMA 测试方法 Testing Methods for linear thermal expansion coefficient of plastics by TMA	以最大 5K/min 测试的 10mm 样品的平均膨胀系数。 Mean expansion coefficient of a 10 mm specimen, measured at maximum of 5 K/min.

3 热固性树脂的基本热效应
Basic thermal effects of thermosets

本节讨论热固性树脂及其原料由热分析观察到的基本热效应。讨论了许多不同的评价方法。

测试是用梅特勒-托利多的 STAR^e 系统和合适的测试仪器完成的。测试和计算曲线一般对程序温度作图。除非另有说明,温度计算值均系样品温度。

这里所呈示的测试提供了关于热固性树脂通常进行的不同类型热分析测定的概况。选择 KU600 粉末涂料(含有基本催化剂的环氧树脂,Ciba-Geigy)作为实例来演示热效应和过程。为了便于各种技术之间的比较,大多数测试是用该材料进行的。

This section discusses the basic effects observed in the thermal analysis of thermosets and their starting materials. A number of different evaluation possibilities are discussed.

The measurements were carried out using the METTLER TOLEDO STAR^e System and the appropriate measuring cells. In general, the measured and calculated curves were plotted against the program temperature. The temperatures resulting from evaluations are based on the sample temperature unless otherwise stated.

The measurements presented here provide an overview of the different types of thermal analysis determinations normally performed on thermosets. KU600 powder coating (epoxy resin with basic catalyst, Ciba-Geigy) was chosen as an example to demonstrate thermal effects and processes. Most measurements were performed with this material in order to facilitate comparisons between individual techniques.

3.1 热效应的 DSC 测量   Measurement effects with DSC

DSC 用来快速测定下列重要的热性能和效应。该技术只需少量样品,典型的仅为几毫克:
- 玻璃化转变温度和比热容曲线
- 固化度
- 固化温度和固化速率(动力学)
- 反应焓
- 后固化
- 添加剂对所述热效应的影响

对于质量控制 DSC 是非常重要的技术,所以许多测试在标准方法中均有所描述。

这里所描述的测试是用 FRS5 传感器进行的。同样也可用 HSS7 传感器完成;有些测试参数必须修改。

DSC is used to rapidly determine the following important thermal properties and events. The technique requires only small sample quantities, typically just a few milligrams:
- Glass transition temperature and specific heat capacity curve
- Degree of cure
- Curing temperature and rate of curing (kinetics)
- Reaction enthalpy
- Postcuring
- Influence of additives on the effects mentioned

Because of the great importance of DSC in quality control, many of these determinations are described in standard methods.

The measurements described here were performed using an FRS5 sensor. They can also be done equally well with the HSS7 sensor; some measurement parameters might have to be modified.

3.1.1 玻璃化转变的测定   Determination of the glass transition

降温时由于协同重排的自由度降低,协同重排被冻结导致较低的比

The freezing of the cooperative rearrangements on cooling results in a lower specific heat capacity, c_p, due to a reduction in the

热容 c_p。因为重排并非在某一精确的温度点冻结,所以比热容的变化发生在某一温度范围。在玻璃态,塑料具有约为 1.5J/gK 的 c_p 值。在玻璃化转变(反玻璃化)处,比热容提高约 0.4J/gK。该变化是相对小的,对于填充材料或仅是部分无定形的材料甚至更小。因此,DSC 测试曲线在玻璃化转变处只显示一个小的变化。然而 DSC 是最常用来测定玻璃化转变的方法,因为对大多数测试来说灵敏度业已足够,而且样品制备容易。关于玻璃化转变性质的进一步讨论可参阅 2.5.3 节"玻璃化转变"。

在玻璃化转变处,比热容发生变化。这引起热流显示一个台阶式增大,产生一个 S 形的 DSC 曲线。描述由该曲线测定特征玻璃化转变温度 T_g 的几个不同的标准方法,为诸如 ASTM E1356、CEI-IEC 1006、DIN 53765、IPC-TM 650、ISO 11357。定义测试和评价参数,以便其结果的比较有意义。

degrees of freedom of the cooperative rearrangements. Since the rearrangements do not freeze at a precise temperature, the change in heat capacity occurs over a temperature range. In the glassy state, plastics have a c_p value of about 1.5 J/gK. At the glass transition (devitrification) the specific heat capacity increases by about 0.4 J/gK. This change is relatively small and is even smaller with filled materials or materials that are only partly amorphous. The DSC measurement curve therefore shows only a small change at the glass transition. DSC, however, is the method most frequently used to determine the glass transition because the sensitivity is perfectly adequate for most determinations and sample preparation is easy.

A further discussion on the nature of the glass transition can be found in Section 2.5.3 "The glass transition".

At the glass transition, the specific heat capacity changes. This causes the heat flow to show a stepwise increase, resulting in an S-shaped DSC curve. Several different standard methods are described to determine a characteristic glass transition temperature T_g from this curve, for example ASTM E1356, CEI-IEC 1006, DIN 53765, IPC-TM 650, ISO 11357. The measurement and evaluation parameters are defined so that the results can be meaningfully compared.

3.1.1.1 玻璃化转变温度的 DSC 测量
Measurement of the glass transition temperature by DSC

目的 **Purpose**	测试固化热固性树脂的玻璃化转变,因此测定玻璃化转变温度的特征值 T_g 和比热容的变化。	To measure the glass transition of a cured thermoset and so determine a characteristic value for the glass transition temperature, T_g, and the change in the specific heat capacity.
样品 **Sample**	固化的 KU600 环氧树脂粉末。	Cured KU600 epoxy resin powder.
条件 **Conditions**	测试仪器:DSC 坩埚: 40μL 铝坩埚,盖钻孔 样品制备: 约 15mg 细粉末称重后装入坩埚。以 10K/min 的升温速率从 25℃ 至 260℃ 加热 5 遍,使粉末完全固化。冷却段不控制。 DSC 测试: 以 10K/min 从 25℃ 升温至 170℃。	Measuring cell:DSC Pan: Aluminum 40μL, with pierced lid Sample preparation: Approx. 15 mg fine powder was weighed into the crucible. The powder was completely cured by heating it 5 times from 25℃ to 260℃ at a heating rate of 10 K/min. The cooling segment was not controlled. DSC measurement: Heating from 25℃ to 170℃ at 10 K/min

气氛：氮气，50 mL/min　　　　　　　　　　Atmosphere：Nitrogen，50 mL/min

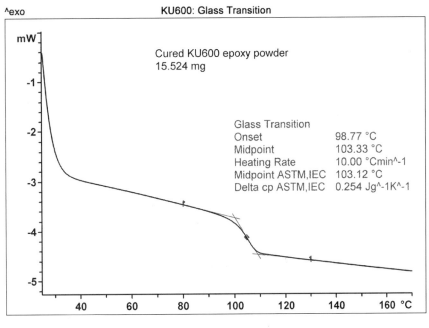

图 3.1　Fig. 3.1

解释　非晶态材料在冷却到足够低的温度时显示玻璃化转变温度。在玻璃化转变温度以上，材料基本上是黏性液体。交联也不改变这种情形。不过，这种限制流动行为的程度，使材料保持着它的形状。在这个温度范围内，材料显示类橡胶的行为，表现液体特有的协同重排。在玻璃化转变温度以下，协同的分子重排被"冻结"。材料处于玻璃态。在玻璃化转变处，分子运动性的变化产生热容曲线上的台阶。

玻璃化转变处曲线的实际形状依赖于样品的热历史和机械历史而变化。如果材料在玻璃化转变温度以下贮存、经历机械应力或被慢慢冷却到玻璃态，由于焓松弛，升温时在玻璃化转变台阶附近经常观察到一个峰。

计算　在玻璃化转变计算中，最重要的量是玻璃化转变温度 T_g 和在玻璃化转变处的热容变化 Δc_p。这些量是通过在热流曲线上高于和低于玻璃化转变的温度处画切线来确

Interpretation　Amorphous materials exhibit a glass transition after cooling to a sufficiently low temperature. Above the glass transition, the material is a more or less viscous liquid. Crosslinking does not change this situation either. However, it restricts flow behavior to such an extent that the material retains its shape. The material exhibits rubbery-like behavior and performs liquid-specific cooperative rearrangements in this temperature range. Below the glass transition temperature, cooperative molecular rearrangements are "frozen". The material is in a glassy state.

The change in molecular mobility causes a step in the heat capacity curve at the glass transition.

The actual shape of the curve at the glass transition varies depending on the thermal and mechanical history of the sample. If the material is stored below the glass transition temperature, undergoes mechanical stress, or is cooled slowly to the glassy state, a peak is often observed next to the glass transition step on heating. The peak is due to enthalpy relaxation.

Evaluation　The most important quantities in the evaluation of the glass transition are the temperature of the glass transition, T_g, and the change in the heat capacity at the glass transition, Δc_p. These quantities are determined by drawing tangents on the heat flow curve at temperatures above and below the glass

定的。必须仔细画切线,因为它们影响计算的质量和重复性。

玻璃化转变温度是在测试曲线上的台阶的特征温度。它可以按照不同的标准方法来确定,例如采用台阶半高处的温度或切线之间角平分线与测试曲线的交点(见下节)来确定。

在上图中,固化的 KU600 环氧树脂的玻璃化转变温度是使用 ASTM 标准方法测定的。得到的值 103.1℃ 与角平分线交点(中点)测定的 T_g 几乎完全相同。

用 ASTM 标准测定的玻璃化转变处的热容变化 Δc_p 是 0.254 $Jg^{-1}K^{-1}$。与无填充聚合物的典型台阶变化(约 0.4 $Jg^{-1}K^{-1}$)相比,其值相对较小,表明在环氧树脂体系中填料的存在。填料含量能用 TGA 测定(39% 填料,见 3.2 节),如果已知未填充材料的值,也可通过玻璃化转变时 Δc_p 值的准确测量来测定。

transition. The tangents must be carefully placed because they affect the quality and reproducibility of the evaluation.

The glass transition temperature is a characteristic temperature for the step in the measurement curve. It can be determined according to various standard methods, for example as the temperature at half the step height or as the point of intersection of the bisector of the angle between the tangents with the measurement curve (see next section).

In the above diagram, the glass transition temperature of cured KU600 epoxy resin was determined using the ASTM standard method. The value of 103.1℃ obtained is almost exactly the same as the T_g determined from the point of intersection of the angle bisector (midpoint).

The change in the heat capacity at the glass transition, Δc_p, was determined to be 0.254 $Jg^{-1}K^{-1}$ using the ASTM standard. The relatively small value compared to the typical step change of an unfilled polymer (approx. 0.4 $Jg^{-1}K^{-1}$ at 100℃) indicates the presence of filler in the epoxy system. The filler content can be determined by TGA (39% filler, see Section 3.2) or by accurate Δc_p measurement during the glass transition if the value for the unfilled material is known.

结论 用 DSC 能快速测试和计算玻璃化转变。该材料性能对于质量保证和新产品开发是非常重要的。

Conclusiong The glass transition can be rapidly measured and evaluated by DSC. This material property is very important for quality assurance and the development of new products.

3.1.1.2 用 DSC 计算玻璃化转变的方法
Evaluation possibilities for the glass transition by DSC

目的	研究贮存对玻璃化转变的影响和描述计算玻璃化转变的各种方法。
Purpose	To study storage effects on the glass transition and describe the various methods for evaluating the glass transition.
样品	在室温下贮存了几年的未固化的 KU600 环氧树脂粉末(原始样品)。
Sample	Uncured KU600 epoxy resin powder stored for several years at room temperature (original sample).
条件	测试仪器:DSC
Conditions	**Measuring cell**:DSC
	坩埚:40μL 铝坩埚
	Pan:Aluminum 40μL
	样品制备:
	Sample preparation:
	约 15mg 细粉末称重后放入坩埚。样品以 10K/min 的升温速率从 30℃ 升温到 90℃。在 90℃,用自动进样器在几秒内取出样品
	Approx. 15 mg fine powder was weighed into the crucible. The sample was heated from 30 to 90℃ at a heating rate of 10 K/min. At 90℃, the sample changer removed the sample within a

放到冷盘上。产生平均降温速率约 2000K/min 的可重复骤冷。

骤冷后的样品或者被立即测试，或者在以 10K/min 从 30℃ 至 90℃ 测试前在室温（RT）下退火 2 天。

DSC 测试：

以 10K/min 从 30℃ 升温至 90℃

气氛：氮气，50 mL/min

few seconds and placed the crucible on the cold tray. This resulted in reproducible shock cooling at a mean cooling rate of about 2000 K/min. The shock-cooled samples were either immediately measured or first annealed at room temperature (RT) for 2 days before measuring from 30 to 90℃ at 10 K/min.

DSC measurement：

Heating from 30℃ to 90℃ at 10K/min

Atmosphere：Nitrogen，50 mL/min

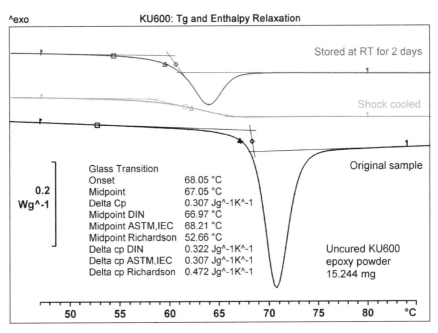

图 3.2　Fig. 3.2

图 3.2 所示为三条 DSC 升温曲线：原始样品曲线、骤冷后立即测试的第二次升温、最后为骤冷后在室温（RT）下贮存两天的样品的升温曲线。计算结果汇总在表 3.1 中。

Figure 3.2 shows three DSC heating curves: the curve of the original sample, the second heating run recorded immediately after shock cooling, and finally the heating curve of a sample that had been shock cooled and then stored at room temperature (RT) for two days. The evaluation results are summarized in Table 3.1.

表 3.1　按照不同方法计算的中点温度（T_g）和 Δc_p。

Table 3.1　Midpoint temperatures (T_g) and Δc_p according to different evaluation methods.

方法 Method		原始样品 Original sample	骤冷后样品 Shock cooled	2 天后样品 After 2 days	
STAR^e	T_g	67.1	62.1	59.5	℃
	Δc_p	0.307	0.250	0.240	$Jg^{-1}K^{-1}$
DIN	T_g	67.0	62.0	59.5	℃
	Δc_p	0.322	0.251	0.248	$Jg^{-1}K^{-1}$
ASTM，IEC	T_g	68.2	62.0	60.0	℃
	Δc_p	0.307	0.239	0.237	$Jg^{-1}K^{-1}$
Richardson	T_g	52.7	61.3	53.0	℃
	Δc_p	0.472	0.240	0.255	$Jg^{-1}K^{-1}$

解释 玻璃化转变曲线的形状与样品的热历史有关。如果样品的升温速率大于样品曾被冷却的速率,则玻璃化转变就会伴随一个焓松弛峰。在玻璃化转变温度范围内或以下温度贮存的样品也会产生焓松弛峰。该效应可在图 3.2 中的黑色曲线和红色曲线上观察到。该图也显示峰的大小随着室温下贮存时间的增加而增大。吸热峰正好等于由于在玻璃化转变下贮存或退火时的松弛而产生的焓的降低。

玻璃化转变的特征量是玻璃化转变温度 T_g 和玻璃化转变处的台阶高度 Δc_p。在 STAR^e 软件中提供了各种标准方法。除了 STAR^e 方法本身,包括 DIN 51007(DIN)、ASTM E 1356 和 IEC 1006 (ASTM, IEC) 的计算方法以及按照 Richardson 测定玻璃化转变假想温度的方法(在 DIN 51007 也有描述)。

如从表中可看到的,得到的玻璃化转变温度值和台阶高度与测定所使用的方法有关。如果玻璃化转变没有焓松弛峰,用 STAR^e、ASTM 和 DIN 方法测定的玻璃化转变温度几乎是相同的。用 Richardson 方法测定的值略低。玻璃化转变处台阶高度的比较显示,c_p 值能分为两组(STAR^e 和 DIN 方法:0.25 Jg^{-1}K^{-1};ASTM 和 Richardson 方法:0.24 Jg^{-1}K^{-1})。

如果出现焓松弛峰,则从不同的计算方法得到的各个 T_g 和 Δc_p 的差值增大。在下面描述的各种方法中,红色曲线被用作例子来说明这些差别是如何产生的,并对特定应用选择最合适的计算方法提供根据。

Interpretation The shape of the glass transition curves depends on the thermal history of the sample. If the heating rate is greater than the rate at which the sample was cooled, the glass transition is accompanied by an enthalpy relaxation peak. Storage of the sample at temperatures in the range of the glass transition or below can also give rise to an enthalpy relaxation peak. This effect can be seen in the black and red curves in the first diagram. The diagram also shows that the peak size increases with increasing storage time at room temperature. The endothermic peak corresponds exactly to the enthalpy decrease due to relaxation during storage or annealing below the glass transition.

The characteristic quantities of the glass transition are the glass transition temperature, T_g, and the step height at the glass transition, Δc_p. Various standard methods are available in the STAR^e software. Besides the STAR^e procedure itself, these include the DIN 51007 (DIN), ASTM E 1356 and IEC 1006 (ASTM, IEC) evaluation procedures as well as the determination of the fictive temperature at the glass transition according to Richardson (also described in DIN 51007).

As can be seen from the diagram, the values obtained for the glass transition temperature and the step height depend on the method used for the determination. If the glass transition does not have an enthalpy relaxation peak, the glass transition temperatures determined by the STAR^e, ASTM and DIN procedures are practically the same. The value determined by the Richardson method is slightly but not significantly lower. A comparison of the step heights at the glass transition shows that the c_p values can be divided into two groups (STAR^e and DIN methods: 0.25 Jg^{-1}K^{-1}; ASTM and Richardson methods: 0.24 Jg^{-1}K^{-1}).

If an enthalpy relaxation peak occurs, the differences between the individual T_g and Δc_p values obtained from the different evaluation methods increase. In the various methods described below, the red curve is used as an example to show how these differences arise and to provide a basis for choosing the most suitable evaluation method for a particular application.

玻璃化转变温度的计算: **Evaluation of the glass transition temperature:**

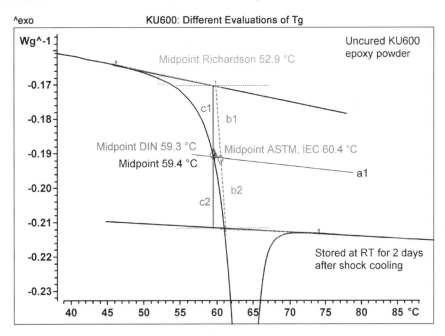

图 3.3　Fig. 3.3

STAR^e 方法(＋):
画玻璃化转变前后切线之间的角平分线 a1。该直线与测试曲线的交点为玻璃化转变温度(中点)。

DIN 方法(Δ):
玻璃化转变温度(DIN 中点)是曲线上比热容变化为总变化一半处(c1＝c2)的温度。c1 是玻璃化转变前的切线与测试曲线之间的距离,c2 是玻璃化转变后的切线与测试曲线之间的距离。

ASTM 方法(◇):
在测试曲线上玻璃化转变区域的拐点作切线。玻璃化转变温度是拐切线起始点和终止点的中点(b1＝b2)。

Richardson 方法(□):
测定外推基线和测试曲线之间的面积。各个面积在图中显示为阴影,并标记为 A1、A2 和 A3。面积 A1 的最高温度等于面积 A2 的最低温度。当 A1＋A3＝A2 时,该温度定义为假想玻璃化转变温度。

除了上面的特征玻璃化转变温度的定义,实际上也经常使用其他的计

STAR^e procedure (＋):
The bisector, a1, of the angle between the tangents above and below the glass transition is drawn. The point of intersection of this line with the measured curve is the glass transition temperature (midpoint).

DIN procedure (Δ):
The glass transition temperature (midpoint DIN) is the temperature on the curve where the change of the specific heat capacity is half of the total change (c1 = c2). c1 is the distance between the measured curve and the tangent below the glass transition. c2 is the distance between the measured curve and the tangent above the glass transition.

ASTM procedure (◇):
The tangent is drawn at the point of inflection in the region of the glass transition on the measured curve. The glass transition temperature is the mid-point between onset and endset of the inflectional tangent (b1 = b2).

Richardson procedure (□):
Areas are determined between the extrapolated tangents and the measured curve. The individual areas are shown shaded in the diagram and labeled A1, A2 and A3. The highest temperature of the area A1 is identical to the lowest temperature of the area A2. When A1 ＋ A3 = A2, this temperature is defined as the fictive glass transition temperature.

Besides the above definitions of a characteristic glass transition temperature, in practice other evaluation procedures are also

算方法,这些方法涉及利用曲线的形状如拐点(微分曲线上的峰)或起始温度(切线的交点)。

often used. These procedures involve the use of the curve shape such as the point of inflection (peak in the derivative curve) or the onset temperature (point of intersection of the tangents).

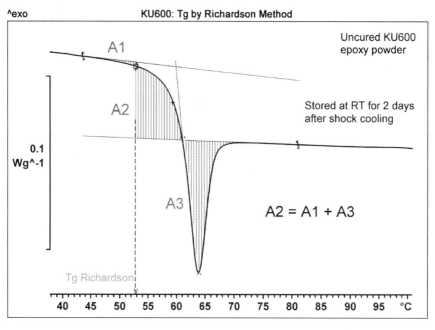

图 3.4　Fig. 3.4

玻璃化转变处台阶高度的计算:
$STAR^e$ 方法:
作拐点处的切线。该拐切线与玻璃化转变前后画的两条外推切线相交。台阶高度 Δc_p 由交点之间的热流差计算:

Evaluation of the step height at the glass transition:
$STAR^e$ procedure:
The tangent at the point of inflection is drawn. This inflectional tangent intersects the two extrapolated tangents drawn above and below the glass transition. The step height, Δc_p, is calculated from the difference in heat flow between the points of intersection:

$$\Delta c_p = \left| \frac{\Phi_u - \Phi_o}{m\beta} \right|$$

Φ_u 和 Φ_o 是外推切线与玻璃化转变前后的拐切线的交点的热流,m 是样品质量,β 是升温速率。

DIN 方法:
玻璃化转变处的台阶高度与 $STAR^e$ 方法中一样测定,但是用测试曲线在 DIN 玻璃化转变温度点的切线代替拐切线。该切线在图中以红色虚线表示。

ASTM 方法:
计算 ASTM 玻璃化转变温度点外推切线之间的距离。

Richardson 方法:
计算 Richardson 玻璃化转变温度点外推切线之间的距离。

Φ_u and Φ_o are the heat flows at the points of intersection of the extrapolated tangents with the inflectional tangent above and below the glass transition temperature, m is the sample mass and β the heating rate.

DIN procedure:
The step height is determined at the glass transition as in the $STAR^e$ procedure, but instead of the inflectional tangent, the tangent of the measured curve at the DIN glass transition temperature is used. The tangent is shown in the diagram as a dashed red line.

ASTM procedure:
The distance between the extrapolated tangents at the ASTM glass transition temperature is calculated.

Richardson procedure:
The distance between the extrapolated tangents at the Richardson glass transition temperature is calculated.

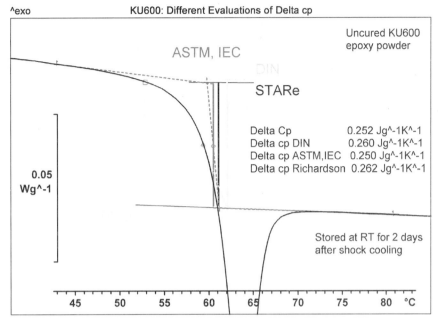

图 3.5　Fig. 3.5

说明　Richardson 玻璃化转变温度描述了测试前材料的实际状态。如果化学反应或物理过程（例如结晶、蒸发）发生在玻璃化转变区域，或者如果与玻璃化转变相重叠，就无法测定。

使用其他方法得到的玻璃化转变温度也受曲线形状即测试条件的影响。因此为了对测试做比较，对 T_g 测定使用相同的测试条件和计算方法是重要的。

玻璃化转变处的台阶高度 Δc_p 用 ASTM 和 Richardson 方法测定。因为得到的两个玻璃化转变温度不同，所以这些值会有所不同。这些值可能直接与材料的性能如填料含量有关。不可能用其他的方法（STAR^e 和 DIN）测定 Δc_p，因为那样测定的台阶高度与玻璃化转变的宽度（切线的斜率）有关。

因此，本手册所述的实验，都用 ASTM 方法测定 Δc_p（除非特别指明）。

结论　说明用于玻璃化转变的测量和定量计算方法是重要的。

Comments　The Richardson glass transition temperature describes the actual state of the material before the measurement. It cannot be determined if chemical reactions or physical processes (such as crystallization, evaporation) take place in the glass transition region or if glass transitions overlap.

The glass transition temperature obtained using the other methods is also influenced by the shape of the curve, i.e. by the measurement conditions. For comparative measurements, it is therefore important to use the same measurement conditions and evaluation method for the T_g determination.

In the ASTM and Richardson methods, the step height, Δc_p, at the glass transition temperature is determined. Differences in these values can arise because the two glass transition temperatures differ. The values can be directly related to properties of the materials such as filler content. This is not possible with the other methods for determining Δc_p (STAR^e and DIN) because the step heights determined then depend on the width of the glass transition (slope of the tangents).

The ASTM method has therefore been used to determine Δc_p in the experiments described in this booklet (unless otherwise noted).

Conclusions　It is important to state the methods used for the measurement and evaluation of quantities determined at the glass transition.

3.1.1.3 样品预处理对玻璃化转变的影响
Influence of sample pretreatment on the glass transition

玻璃化转变是与协同分子重排有关的物理现象。如果一个过冷液体冷却到足够低的温度,则协同分子重排在玻璃化转变处冻结。生成的玻璃态具有固体的力学性能,但具有液体的结构。这个玻璃化特征温度也称为玻璃化假想温度,该假想温度依赖于降温和贮存条件。降温速率越慢,假想温度越低。假想温度是热力学玻璃化转变温度。在 STAR^e 软件中用于测定该玻璃化转变温度的方法是 Richardson 方法。

The glass transition is a physical phenomenon that has to do with cooperative molecular rearrangements. If a supercooled liquid is cooled to a sufficiently low temperature, cooperative molecular movement freezes at the glass transition. The resulting glass has the mechanical properties of a solid but the structure of the liquid. This characteristic temperature of vitrification is also known as the fictive temperature of the glass. The fictive temperature depends on cooling and storage conditions. The slower the cooling rate, the lower the fictive temperature. The fictive temperature is the thermodynamic glass transition temperature. The method used to determine this glass transition temperature in the STAR^e software is the Richardson method.

目的 说明冷却速率如何影响热固性树脂的玻璃化转变。由于玻璃化转变是一个松弛过程,所以 DSC 曲线的形状和特征温度与降温和升温速率有关。这是比较了不同计算方法的结果。

Purpose To show how the cooling rate influences the glass transition of a thermoset. Since the glass transition is a relaxation process, the shape of the DSC curve and the characteristic temperature depend on the cooling and heating rates. The results of the different evaluation methods are compared.

样品 固化的 KU600 环氧树脂粉末。
Sample Cured KU600 epoxy resin powder.

条件 测试仪器:DSC
Conditions
坩埚:
40μL 铝坩埚
样品制备:
12.357mg 环氧粉末被完全固化。样品从 150℃冷却至 60℃,即以 0.1 至 40K/min 之间的降温速率冷却到玻璃态。
为了获得最快的降温速率(骤冷),用镊子在 150℃将样品从炉中取出,立即放在平的金属表面。产生的降温速率约为 50K/s(即 3000K/min)。
DSC 测试:
升温速率始终是 10K/min。
气氛:50mL/min 氮气

Measuring cell:DSC
Pan:
Aluminum 40μL
Sample preparation:
12.357 mg of the epoxy powder was completely cured. Samples were cooled from 150℃ to 60℃, i.e. to the glassy state at cooling rates between 0.1 and 40 K/min.
To achieve the fastest possible cooling rate (shock cooling), the sample was removed from the furnace with tweezers at 150℃ and immediately placed on a flat metal surface. The resulting cooling rate was about 50 K/s (i.e. 3000 K/min).
DSC measurement:
The heating rate was always 10 K/min.
Atmosphere:50 mL/min nitrogen

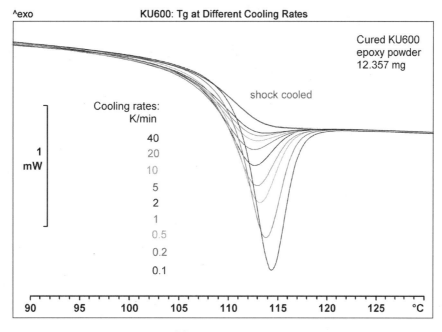

图 3.6　Fig. 3.6

解释　图 3.6 中的 DSC 曲线表明，松弛效应随着降温速率提高而变小。这也影响计算，实际上影响拐切线的位置。如用骤冷，则观察不到松弛峰。

此外曲线还表明，玻璃化转变处的台阶易与焓松弛峰重叠。松弛峰与样品历史和测试条件有关。在所描述的实验中，如果降温速率大于升温速率（10K/min），就几乎不出现峰。升温速率比降温速率越大，松弛峰就越大。

如上所述，预期在较低降温速率下玻璃化转变处的假想温度较低。从玻璃化转变的理论考虑，可预期降温速率对数与玻璃化转变之间几乎为线性的关系。如果用 Richardson 方法计算玻璃化转变温度，这个关系得到了证实（见表 3.2 和图示）。用其他计算方法，则结果受曲线实际形状的影响。在没有焓松弛峰发生的情况下，按 STAR^e、ASTM 和 Richardson 方法测定的玻璃化转变温度显示相似的升温速率依赖性。当松弛峰增大时，ASTM 玻璃化转变温度移向更高值。所有依赖于曲

Interpretation　The DSC curves in Figure 3.6 show that the relaxation effect becomes smaller with increasing cooling rate. This also influences the evaluation and in particular the position of the inflectional tangent. With shock cooling, no relaxation peak is observed.

The curves also show that the step in the glass transition and the enthalpy relaxation peak can easily overlap. The relaxation peak depends on the sample history and the measurement conditions. In the experiment described, practically no peak occurs if the cooling rate is greater than the heating rate (10 K/min). The relaxation peak becomes larger the greater the heating rate is in comparison to the cooling rate.

As discussed above, one expects the fictive temperature at the glass transition to be lower at lower cooling rates. From theoretical considerations of the glass transition, an almost linear relationship between the logarithm of the cooling rate and the glass transition temperature is expected. This relationship is confirmed if the glass transition temperature is evaluated using the Richardson method (see Table 3.2 and the graphical presentation). With the other evaluation methods, the result is influenced by the actual shape of the curve. In cases in which no enthalpy relaxation peak occurs, the glass transition temperatures determined according to the STAR^e, ASTM and Richardson methods show a similar dependence on the heating rate. As the relaxation peak increases, the ASTM glass transition temperature shifts to higher values. All the methods

线形状的用来测定玻璃化转变的方法(例如半台阶高法、拐点法等),表明玻璃化转变对升温速率的依赖性与 ASTM 方法是相似的。

used for the determination of the glass transition temperature that depend on the shape of the curve (e. g. half step height method, point of inflection, etc.) show a similar dependence of the glass transition temperature on the heating rate as the ASTM method.

计算 按照 STAR^e、ASTM 和假想温度(Richardson)方法从曲线测定了玻璃化转变温度。在计算中,注意保证玻璃化转变前后的切线在各种情况下相互平行。可以看到,得到的玻璃化转变温度与测试之前的样品降温速率有关(表 3.2)。在下图用得到的玻璃化转变温度与降温速率的对数作图。

Evaluation The glass transition temperatures were determined from the curves according to the STAR^e, ASTM and fictive temperature (Richardson) methods. In the evaluation, care was taken to ensure that the tangents above and below the glass transition were in each case parallel to one another. It can be seen that the glass transition temperatures obtained depend on the cooling rate of the sample prior to the measurement (Table 3.2). In the following diagram, the glass transition temperatures obtained were plotted as a function of the logarithm of the cooling rate.

表 3.2 以 10K/min 测量的玻璃化转变温度和相应的 c_p 与降温速率的关系。冷却速率 400K/min 几乎相当于骤冷。

Table 3.2 Glass transition temperature and corresponding changes of the c_p measured at 10 K/min, as a function of the cooling rate. The cooling rate of 400 K/min corresponds more or less to shock cooling.

冷却速率 Cooling rate K/min	T_g STAR^e ℃	T_g ASTM ℃	T_g Richardson ℃	Δc_p ASTM J/gk	Δc_p Richardson J/gk
0.1	109.3	110.9	100.8	0.255	0.281
0.2	108.6	110.0	101.1	0.255	0.275
0.5	108.0	109.1	102.1	0.247	0.264
1.0	107.7	108.5	102.7	0.249	0.264
2.0	107.4	108.0	103.2	0.243	0.255
5.0	107.6	107.8	104.9	0.241	0.249
10.0	107.8	107.9	105.5	0.244	0.250
20.0	108.1	108.0	106.5	0.245	0.250
40.0	108.2	108.2	106.8	0.249	0.253

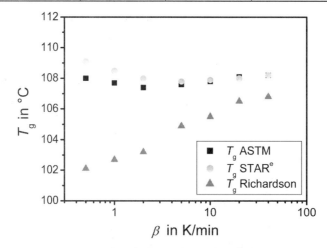

图 3.7 以 10K/min 测量的和用不同方法计算的玻璃化转变温度与降温速率的关系。

Fig. 3.7 Glass transition temperatures measured at 10 K/min as a function of the cooling rate and evaluated using different methods.

结论 无论何时引用玻璃化转变温度时,重要的是均须说明所使用的测定方法。

为了有意义地比较玻璃化转变的测试,样品预处理与计算方法应该是完全一样的。

参考文献 J. E. K. Schawe,接近热玻璃化转变的聚苯乙烯的热松弛的描述,J Polymer Science:Part B:Polymer Physics, Vol. 36,(1998) 2165.

Conclusions Whenever glass transition temperatures are quoted, it is important to state the method used for the determination.

Sample pretreatment and the evaluation procedure should be identical to meaningfully compare glass transition measurements.

Reference J. E. K. Schawe, Description of thermal relaxation of polystyrene close to the thermal glass transition, J Polymer Science:Part B:Polymer Physics, Vol. 36,(1998) 2165.

3.1.1.4 玻璃化转变的 ADSC 测量 Measurement of the glass transition by ADSC

温度调制 DSC(ADSC、IsoStep、TOPEM™)可在不同频率下测试玻璃化转变。这种技术经常用来分离重叠效应。例如,在玻璃化转变处,蒸发效应、焓松弛、相转变和化学反应可能与比热容变化重叠,温度调制能使比热容变化与其他效应分开,因而玻璃化转变温度能够可靠测定。

Temperature-modulated DSC (ADSC, IsoStep, TOPEM™) allows the glass transition to be measured at different frequencies. The technique is often used to separate overlapping effects. For example, at the glass transition, vaporization effects, enthalpy relaxation, phase transitions and chemical reactions can overlap the change in the specific heat capacity. Temperature modulation allows heat capacity changes to be separated from other effects. The glass transition temperature can then be reliably determined.

目的	热固性树脂的 DSC 分析常观察到重叠效应,而这使得难于解释。本实验的目的是说明如何用 ADSC 分离重叠的热容和动力学分量,例如玻璃化转变和焓松弛。
Purpose	Overlapping effects are often observed in the DSC analysis of thermosets. This makes interpretation difficult. The purpose of the experiment is to show how ADSC can be used to separate overlapping heat capacity and kinetic components such as the glass transition and enthalpy relaxation.
样品	未固化的 KU600 环氧树脂粉末。
Sample	Uncured KU600 epoxy resin powder.
条件	测试仪器:DSC
Conditions	坩埚:
	40μL 铝坩埚
	样品制备:称量 11.498mg 细粉末装入坩埚,未经预处理。
	DSC 测试:
	以 1K/min 的平均速率从 30℃升温至 130℃,温度振幅 0.5K,周期 48s。在相同条件下进行三次 ADSC 测试:用空样品坩埚(无盖)和空参比坩埚(无盖)进行空白测试。

Measuring cell:DSC

Pan:

Aluminum 40 μL

Sample preparation:11.498 mg fine powder was weighed into the crucible, no pretreatment.

DSC measurement:

Heating from 30℃ to 130℃ at a mean rate of 1 K/min, temperature amplitude 0.5 K, period 48s. Three ADSC measurement were performed under the same conditions: A blank measurement with an empty sample crucible (without lid) and an empty

用空样品坩埚(带盖)和同一个参比坩埚(无盖)进行校准测试。

A calibration measurement with an empty sample crucible (with lid) and the same empty reference crucible (without lid).

用装好样品的样品坩埚(带盖)和同一个空参比坩埚(无盖)进行样品测试。

A sample measurement with a sample crucible filled with sample (with lid) and the same empty reference crucible (without lid).

气氛:50mL/min 氮气

Atmosphere:50 mL/min nitrogen

介绍 调制DSC(ADSC)的温度程序是由一个线性的升温或降温速率叠加上一个小的周期性温度调制组成的。正弦调制的特征为温度振幅A_T和定义为$2\pi/P$的角频率ω,这里P表示正弦波的周期。一般的温度程序$T(t)$这样给出：

Introduction In alternating DSC (ADSC), the temperature ramp consists of a linear heating or cooling rate overlaid with a small periodic temperature modulation. The sinusoidal modulation is characterized by a temperature amplitude A_T and an angular frequency, ω, defined as $2\pi/P$, where P denotes the period of the sine wave. The general temperature program, T(t), is thus given by:

$$T(t) = T_0 + \beta t + A_T \sin\omega t$$

响应这个温度程序的测试热流(Φ)也是周期性的。由玻璃化转变引起的比热容变化这样的一些效应能够跟上施加的升温速率("可逆"现象),而像焓松弛、结晶、化学反应这样的其他效应则不能("不可逆"现象)。周期性热流信号是同相热流分量和异相热流分量的叠加。同相热流与升温速率同时发生,而异相热流滞后90°。

The measured heat flow (Φ) in response to this temperature program is also periodic. Certain effects such as changes in the specific heat capacity due to a glass transition can follow the applied heating rate ("reversing" phenomena), whereas other effects such as enthalpy relaxation, crystallization, chemical reaction cannot ("non-reversing" phenomena). The periodic heat flow signal is the superposition of an in-phase heat flow component and an out-of-phase component. While the in-phase heat flow coincides with the heating rate, the out-of-phase component lags behind by 90°.

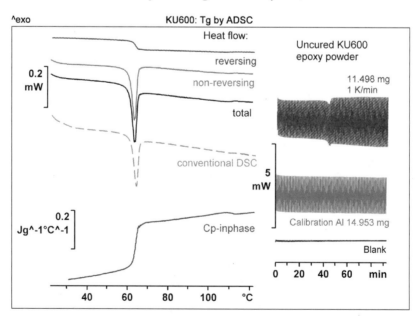

图3.8　Fig.3.8

解释 图右边的图所示为空白测试、校准测试和未固化 KU600 环氧树脂样品的样品测试。参比物质是铝，校准轮中样品坩埚（带盖）和参比坩埚（无盖）之间的质量差作为铝的质量。

如图所见，未固化 KU600 环氧树脂的可逆热流在约 63℃ 呈台阶变化，这可归因于玻璃化转变。由 c_p 同相热容曲线，对玻璃化转变的温度和热容台阶变化的计算分别得到 62.6℃ 和 0.25J/gK 的值。

在与玻璃化转变相同的温度范围，不可逆热流呈现吸热峰，这是由于焓松弛，从峰的积分得到 4.3J/g 的值。

结论 本实例证明 ADSC 能够分离所谓的"可逆"和"不可逆"效应，即玻璃化转变和重叠的松弛。一般而言，可逆效应是那些能够跟得上升温速率变化的效应如热容（即玻璃化转变时的 c_p 变化）。不可逆效应是那些跟不上调制的效应如焓松弛、结晶和化学反应。这对于解释和正确判定测试效应是特别有用的。

Interpretation The diagram on the right of the figure shows the blank, calibration and sample measurements of the uncured KU600 epoxy resin sample. The reference material was aluminum. The mass difference between the sample crucible (with lid) and reference crucible (without lid) in the calibration run was taken as the mass of the aluminum.

As can be seen in the figure, the reversing heat flow of the uncured KU600 epoxy resin shows a step change at approximately 63℃, which can be attributed to the glass transition. Evaluation of the temperature and the step change of the heat capacity of the glass transition yields values of 62.6℃ and 0.25 J/gK respectively using the c_p in-phase heat capacity curve.

The non-reversing heat flow shows an endothermic peak in the same temperature range as the glass transition. This is due to the enthalpy relaxation. Integration of the peak gave a value of 4.3 J/g.

Conclusions The example demonstrates that ADSC is able to separate so-called "reversing" from "non-reversing" effects, that is, the glass transition from the overlapping relaxation. In general, reversing effects are those that can follow the change of heating rate such as the heat capacity (i.e. c_p change during the glass transition). Non-reversing effects are those that do not follow the modulation, such as enthalpy relaxation, crystallization and chemical reactions. This is particularly useful for the interpretation and correct identification of the measured effects.

3.1.2 比热容测定 Determination of the specific heat capacity

目的	描述用 DSC 测定比热容的方法。	
Purpose	To describe the method for the determination of specific heat capacity by DSC.	
样品	完全固化的 KU600 环氧树脂粉末。	
Sample	Fully cured KU600 epoxy resin powder.	
条件 **Conditions**	测试仪器：DSC 坩埚： 40μL 铝坩埚 样品制备： 样品：通过加热若干次至 260℃，55.418mg 样品在坩埚中被完全固化。 蓝宝石参比：两个蓝宝石片放入盖子有小孔的坩埚内。 空坩埚：挑选三对坩埚和盖，质量差	Measuring cell：DSC Pan： Aluminum 40μL Sample preparation： Sample：55.418mg were completely cured in the crucible by heating several times to 260℃. Sapphire reference: two sapphire disks were enclosed in a crucible with a small hole in the lid. Empty crucible: the three crucibles and lids were

尽可能小(相差最大0.4mg);在校准软件中输入坩埚和盖子的重量。

DSC 测试:
以 5K/min 从 60℃升温至 160℃,从开始至结束温度恒温 5min。
气氛:静态空气

selected so that the smallest possible weight differences resulted (here maximum 0.4mg); the weights of crucible ＋ lid were entered in the software for the calculation.

DSC measurement:
Heating from 60℃ to 160℃ at 5 K/min with 5-min isothermal periods at the start and end temperature.
Atmosphere: Static air

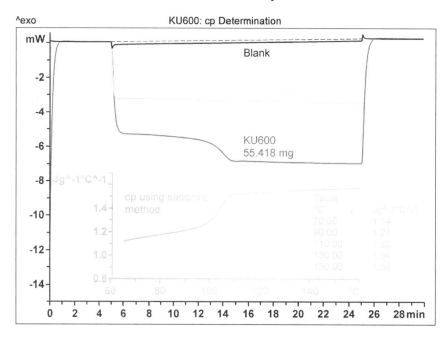

图 3.9　Fig.3.9

解释　作为常规方法,用蓝宝石法(DIN 51007)测定比热容可得到最准确的值。不过需要三次单独的测试,即样品、蓝宝石参比和空坩埚。三条与时间的关系曲线如图所示。样品量较大,以便测得大的信号(吸热的 DSC 偏移)。用相对低的升温速率 5K/min,以使样品中可能的温度梯度最小化。

比热 c_p 对样品温度作图,能清晰地看到 90℃至 110℃之间玻璃化转变处大约 0.31 $Jg^{-1}K^{-1}$ 的 c_p 增大。

计算　可不用蓝宝石作参比,根据样品和空白曲线之间的热流差($\Phi_{sample}-\Phi_{blank}$)按照下列方程直接测定比热容:

Interpretation　As a conventional method, the determination of the specific heat capacity using the sapphire method (DIN 51007) yields the most accurate values. Three separate measurements are however required, namely the sample, the sapphire reference and the empty crucible. The figure shows the three curves as a function of time. The sample size was relatively large in order to produce a large signal (an endothermic DSC deflection). A relatively low heating rate of 5K/min was used to minimize any possible temperature gradients in the sample.

The specific heat capacity, c_p, is plotted as a function of the sample temperature. The c_p increase of about 0.31 $Jg^{-1}K^{-1}$ can be clearly seen at the glass transition between 90℃ and 110℃.

Evaluation　The direct determination of the heat capacity without the sapphire reference is based on the difference of the heat flow between sample and the blank curve ($\Phi_{sample}-\Phi_{blank}$) according to the equation:

$$C_p = \frac{\Phi_{sample} - \Phi_{blank}}{M_0 \beta}$$

式中 Φ 是热流，m_0 是样品质量，β 是升温速率。

TOPEM™ 也可直接测量准等温条件下的比热容。经过热流校准后仅仅一次测试就足够。

表 3.3 汇总了从三种测试和计算方法得到的热容。比较表明，用简单测试会有 10% 的偏差。TOPEM™ 和蓝宝石方法得到几乎相同的值。

where Φ is the heat flow, m_0 the sample mass and β the heating rate.

A TOPEM™ measurement also directly provides the specific heat capacity for quasi-isothermal conditions. After heat flow calibration just one measurement is sufficient.

Table 3.3 summarizes the heat capacities obtained from the three measurement and evaluation methods. The comparison shows that with simple measurements deviations of 10% are possible. The TOPEM™ and sapphire methods yield practically the same values.

表 3.3 用三种不同方法测定的完全固化 KU600 环氧树脂的热容值

Table 3.3 Heat capacity values of the fully cured KU600 epoxy resin determined using three different methods.

温度 Temperature ℃	蓝宝石法 Sapphire J/gK	直接法 Direct J/gK	TOPEM™法 TOPEM™ J/gk
70	1.14	1.13	1.15
90	1.21	1.19	1.24
110	1.52	1.49	1.51
130	1.54	1.51	1.53
150	1.56	1.53	1.56

结论 在再现性良好的条件下，按照蓝宝石方法进行的 DSC 测试能在 5% 偏差内测定比热容。TOPEM™ 也得到准确值且只需一次测试，但热流必须准确校正。对于热容测试要用相对大的样品。如果材料的导热性差，则需要相对低的升温速率以使样品中的温度分布保持尽可能的均一。

Conclusions The specific heat capacity can be determined with a deviation of 5% from DSC measurements performed according to the sapphire method, provided good measurement reproducibility is obtained. TOPEM™ also yields accurate values and only one measurement is necessary. The heat flow must however be accurately adjusted. Relatively large samples are used for heat capacity measurements. If the thermal conductivity of the materials is poor, the heating rate should be relatively low so that temperature distribution in the sample remains as uniform as possible.

3.1.3 用 DSC 测试的固化反应 The curing reaction measured by DSC

本节讨论使用 DSC 分析固化反应和产物的性能。下面的体系用作代表性树脂：

官能团化学计量比 1∶2 的双酚 A 二缩水甘油醚型环氧树脂 DGEBA(壳牌化学，Epikote 828)和硬化剂 4,4'-二氨基二苯甲烷 DDM(Aldrich)。两个组分都加热到 120℃，20s 混合好，然后快速冷却到室温。制备后，树脂

In this section, the use of DSC for the analysis of curing reactions and the resulting material properties is discussed. The following system was used as a typical resin:

Epoxy resin based on diglycidyl ether of bisphenol A, DGEBA (Shell Chemical, Epikote 828) and 4,4'-diaminodiphenylmethane, DDM (Aldrich) as hardener in the stoichiometric ratio of the functional groups (1∶2). Both components were heated to 120℃ for 20s, mixed well and then rapidly cooled to room temperature. After preparation, the resin was stored at −30℃.

贮存在-30℃。

本节先讨论动态和等温固化反应的DSC曲线的外观和可得知的结果，然后论证固化度与玻璃化转变温度之间的关系。

然后，根据动态和等温测试介绍动力学计算。动力学描述化学反应发生有多快，例如固化反应，由于反应的每一个官能团释放一定量的热，所以测量的DSC曲线与给定条件下的反应速率成正比。分析仅从若干DSC测试得到的动力学数据，便可预测在其他条件下具有怎样的固化反应行为。

This section first discusses the appearance of the DSC curve of dynamic and isothermal curing reactions and the results that can be obtained. The relationship between the degree of cure and the glass transition is then demonstrated.

After this, kinetic evaluations are presented based on dynamic and isothermal measurements. The kinetics describe how rapidly a chemical reaction occurs, for example the curing reaction. Since a certain amount of heat is released for every functional group that reacts, the measured DSC curve is proportional to the reaction rate under the given conditions. An analysis of the kinetic data obtained from just a few DSC measurements allows predictions to be made about how a curing reaction behaves under other conditions.

3.1.3.1 动态固化：第一次和第二次升温测量
Dynamic curing: first and second heating measurements

目的 说明如何能够从各个DSC测试快速测定树脂体系的重要性能，这些包括由于固化产生的玻璃化转变温度的改变、反应焓、反应发生的温度范围、转化率与温度关系的曲线。所有这些都需要对同一个样品进行两次升温测试，就如本实例所示。

Purpose To show how important properties of a resin system can be rapidly determined from individual DSC measurements. These include the change of the glass transition temperature due to curing, the reaction enthalpy, the temperature range in which the reaction occurs, and the conversion curve as a function of temperature. All that is necessary is to perform two heating measurements on the same sample, as the example shows.

样品 以活性基团化学计量比混合的DGEBA和DDM环氧树脂体系。

Sample Epoxy resin system from DGEBA and DDM mixed in the stoichiometric ratio of the reactive groups.

条件 测试仪器：DSC

Conditions

坩埚：
40μL铝坩埚，盖钻孔

Measuring cell: DSC

Pan: Aluminum 40μL with pierced lid

样品制备：
称量1.984mg新制备的树脂混合物，放入坩埚，立即开始测试。

Sample preparation: 1.984 mg of the freshly prepared resin mixture was weighed into the crucible and the measurement immediately started.

DSC测试：
两次测试都以10K/min从-40℃升温至220℃。

DSC measurement: Heating from -40℃ to 220℃ at 10 K/min for both runs.

气氛：氮气，50mL/min

Atmosphere: Nitrogen, 50mL/min

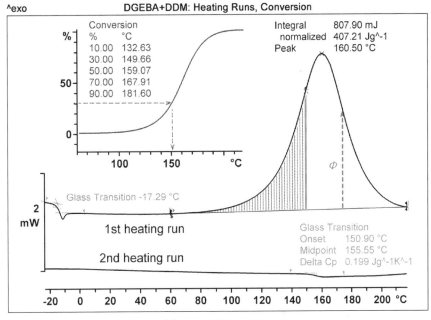

图 3.10　Fig. 3.10

解释　未固化环氧树脂的第一次升温在约-17℃显示一个伴有焓松弛的玻璃化转变。

然后发生放热的固化反应：反应慢慢开始，在峰温 160.5℃时达到最大反应速率。然后反应慢下来，因为可供反应的未反应物质变少了。

用直线基线对峰积分，得到固化热 407.2 J/g。这是完全固化的参比值（即体系的最大反应焓）。

反应速率 r 等于用总反应焓 Δh_{tot} 对单位时间内产生的热量 Φ（见图中箭头）进行归一化：

$$r = \frac{\Phi}{\Delta h_{tot}}$$

由部分积分计算得到反应转化率与温度关系的曲线，如插图所示。它能用来估测反应过程。例如：如果在 149.7℃达到 30%的转化率（见虚线箭头），那么到达该温度时已经释放了 30%的反应焓（见相应的整个表面积的阴影面积）。

固化反应后，材料的玻璃化转变移至 155.6℃，如第二次升温的 DSC 曲线所示。在该高温下，Δc_p 台阶仅为 0.20 J/gK。这主要是由于玻

Interpretation　The 1st heating run of the uncured epoxy resin shows the glass transition associated with enthalpy relaxation at approximately -17℃.

The exothermic curing reaction then takes place: the reaction begins slowly and reaches a maximum reaction rate at the peak temperature of 160.5℃. The reaction then slows down because less unreacted material is available to react.

Integration of the peak using a straight baseline yields a heat of curing of 407.2 J/g. This is the reference value for complete curing (i.e. maximum reaction enthalpy of the system).

The reaction rate, r, corresponds to the amount of heat Φ generated per unit of time (see the arrow in the diagram) normalized with respect to the total reaction enthalpy, Δh_{tot}:

$$r = \frac{\Phi}{\Delta h_{tot}}$$

The conversion curve of the reaction was calculated as a function of temperature by partial integration and is shown in the inserted diagram. It can be used to estimate the course of the reaction. For example, if 30% of the conversion is reached at 149.7℃ (see the dashed arrows), then 30% of the reaction enthalpy has been released up to this temperature (see hatched area in relation to the entire surface area).

After the curing reaction, the glass transition of the material shifts to 155.6℃, as the DSC curve of the 2nd heating run shows. At this high temperature, the Δc_p step is only 0.20 J/gK. This is mostly due to the temperature dependence of the heat capacity in the glassy state, whose value increases relatively strongly with temperature. In contrast, the heat capacity of the liquid (or the rubbery state) is almost constant, as shown schematically in Figure 3.11.

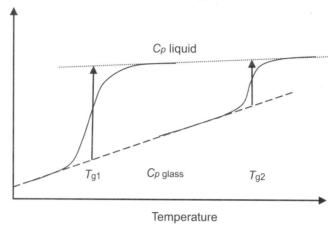

Fig. 3.11 Temperature dependence of the specific heat capacity.

The 2nd heating run shows a slight exothermic tendency at higher temperatures after the glass transition. This indicates further crosslinking.

Evaluation The evaluation of the conversion curve was performed using the DSC Conversion evaluation program. The conversion, $\alpha(T)$, is calculated from the heat flow curve according to the equation:

$$\alpha(T)=\frac{\Delta h_{part}}{\Delta h_{tot}}$$

The partial area of the peak from the start temperature of integration to the actual temperature is Δh_{part}. The total area of the peak is Δh_{tot}.

The glass transition temperature of the fresh resin is $-17.3℃$ (evaluated according to Richardson) with a heat capacity change of 0.528 J/gK. The cured system has a glass transition temperature of 155.6℃ and a heat capacity change of 0.199 J/gK.

Conclusions The DSC measures the temperature range, the reaction rate, and the enthalpy of the curing reaction. A start temperature of 60℃ was chosen for the integration of the reaction peak, a temperature at which practically no exothermic effect is apparent. This does not mean, however, that no reaction occurs. It is just so weak that it cannot be detected with

被检测到。因此,当计算反应峰时,选择反应开始时的基线十分重要,使低转化率值时的反应速率尽可能地与实际相符。特别对于动力学计算,这具有重要意义。

转化率曲线能用来估测反应过程以及优化与温度和速率有关的交联体系。曲线也能用于动力学分析。

the small sample mass. The choice of the baseline at the beginning of the reaction is therefore very important when evaluating reaction peaks so that the reaction rate at low values of conversion corresponds as far as possible to reality. This has important implications especially for kinetic evaluations.

Conversion curves can be used to estimate the course of a reaction as well as to optimize the crosslinking system with respect to temperature and rate. The curves can also be used for kinetic analysis.

3.1.3.2 等温固化的 DSC 测量　Isothermal curing by DSC

目的　研究热固性体系的等温固化行为并直接计算等温转化率曲线。
Purpose　To study the isothermal curing behavior of a thermosetting system and calculate the isothermal conversion curve directly.

样品　以活性基团化学计量比混合的由 DGEBA 和 DDM 组成的环氧树脂体系。
Sample　Epoxy resin system consisting of DGEBA and DDM mixed in the stoichiometric ratio of the reactive groups.

条件　测试仪器:DSC
Conditions　
坩埚:
40μL 铝坩埚
样品制备:
称量 2.87mg 的树脂混合物放入坩埚,立即开始测试。
DSC 测试:
在 160℃恒温 80min。
在反应非常慢的温度下将坩埚放入 DSC 中,然后以尽可能快的升温速率将样品升温至等温测试温度。例如,从 60℃ 开始,接着以 200K/min。这个方法甚至可测试相当快的反应。

可选的另外方法:可将样品在反应温度直接放入。不过,测试必须随后立刻开始,以便观测到发生在升温阶段的反应。如果使用自动进样器,这个方法仅适用于慢反应。

气氛:氮气,50mL/min

Measuring cell:DSC
Pan:
Aluminum 40μL
Sample preparation:
2.87 mg of the resin mixture was weighed into the crucible and the measurement immediately started.
DSC measurement:
Isothermal at 160℃ for 80 min.
This was done by inserting the crucible into the DSC at a temperature at which the reaction is very slow and then heating the sample to the isothermal measurement temperature at the highest possible heating rate. For example, start at 60℃ followed by heating at 200 K/min. This method allows even relatively fast reactions to be measured.

Alternative method: The sample can be inserted directly at the reaction temperature. The measurement must however be started immediately afterward so that reactions that occur during the heating phase are measured. This method is only suitable for slow reactions if the sample robot is used.

Atmosphere:Nitrogen,50mL/min

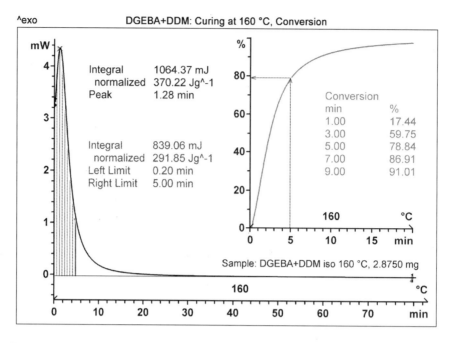

图 3.12　Fig 3.12

解释　图表示等温部分测试的 DSC 曲线（未包含升温阶段）。因为测试所选温度下的反应相当快，所以开始时的 DSC 信号表明业已放热，在半个小时后再接近基线，这表明交联反应已经结束。反应可能完全结束或者由于其他效应而停止，例如由于玻璃化。仅仅根据测试曲线是不可能区分这两种情形的。

基线的选择影响反应焓的计算。对于实际上完全的反应，热流信号逐渐逼近水平基线。因此可用"右水平"基线。不过，这个方法并未考虑样品升温到等温温度时反应产生的热。

反应的转化率曲线表示在插图中，是用部分积分计算的，表示为与时间的关系。可用其来估测反应的进程。开始时反应以最大速率进行，随着反应时间增加而慢下来。例如固化 5min 后达到 78.8% 的转化率，如红箭头所指示的情况。这与积分至 5min 的 DSC 曲线的百分面积（阴影面积）是相同的。

Interpretation　The diagram shows the DSC curve of the isothermal part of the measurement (without the heating phase). Since the reaction is relatively fast at the temperature chosen for measurement, the DSC signal is already exothermic at the beginning and approaches the baseline again after half an hour. This indicates that the crosslinking reaction has finished. It can stop completely or come to a stop due to other effects, for example, through vitrification. It is not possible to distinguish between these two situations based on the measurement curve alone.

The choice of baseline influences the evaluation of the reaction enthalpy. For a reaction that is practically complete, the heat flow signal asymptotically approaches a horizontal line. The "Horizontal right" baseline can therefore be used. This method, however, does not take into account the heat produced in the reaction while the sample is heated to the isothermal temperature.

The conversion curve of the reaction is shown in the inserted diagram and was calculated as a function of time using partial integration. It can be used to estimate the course of the reaction. The reaction proceeds at a maximum rate at the beginning and slows down with increasing reaction time. For example, a conversion of 78.8% is achieved after curing for 5min, as the red arrows indicate. This is the same as the percentage area of the DSC curve integrated up to 5min (hatched area).

计算 用 DSC 转化率计算程序进行转化率曲线的计算。

转化率 $\alpha(t)$ 是用下列方程从热流曲线计算的:

$$\alpha(t)=\frac{\Delta h_{part}(t)}{\Delta h_{tot}}$$

Δh_{tot} 为峰的总面积。从时间 0 到实际时间 t 峰的部分面积为 Δh_{part}。

结论 DSC 是测试等温固化反应焓并估算指定温度下等温固化达到一定转化率的时间的出色方法。对从等温测试定量测定反应焓,应注意两个重要因素:

① 反应速率不可太快——反应时间应该大于两分钟,否则反应开始时的不确定性太大。这个测试上的问题,可通过选择合适的等温温度来解决。

② DSC 曲线的时间记录必须足够长,否则不能准确地确定反应结束时的水平基线。例如,如果曲线只被积分 30min,则峰面积小 3%。

等温转化率曲线也经常用来检验动力学分析和预测的可靠性。

单个等温测试无法表明反应是否完成或者是否因其他原因而停止。这可通过进行一个动态后固化实验来研究,实验中将等温固化的样品升温至完全高于预料的最大玻璃化转变温度(见下一应用实例)。

Evaluation The evaluation of the conversion curve was performed using the DSC Conversion evaluation program. The conversion, $\alpha(t)$, is calculated from the heat flow curve by means of the following equation:

$$\alpha(t)=\frac{\Delta h_{part}(t)}{\Delta h_{tot}}$$

The total area of the peak is Δh_{tot}. The partial area of the peak from the time 0 to the actual time t is Δh_{part}.

Conclusions DSC is an excellent method to measure the enthalpy of an isothermal curing reaction and to estimate the time for isothermal curing to achieve a certain conversion at a specified temperature. Two important factors should be noted for the quantitative determination of the reaction enthalpy from isothermal measurements:

① The reaction rate must not be too high-the reaction should take longer than two minutes because otherwise the uncertainty at the beginning of the reaction is too great. This measurement problem can be overcome by choosing an appropriate isothermal temperature.

② The DSC curve must be recorded sufficiently long because otherwise the horizontal baseline at the end of the reaction cannot be determined exactly. For example, if the curve is only integrated for 30min, the area is 3% less.

The isothermal conversion curve is also often used to verify the reliability of the kinetics analysis and predictions.

One single isothermal measurement cannot show whether the reaction is complete or whether it has stopped for other reasons. This can be investigated by performing a dynamic postcuring experiment in which the isothermally cured sample is heated to a temperature well above the expected maximum glass transition temperature (see next application).

3.1.3.3 后固化和固化度的 DSC 测量 Postcuring and degree of cure by DSC

目的	说明如何能用 DSC 来研究后固化。	
Purpose	To show how DSC can be used to study postcuring.	
样品	以活性基团化学计量比混合的由 DGEBA 和 DDM 组成的环氧树脂体系。	
Sample	Epoxy resin system consisting of DGEBA and DDM mixed in the stoichiometric ratio of the reactive groups.	
条件	测试仪器:DSC	**Measuring cell**:DSC
Conditions	坩埚:	**Pan**:
	40μL 铝坩埚,盖钻孔	Aluminum 40μL, with pierced lid

样品制备:

样品量约 3mg。

为了得到部分固化的材料,未固化的树脂在 DSC 炉中 100℃等温保持 200min。

DSC 后固化测试::

以 10K/min 从-40℃升温至 230℃。

气氛:氮气,50mL/min

Sample preparation:

Sample size is approximately 3mg.

The uncured resin was held isothermally at 100℃ for 200 min in the DSC furnace in order to obtain partially cured material.

DSC postcuring measurement::

Heating from -40℃ to 230℃ at 10 K/min

Atmosphere:Nitrogen,50mL/min

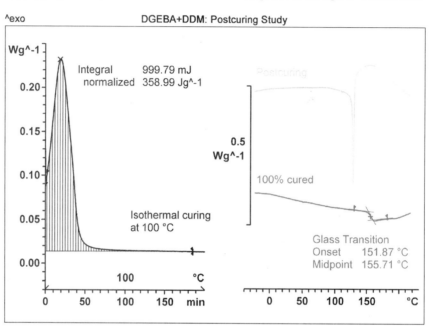

图 3.13　Fig. 3.13

解释　左图所示的第一次测试,是为了制备部分固化的热固性树脂(100℃下等温反应)。

第二次 DSC 测试(后固化),将材料升温到足以高过玻璃化转变温度。在玻璃化转变后随即观察到的放热峰就是部分固化材料的后固化反应。其起始点约在 120℃。固化反应由于玻璃化而停止。玻璃化转变被焓松弛效应的吸热峰所重叠。因为协同分子运动在玻璃化转变温度以上不再被冻结,活性基团自由运动,所以后固化反应立即开始。从而观察不到实际存在的玻璃化转变的 DSC 台阶。然而玻璃化转变温度仍能按下面描述的来测定。另外,可选用温度调制 DSC 来测量。

后固化使玻璃化转变温度从 120℃

Interpretation　The purpose of the first measurement shown in the left diagram was to prepare a partially cured thermoset (isothermal reaction at 100℃).

In the second DSC measurement (postcuring), the material was heated to well above the glass transition temperature. The postcuring reaction of this partially cured material is observed as an exothermic peak directly after the glass transition. This has an onset at about 120℃. The curing reaction stopped due to vitrification. The glass transition is overlapped by the endothermic peak of the enthalpy relaxation effect. Since cooperative molecular movements are no longer frozen-in above the glass transition temperature and the reactive groups are free to move, the postcuring reaction begins immediately. This makes it difficult to see the actual DSC step of the glass transition. The glass transition temperature can however still be determined as described below. Alternatively, temperature-modulated DSC can be used.

The postcuring increases the glass transition temperature from about 120℃ to 155.7℃ (100% cured curve).

Evaluation The degree of cure (or conversion) is an important quality control parameter of partially cured material such as coating materials and prepregs with respect to their manufacture, further processing, and end-product properties. The degree of cure can be determined according to the following equation:

$$\alpha = \frac{\Delta h_0 - \Delta h_p}{\Delta h_0}$$

where Δh_0 is the specific enthalpy of the curing reaction of the pure unreacted resin and Δh_p is the specific enthalpy of the postcuring. The determination of Δh_0 of the DGEBA/DDM system (407.2 J/g) is described in Section 3.1.3.1.

Δh_p can be determined from isothermal or dynamic postcuring experiments. To determine the enthalpy of postcuring a straight baseline was chosen so that its course or slope corresponded to the curve of the fully cured material (see Fig. 3.14). Comparison of the measured reaction enthalpy (47.3 J/g) with the total reaction enthalpy (407.2 J/g) gives a degree of cure of 88.4% (11.6% postcuring enthalpy).

The isothermal reaction at 100℃ yields 359 J/g, which corresponds to a conversion of 88.2%. In this curing experiment, the isothermal and dynamically measured reaction enthalpies add up very well to the total possible reaction enthalpy (see Section 3.1.3.1). Depending on the resin system, this is not always the case, especially if the course of the chemical processes differ depending on temperature.

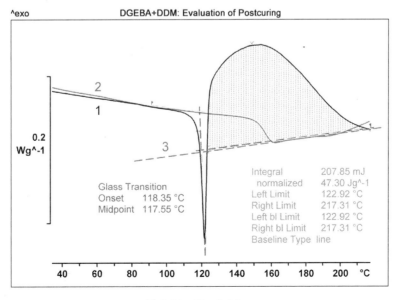

图 3.14 Fig. 3.14

后固化焓和玻璃化转变温度测定的各个计算步骤示于 3.14 图。

The individual evaluation steps for the determination of the postcuring enthalpy and the glass transition temperature are shown in Fig. 3.14.

如有必要可将部分固化的样品(1)和完全固化的样品(2)的 DSC 曲线垂直移动,以使它们在玻璃化转变前和后固化后实际上位于相互的顶部。直线 3 描述完全固化后玻璃化转变以上的 DSC 曲线的行进方向。因此它代表积分的基线和确定 T_g 的切线。积分确定为图 3.14 中所示的虚线面积。基线为与直线 3 一致的直线。精确定义了起始(曲线 3 和 1 的交点,122.9℃)和终止(曲线 1 和直线 3 的交迭处,217.3℃)的直线("直线"类型)常常就已足够。

The DSC curves of the partially cured sample (1) and the fully cured sample (2) are shifted vertically (if necessary) so that they lie practically on top of one another before the glass transition and after postcuring. Line 3 describes the course of the DSC curve after complete curing above the glass transition. It therefore represents the baseline for the integration and the tangent for the T_g determination. The integration is determined as the dashed area shown in Figure 3.14. The baseline is a line corresponding to Line 3. Often a straight line (Type "Line") with precisely defined start (point of intersection of curves 3 and 1, at 122.9℃) and end (overlap of curve 1 and line 3; at 217.3℃) is adequate.

测定玻璃化转变温度的切线在转变前与曲线 1 和 2 平行,在转变后与直线 3 平行。

The tangents for the determination of the glass transition temperature are drawn parallel to curves 1 and 2 before the transition and parallel to line 3 after the transition.

结论 易于用 DSC 来表征部分固化材料如预浸料和涂层材料的后固化过程。这对于产品的质量控制和过程优化是非常重要的。
后固化焓可从动态升温实验的反应峰的积分测定。固化度(即转化率)可由后固化焓和未固化材料的固化焓计算。
后固化通常使得玻璃化转变温度提高。这可作为判断质量的标准(例如 IPC_TM_650 No 2.4.25 中的"固化因子")。

Conclusions The postcuring process of partially cured materials such as prepregs and coating materials can easily be characterized by DSC. This is very important for the quality control of products and process optimization.
Postcuring enthalpies can be determined from the integral of the reaction peaks in the dynamic heating runs. The degree of cure (or conversion) can be calculated from the postcuring enthalpy and the enthalpy of curing of the uncured material.
Postcuring usually leads to an increase in the glass transition temperature. This is used as a quality criterion (e.g. "cure factor" in the IPC_TM_650 No 2.4.25) standard.

3.1.3.4 玻璃化转变与转化率的关系　Glass transition as a function of the conversion

目的	玻璃化转变与固化度的关系经常用作测定产品固化质量的简易方法。本实例旨在说明如何测定反应转化率(固化度)和玻璃化转变温度之间的关系。
Purpose	The dependence of the glass transition on the degree of cure is often used as a simple method to determine the quality of a product with regard to curing. The purpose of this example is to show how the relationship between the reaction conversion (degree of cure) and the glass transition temperature can be determined.
样品	DGEBA 和 DDM 组成的环氧树脂体系,按活性基团化学计量比混合。
Sample	Epoxy resin system consisting of DGEBA and DDM mixed in the stoichiometric ratio of

条件 Conditions	the reactive groups. 测试仪器：DSC 坩埚：40μL 铝坩埚,盖钻孔 样品制备：样品量约 3mg DSC 测试： 在 100℃恒温不同的时间,接着以 10K/min 从－40℃升温到 230℃ 气氛：氮气,50ml/min	Measuring cell：DSC Pan：Aluminum 40μL, with pierced lid Sample preparation：sample size is approx 3mg DSC measurement： Isothermal at 100℃ for different times followed by heating from －40℃ to 230℃ at 10 K/min Atmosphere：Nitrogen，50 ml/min

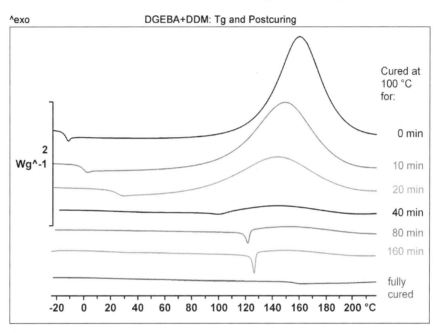

图 3.15　Fig. 3.15

解释 图示为一系列样品的后固化曲线,样品已在 100℃经不同时间进行部分固化,以得到不同的固化度。之后立即将样品快速冷却至－40℃并测试。系列曲线表明放热后固化峰如何随着固化时间而减小。相应的玻璃化转变温度 T_g 随着反应时间的延长而提高,但不超过 125℃（材料玻璃化,见本书中的相关章节）。在刚好低于玻璃化转变的 100℃固化了 80 和 160min 的样品显示明显的焓松弛,这是在升温测试所观察到的玻璃化转变处的吸热峰。只有在较高温度下后固化才导致完全交联,达到可能的最高玻璃化转变温度 155℃。为此,必须将热固性材料加热到至少 170℃。

Interpretation The diagram shows the DSC postcuring curves of a series of samples that had been partially cured at 100℃ for different times in order to obtain different degrees of cure. Immediately after, the samples were cooled rapidly to －40℃ and measured. The series of curves shows how the exothermic postcuring peak decreases with increasing curing time. The corresponding glass transition temperature, T_g, increases with longer reaction times but does not exceed 125℃ (the material vitrifies, see the corresponding chapter in the book). The samples cured for 80 and 160 min at 100℃, a temperature just below the glass transition, show a pronounced enthalpy relaxation. This is observed in the heating measurement as an endothermic peak at the glass transition. Only postcuring at a higher temperature leads to complete crosslinking and the maximum possible glass transition temperature of 155℃. To achieve this, the thermosetting material has to be heated to at least 170℃.

计算 可以上例中描述的相同方法给曲线确定固化度 α 和相应的 T_g。

固化度和玻璃化转变温度之间的关系显示在下图。曲线能用模型方程和通过曲线拟合确定的参数来描述。这使得在常规分析中从 T_g 值测定固化度成为可能。

DiBenedetto 方程经常被用来描述 T_g 和 α 之间的函数关系：

T_{g0} 是未固化树脂的玻璃化转变温度（-15℃）；T_{g1} 是完全固化体系的玻璃化转变温度（155℃）；λ 是拟合参数。

Evaluation The degree of cure, α, and the corresponding T_g can be determined for the curves in the same way as described in the previous example.

The relationship between the degree of cure and the glass transition temperature is shown in the following diagram. The curve can be described by model functions and their parameters determined by curve fitting. This allows the degree of cure to be determined in a routine analysis from the value of T_g.

The DiBenedetto equation is often used to describe the functional relationship between T_g and α:

$$T_g(\alpha) = \frac{\lambda\alpha(T_{g1} - T_{g0})}{1-(1-\lambda)\alpha} + T_{g0}$$

T_{g0} is the glass transition temperature of the unreacted resin (-15℃); T_{g1} is the glass transition temperature of the fully cured system (155℃); λ is a fit parameter.

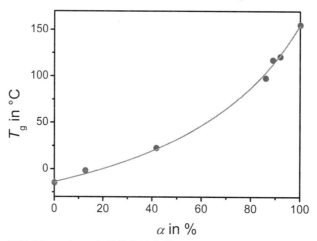

图 3.16 根据 DiBenedetto 方程拟合的曲线，$T_{g0} = -14$℃、$T_{g1} = 155$℃、$\lambda = 0.37$。

Fig. 3.16 The fit curve according to the DiBenedetto equation with $T_{g0} = -14$℃, $T_{g1} = 155$℃ and $\lambda = 0.37$.

结论 对于热固性树脂，玻璃化转变温度 T_g 随着固化度即交联度 α 而增高。这种关系可以经验为主或用 DiBenedetto 方程来描述。这经常用于质量控制中，通过简单测定玻璃化转变温度来检查固化的完全程度。

如果样品存在玻璃化，则 α 和 T_g 的测定要困难一些，而用上例中描述的方法或用 TMDSC 进行测定。

Conclusions With thermosetting resins, the glass transition temperature, T_g, increases with the degree of cure or crosslinking, α. This relationship can be described empirically or by the DiBenedetto equation. This is often used in quality control to check the completeness of curing by means of simple determinations of the glass transition temperature.

If vitrified samples are present, the determination of α and T_g is somewhat more difficult and is performed using the method described in the previous example or by TMDSC.

参考 J. E. K. Schawe, 固化动力学的等温动力学中的化学和扩散控制的描述，Thermochimica Acta 388 (2002) 299-312

Reference J. E. K. Schawe, A description of chemical and diffusion control in isothermal kinetics of cure kinetics, Thermochimica Acta 388 (2002) 299-312

3.1.3.5 固化速率和动力学的等温测量
Rate of cure and kinetics, isothermal measurements

目的	用等温 DSC 测试说明温度对固化反应速率和反应时间的影响。显示如何用简单动力学模型估算反应活化能。
Purpose	To demonstrate the influence of temperature on the rate and duration of the curing reaction using isothermal DSC measurements. To show how the activation energy of the reaction can be estimated by means of simple kinetic models.
样品	DGEBA 和 DDM 组成的环氧树脂体系,按活性基团化学计量比混合。
Sample	Epoxy resin system consisting of DGEBA and DDM mixed in the stoichiometric ratio of the reactive groups.

条件　　测试仪器:DSC　　　　　　　　　　　　**Measuring cell**:DSC
Conditions　坩埚:　　　　　　　　　　　　　　　　　　　　**Pan**:
　　　　　40μL 铝坩埚,盖钻孔　　　　　　　　　　Aluminum 40μL, with pierced lid
　　　　　样品制备:　　　　　　　　　　　　　　　　　**Sample preparation**:
　　　　　样品量在 2 到 3mg 之间。　　　　　　　　Sample size is between 2 and 3 mg.
　　　　　DSC 测试:　　　　　　　　　　　　　　　　　**DSC measurements**::
　　　　　在 160℃、140℃ 和 120℃ 下恒温 60min。　Isothermal at 160℃, 140℃ and 120℃ for 60 min.
　　　　　在反应很慢的温度下将样品放入 DSC。然后快速升温(例如 200K/min)至等温温度。该方法甚至可测试相当快的反应。　The sample was inserted into the DSC at a temperature at which the reaction is very slow. It was then heated rapidly to the isothermal temperature (e.g. at 200 K/min). This method allows even relatively fast reactions to be measured.
　　　　　气氛:氮气,50mL/min　　　　　　　　　　**Atmosphere**:Nitrogen, 50 mL/min

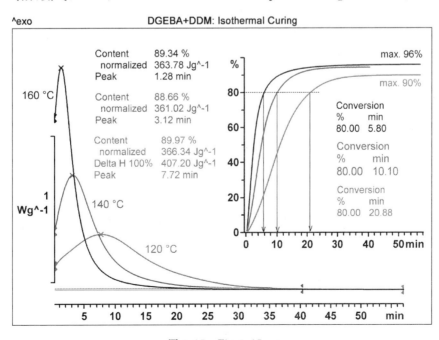

图 3.17　Fig. 3.17

Interpretation As expected, a comparison of the DSC curves shows that the reaction rate is higher at higher temperature. The DSC peak at 160℃ is higher and narrower (i.e. it does not last so long) than in the other two measurements. It follows that the conversion curves at higher temperatures are shifted to shorter times. In the measurement at 120℃, it can be seen that the highest reaction rate is reached at 7.7 minutes and not at the beginning of the measurement. This indicates that the reaction is self-accelerating. The conversion curves therefore exhibit an S-shape. For process design, the reaction time for a certain conversion can be read off directly. For example at 120℃, it takes 20.9 minutes to achieve 80% conversion. A comparison with the dynamic measurement (Section 3.1.3.1) shows that the reaction enthalpies measured in these isothermal measurements are only about 89% of the theoretically possible $\Delta h_r = 407.2 \text{J/g}$. This indicates that curing is incomplete and has stopped due to vitrification. Postcuring experiments showed small exotherms, so that the total conversion α_{iso} must be corrected using the isothermal experiment results (incl. heating) for each temperature by means of the following equation:

$$\Delta h_r = \Delta h_{heating} + \Delta h_{iso} + \Delta h_{postcuring} = 407 \text{J/g}$$
$$\alpha_{iso} = (\Delta h_{heating} + \Delta h_{iso})/\Delta h_r$$

α_{iso} results are from J. E. K. Schawe, according to the literature reference at the end of this example.

Table 3.4 Correction values for the conversion calculation

T_{iso} ℃	Δh_{iso} J/g	Δh_{post}curing J/g	$\Delta h_{heating}$ J/g	$\Delta \alpha_{iso}$ %
160	363	16	28	96
140	361	24	22	94
120	366	40	1	90

The amount of heat not measured before reaching the isothermal temperature, $\Delta h_{heating}$, depends on the temperature, i.e. on the reaction rate and (as shown in Table 3.4) cannot be neglected, especially at higher temperatures.

计算 动力学描述 A：

Evaluation Kinetics description A：

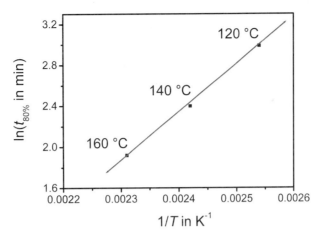

图 3.18　Fig. 3.18

上图所示为简单的动力学描述，对 80% 转化率的时间（时间 t 的自然对数）表示为与温度的倒数 [$1/T$ (Kelvin]的关系。这个关系对应于下面的动力学描述：

A simple kinetics description is shown in the above diagram. The time (natural logarithm of the time t) to 80% conversion is displayed as a function of the inverse reaction temperature $1/T$ (in Kelvin). This relationship corresponds to the following kinetic description：

$$\ln t = K + \frac{E_a}{RT}$$

K 是一个常数。

从图中的斜率可计算该反应的活化能：

K is a constant.

From the slope in the diagram, the activation energy for this reaction can be calculated：

$$E_a = 38.5 \text{kJ/mol}$$

不同温度下达到 80% 转化率的时间可直接从图 3.17 得到。

动力学描述 B：

更多的动力学参数能用动力学计算程序来计算。在最简单的模型中应用了 n 级动力学，其中起始物以简单方式直接反应至最终产物，见 2.3.3 节。

The reaction time to reach 80% conversion for different temperatures can be directly taken from Figure 3.17.

Kinetics description B：

Further kinetic parameters can be calculated using the kinetics evaluation programs. In the simplest model, n^{th} order kinetics is applied in which the starting product reacts in a simple way directly to the end product, see Section 2.3.3.

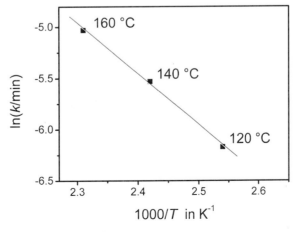

图 3.19　Fig. 3.19

Arrhenius 图（以 1/T 为函数的 ln$k$）从三个等温测试得到。各个速率常数用 n 级动力学程序从三条 DSC 曲线测定，这里只使用了高于 50% 转化率的测量值（自加速行为较轻）。

依靠过三点的线性回归，从斜率确定活化能 E_a=41.3kJ/mol，从截距确定 lnk_0=6.45。三个反应级数确定的平均值为 1.2。如果用可利用的软件动态计算三个等温测试，则三个动力学参数都能直接测定。如果用较低转化率下的测试值，则得到完全不同的动力学参数。这指出该动力学描述对于复杂的固化反应过于简单，因此几乎不能提供可靠的预测。

结论 等温 DSC 测试直接显示特定温度下的固化反应有多快。这可用来比较不同的树脂体系。还能对体系关于温度和时间优化生产工艺。此外必须考虑升温速率、热传递行为和涉及的热量。

用简单的模型通过若干参数能描述动力学行为。这些能用来对其他反应条件作出预测，例如更高温度下的更快速固化，即预测难以测试条件下的反应。然而，由于使用含促进剂和稳定剂的多组分体系，因而交联反应通常是复杂的。这意味着简单的动力学描述只能应用到非常有限的程度，必须小心对待结果。对于预测尤其如此。

非模型动力学提供了更好的可能性。

参考文献 J. E. K. Schawe，固化动力学的等温动力学中化学和扩散控制的描述，Thermochimica Acta 388（2002）299—312

The Arrhenius diagram (ln k as a function of 1/T) follows from the three isothermal measurements. The individual rate constants were determined from the three DSC curves using the nth order kinetics program, whereby only measured values above 50% conversion were used (self-accelerating behavior is weighted less).

By means of linear regression through three points, the activation energy E_a = 41.3 kJ/mol is determined from the slope and ln k_0 = 6.45 from the intercept. The mean value determined for the three reaction orders n is 1.2. If the three isothermal measurements are kinetically evaluated using the available software, all three kinetic parameters can be directly determined. If measured values at lower conversions are used, quite different kinetic parameters are obtained. This is an indication that this kinetic description is too simple for complex curing reactions and can therefore hardly provide reliable predictions.

Conclusions Isothermal DSC measurements show directly how fast a curing reaction is at a particular temperature. This can be used to compare different resin systems. Alternatively, a production process can be optimized for a system with regard to temperature and time. The heating rate, heat transfer behavior and the amounts involved must also be taken into account.

The kinetic behavior can be described by a few parameters using simple models. These can be used to make predictions for other reaction conditions, e.g. for faster curing at higher temperatures, that is, for conditions that are difficult to measure. The crosslinking reactions are however usually complex due to the use of multicomponent systems with accelerators and stabilizers. This means that the simple kinetic description can only be applied to a very limited extent and the results must be treated with care. This is particularly the case with predictions.

Model free kinetics offers better possibilities.

Reference J. E. K. Schawe, A description of chemical and diffusion control in isothermal kinetics of cure kinetics, Thermochimica Acta 388 (2002) 299—312

3.1.3.6 固化速率的动态测量 Curing rate, dynamic measurements

目的
Purpose
说明升温速率对固化反应和最大反应速率的影响。
To demonstrate the influence of heating rate on the curing reaction and the maximum reaction rate.

样品
Sample
DGEBA 和 DDM 组成的环氧树脂体系,按活性基团的化学计量比混合。
Epoxy resin system consisting of DGEBA and DDM mixed in the stoichiometric ratio of the reactive groups.

条件
Conditions

测试仪器:DSC
坩埚:
40μL 铝坩埚,盖钻孔
样品制备:
样品量 2 到 12mg。
DSC 测试:
以三个不同的升温速率从 −40℃ 加热至 240℃。
气氛:氮气,50 ml/min

Measuring cell:DSC
Pan:
Aluminum 40μL, with pierced lid
Sample preparation:
Sample masses of 2 to 12 mg.
DSC postcuring measurement:
Heating from −40℃ to 240℃ at three different heating rates.
Atmosphere:Nitrogen, 50 ml/min

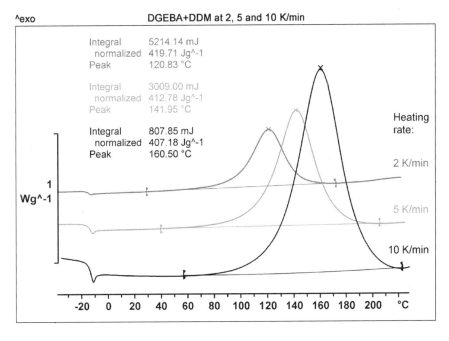

图 3.20 Fig. 3.20

解释 DSC 曲线的比较表明,在较高的升温速率下反应向较高温度移动。这是反应动力学的结果。反应在玻璃化转变温度以上开始,但是仍然非常慢。随着温度升高,反应变得越来越快。这就如 DSC 信号的增大所示。

Interpretation A comparison of the DSC curves shows that the reaction is shifts to higher temperatures at higher heating rates. This is a result of reaction kinetics. The reaction begins above the glass transition temperature, but is still very slow. With increasing temperature, the reaction becomes faster and faster. This is shown by the increasing DSC signal.

计算	反应焓随着升温速率的提高而下降。这可以通过考虑反应时间来解释。在低升温速率下,有更多的时间供活性基团"扩散",因此交联更完全,从而反应焓更大。这也可由最终的玻璃化转变温度明显看到。
Evaluation	The reaction enthalpies decrease with increasing heating rates. This can be explained by considering the reaction time. At low heating rates, more time is available for the reactive groups to "diffuse". Crosslinking is therefore more complete, i. e. a larger reaction enthalpy results. This is also apparent from the resulting glass transition temperatures.
结论	DSC 曲线反应峰的位置取决于升温速率。这与反应动力学有关,这就是为什么这样的曲线被用于动力学计算的原因。然而,特定温度下等温固化反应的持续时间实际上是不能从一次动态 DSC 测试估算的。
Conclusions	The location of the reaction peak on the DSC curve depends on the heating rate. This has to do with the kinetics of the reaction, which is the reason why such curves are used for kinetic evaluations. The duration of an isothermal curing reaction at a particular temperature cannot however in practice be estimated from one dynamic DSC measurement.

3.1.3.7 动力学计算和预测 Kinetic evaluations and predictions

目的	说明不同加热速率下的 DSC 测试如何用于动力学计算。本应用实例呈现了由不同计算方法得到的结果并说明如何来检验预测。
Purpose	To show how DSC measurements at different heating rates can be used for kinetic evaluations. The application shows the results that can be obtained from the different evaluation methods and how the predictions can be checked.
样品	DGEBA 和 DDM 组成的环氧树脂体系,按活性基团的化学计量比混合。测试曲线来自上例。
Sample	Epoxy resin system consisting of DGEBA and DDM mixed in the stoichiometric ratio of the reactive groups. The measured curves originate from the previous examples.

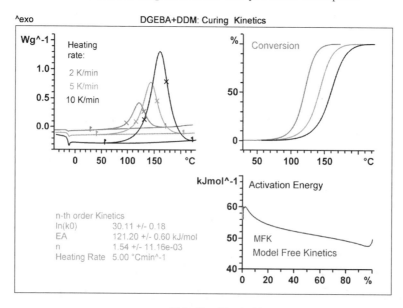

图 3.21   Fig. 3.21

计算 可用动力学程序来分析三条动态测试的 DSC 曲线。可用 n 级动力学方法单独计算每个测试,或可用非模型动力学(MFK)一起计算三条曲线。

在两类计算方法中,正确选择基线是非常重要的。在 MFK 中,先计算转化率曲线。如图所示,在较高加热速率下,转化率曲线向较高温度移动。MFK(和高级 MFK)从这些转化率曲线计算出一条活化能与转化率关系的新曲线。随着转化率增加,活化能 E_a 曲线清楚地表明,在整个固化反应中,E_a 并不如同人们经常认为的那样保持为常数。这说明反应包含复杂过程。此外明显的事实是,由 n 级动力学得到的活化能显著大于从等温实验测定的(见 3.1.3.5 节)或如 MFK 的 E_a 曲线所示的活化能。

预测

动力学计算的目的是为了得到反应行为的准确描述,从而能用它来预测其他条件下的行为。例如,预测在特定的等温条件下转化率与时间的关系,或作出对特定转化率的时间-温度图(等转化率图)。二者都能用非模型动力学或 n 级动力学计算。

图 3.22 表示两种可能的预测:

1. 160℃等温下转化率与时间的关系(图左)。

2. 对于 70% 转化率的等转化率曲线(图右)。

例如在左图中可见,根据 MFK 预测曲线(红色箭头),在 160℃下,对于 70% 转化率需要 3.6min 的反应时间。而根据 n 级动力学的预测,似乎仅需 1min。

例如在右图中,对 300min 反应时间后(黑色箭头)要达到 70% 转化率,可读出必需的反应温度为 55.5℃。

Evaluation The three dynamically measured DSC curves can be analyzed using kinetics programs. Each measurement can be individually evaluated with the nth order kinetics program or all three curves can be evaluated together using model free kinetics (MFK).

In both types of evaluation, the correct choice of baselines is very important. In MFK, the conversion curves are first calculated. As the diagram shows, these conversion curves shift to higher temperatures at higher heating rates. MFK (and advanced MFK) calculates a new curve from these conversion curves that shows how the activation energy changes with conversion. The curve of the activation energy E_a with increasing conversion clearly shows that E_a does not remain constant during the entire curing reaction as is often assumed. This indicates that the reaction includes complex processes. This is also apparent from the fact that n^{th} order kinetics yields an activation energy that is significantly larger than that determined from the isothermal experiments (see Section 3.1.3.5) or that shown by the E_a curve of MFK.

Predictions

The aim of kinetic evaluations is to obtain an accurate description of the reaction behavior so that it can be used to predict behavior under other conditions. For example, to predict the conversion curve as a function of time under isothermal conditions at a particular temperature or the time-temperature diagram (Iso-Conversion plot) for a particular conversion. Both can be calculated using model free kinetics or n^{th} order kinetics.

The Figure 3.22 shows two possible predictions:

1. The conversion as a function of time for an isothermal temperature of 160℃ (left diagram).

2. The iso-conversion curve for 70% conversion (right diagram).

In the left diagram, for example, it can be seen that at 160℃ a reaction time of 3.6 min is needed for a conversion of 70% if the MFK prediction curve is used (red arrow). Based on the predictions of nth order kinetics, only 1 min would appear to be necessary.

In the right diagram, for example, the reaction temperature can be read off that is necessary to achieve 70% conversion in 300 min reaction time (black arrow)—in this case 55.5℃.

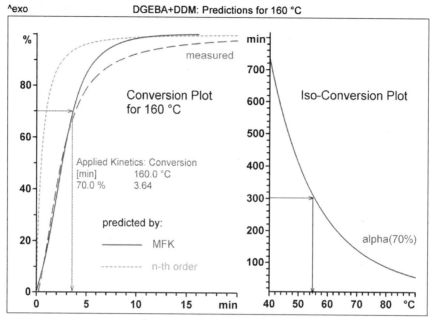

图 3.22 Fig. 3.22

检验预测

通过进行 160℃ 等温 DSC 测试来检验预测,由此计算的转化率曲线也绘于上图中(灰色虚线,见 3.1.3.2 节中的实例)。

转化率与时间关系曲线的比较(MFK 预测/等温 DSC 测试)表明,高达约 70% 转化率,MFK 的预测非常好。之后,测试转化率低于预测转化率。与之不同,基于一次动态测试的 n 级动力学所预测的时间要短得多。这与测试不相符,因为固化过程非常复杂,n 级动力学仅用三个动力学参数不能充分地描述该化学反应。

Checking the predictions

The predictions were checked by performing an isothermal DSC measurement at 160℃. The conversion curve calculated from it was also plotted in the above diagram (dashed gray line, see the example in Section 3.1.3.2).

A comparison of the conversion curves (MFK prediction/isothermal DSC measurement) as a function of time shows that MFK gives a very good prediction up to about 70% conversion. Afterward the measured conversion is lower than that predicted. In contrast, n^{th} order kinetics predicts much shorter times based on one dynamic measurement. This does not agree with the measurement because the curing process is a very complex chemical reaction that cannot be adequately described by just three kinetic parameters.

表 3.5 160℃ 等温反应温度下 70% 转化率的预测和测试
Table 3.5 Predictions and measurement of a conversion of 70% at an isothermal reaction temperature of 160℃.

方法 Method	时间,min Time in min
MFK 预测 MFK prediction	3.6
n 级预测(从 5K/min 动态曲线) n^{th} order prediction (from 5 K/min)	1.0
等温 DSC 测试(实验检验预测) Isothermal DSC measurement (to check the prediction experimentally)	3.8

结论 能用不同的方法计算一条或几条固化反应DSC曲线的动力学。动力学通常用来描述反应速率并对其他温度下的行为作预测。由非模型动力学（MFK）可得到最可靠的结果。MFK需要以不同升温速率进行的若干次测试（用高级MFK，等温测试也可）。

应尽可能通过进行独立的实验来检验预测。在尽可能接近期望的条件但仍然可进行热流测量的温度，用等温DSC测试来检验是最方便的。

选择其他实验条件（例如低温）可能改变化学反应的进程，例如由于玻璃化。下述观察可得到有关这种改变的迹象：

- E_a 曲线的形状。
- 根据不同升温速率的测试曲线计算得到的不同结果。

实际上，对于通常复杂的固化反应的可靠动力学描述，总是需要若干个不同的DSC测试。等温和动态测试的结合并使用高级MFK方法通常可得到最可靠的结果。

Conclusions The kinetics of one or several DSC curves of a curing reaction can be evaluated by different methods. This is mostly used to describe the reaction rate and to make predictions about behavior at other temperatures. The most reliable results are obtained from model free kinetics (MFK). This needs several measurements performed at different heating rates (with advanced MFK, also isothermal measurements).

The predictions should as far as possible be checked by performing independent experiments. This is easiest to do using an additional isothermal DSC measurement at a temperature that is as close as possible to the desired conditions but that still provides a measurable heat flow.

The choice of other experimental conditions (e. g. low temperature) can change the course of the chemical reaction, for example through vitrification. Indications of such changes can be given by following observations:

- the shape of the E_a curve
- different results based on the evaluation of curves measured at different heating rates.

In practice, a reliable kinetic description of the normally complex curing reaction always requires several different DSC measurements. A combination of isothermal and dynamic measurements and use of the advanced MFK method usually yields the most reliable results.

3.1.4 玻璃化转变和后固化的分离（TOPEM™法）
Separation of the glass transition and postcuring (TOPEM™)

目的	说明TOPEM™（温度调制DSC方法）如何能分离不同的效应从而测定玻璃化转变温度 T_g 和后固化焓 Δh_{post}。后固化反应通常开始于刚刚超过玻璃化转变温度。常规DSC测试往往难以甚至不可能测定玻璃化转变温度和后固化焓。
Purpose	To demonstrate how TOPEM™ (a temperature-modulated DSC method) can separate different effects and so allow the determination of the glass transition temperature, T_g, and the postcuring enthalpy, Δh_{post}. A postcuring reaction usually begins just above the glass transition. In conventional DSC measurements, this often makes it difficult or even impossible to determine the glass transition temperature and the postcuring enthalpy.
样品	由碳纤维和环氧树脂制成的具有很高玻璃化转变温度的层压复合材料。
Sample	Laminated composite material made of carbon fibers and epoxy resins with a very high glass transition temperature.

条件 Conditions	测试仪器：DSC	Measuring cell：DSC
	坩埚：	Pan：
	40 μL 铝坩埚	Aluminum 40 μL
	样品制备：	Sample preparation：
	从一块复合材料板上锯下的重19.326 mg（常规 DSC）和 24.188 mg（TOPEM™）的试片。	Pieces weighing 19.326 mg (conventional DSC) and 24.188 mg TOPEM™ were sawn out of a sheet of the composite.
	DSC 测试：	DSC measurement：
	以 2 K/min 从 50 ℃ 升温至 280 ℃。	Heating from 50 ℃ to 280 ℃ at 2 K/min.
	TOPEM™：脉冲高度 0.5 K，脉冲宽度 15 到 30 s。	TOPEM™：Pulse height 0.5 K, pulse width 15 to 30 s.
	气氛：氮气，50 mL/min	Atmosphere：Nitrogen, 50 mL/min

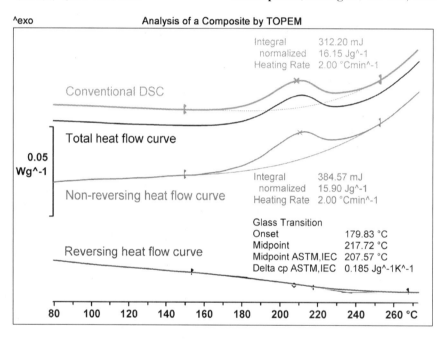

图 3.23　Fig 3.23

解释　常规 DSC 测试（最上面曲线）表明，后固化反应为一放热峰，与放热分解过程开始相互重叠。曲线上观察不到可用来测定 T_g 的台阶。

Interpretation　The conventional DSC measurement (upper curve) shows the postcuring reaction as an exothermic peak which is overlapped by the beginning of the exothermic decomposition process. A step in the curve that could be used to determine the T_g is not apparent.

计算　TOPEM™ 测试和计算产生三条曲线：
1. 可逆热流（红色曲线）呈现一个由于玻璃化转变产生的台阶。用 ASTM 方法计算得到玻璃化转变温度 $T_g = 208.6$ ℃。
2. 不可逆热流，它表示后固化反应和分解。

Evaluation　The TOPEM™ measurement and evaluation yields three curves：
1. The reversing heat flow (red curve) shows a step due to the glass transition. Evaluation using the ASTM method gave a glass transition temperature, $T_g = 208.6$ ℃.

2. The non-reversing heat flow, which shows the postcuring reaction and decomposition.

3. 作为二者之和的总热流,相当于常规 DSC 曲线。
在重复性范围内,两种测试的后固化焓相同。

结论 后固化焓可用常规 DSC 测定,但是常常只能用 TMDSC 方法(例如 TOPEM™)才能观察到伴随的玻璃化转变。

3. The total heat flow as the sum of the two others. This corresponds to the conventional DSC curve.
The postcuring enthalpy is the same for both measurements, within the limits of reproducibility.

Conclusions The postcuring enthalpy can be determined by conventional DSC, but the accompanying glass transition can often only be observed using a TMDSC method (e. g. TOPEM™).

3.1.5 紫外光固化的 DSC 测量 UV curing measured by DSC

目的 本应用的目的是说明如何能用 DSC 来研究紫外光固化体系。
在紫外光固化中,出现三个主要问题:
1. 样品必须对紫外光曝光多久才能达到足够的固化度即交联度?
2. 温度和光强度对固化过程有什么影响?
3. 交联反应的最优化参数是什么?
本节描述光量热技术(这里是 UV-DSC)。上述问题将在本书后面的其他章节里详加讨论。

Purpose The purpose of this application is to show how UV curing systems can be investigated by DSC.
In UV curing, three main questions arise:
1. How long does the sample have to be exposed to UV light to achieve an adequate degree of cure or crosslinking?
2. What influence do the temperature and light intensity have on the curing process?
3. What are the optimum parameters for the crosslinking reaction?
The following section describes the technique of photocalorimetry (here UV-DSC). The questions outlined above will be addressed in more detail later.

样品 所测试的样品是商品粉末涂料,作为透明底漆主要用于中密度纤维板(MDF)和刨花板。

Sample The sample investigated was a commercially available powder coating material that is used as a transparent primer especially on medium density fiberboard (MDF) and chipboard.

条件
测试仪器:
带光量热附件的 DSC
坩埚:
40μL 铝坩埚,不带盖
样品制备:
将粉末样品(8.42mg)均匀地撒在坩埚底部,形成 0.8mm 厚的一层。

DSC 测试:
样品在固化温度下等温保持 25min。在该时间段,样品对设定强度的紫外光曝光 15min。

Conditions
Measuring cell:
DSC with photocalorimetry accessory
Pan:
Aluminum 40μL, without lid
Sample preparation:
The powder sample (8.42 mg) was spread evenly over the bottom of the crucible, forming a layer about 0.8 mm thick.

DSC measurement:
The sample was held isothermally for 25 minutes at the curing temperature. During this time interval, the sample was exposed UV light of a

defined intensity for 15 minutes.

Atmosphere: Static air

Design of the UV-DSC The DSC was equipped with an accessory that allowed the sample to be exposed to UV light. The design and geometry of the photocalorimeter system is shown in Figure 3.24.

The light source used was a Hamamatsu "Lightningcure 200" system with a built-in mercury-xenon lamp. This light source emits UV light mainly in the so-called UV-A region (315～400 nm). The UV light emitted in the 290～315 nm region is referred to as UV-B light. UV light of still shorter wavelengths is called UV-C light.

Other light sources can be used for other applications, e. g. halogen light.

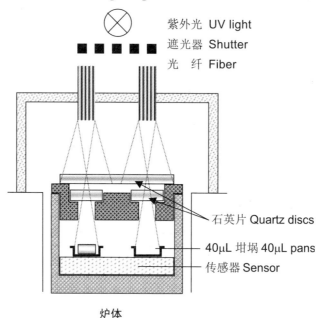

Fig. 3.24 Optical setup for projecting UV light into the DSC

Determination of the light intensity The light intensity on the sample side was determined using carbon black to ensure that the light was 100% absorbed. The carbon black must cover the bottom of the crucible as evenly as possible. The DSC cell is held isothermally at any desired temperature. After constant equilibrium heat flow has been reached, the light source is switched on for a few minutes while the reference side remains covered. A step evaluation of this curve yields the heat flow generated through the UV light absorption by the sample (Fig. 3.25). Assuming a homogeneous distribution of light over the entire area of the crucible, the light intensity can be calculated from the ratio of the step height to the area of the crucible. Figure 3.25 shows the measurement curves for the

的测试曲线。 three incident light intensities used here.

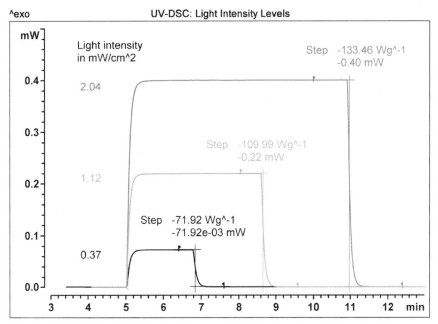

图 3.25 从 DSC 曲线测定入射光强度。(光源遮光器是手动开关的。)
Fig. 3.25 Determination of the incident light intensity from DSC curves.
(The shutter of the light source was opened and closed manually.)

入射光强度 I 从下式计算： The incident light intensity, I, is calculated from the equation
$$I = 台阶高度/\pi r^2$$ $$I = \text{step height}/\pi r^2$$

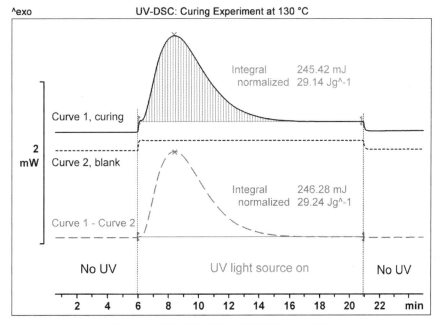

图 3.26 UV-DSC 实验反应进程和焓的测定。
Fig. 3.26 Determination of the course of the reaction and the enthalpy in UV-DSC experiments.

用 STEP 程序计算的台阶高度 mW 如图所示。40μL 铝坩埚底部的内半径为 0.25cm。

The step height in mW is calculated as shown in the diagram using the STEP program. The internal radius of the bottom of the 40μL aluminum crucible is 0.25 cm.

The incident light intensity can also be measured with a power meter.

Interpretation The sample was held isothermally at 130℃ for 25 min. After 6 min the shutter of the light source was opened for 15 min and the sample cured with UV light. The reaction begins as soon as the light is switched on and takes place in the same way as an isothermal curing process, that is, it dies down after a certain time because the possible reactants have been used up, i.e. undergone conversion (Curve 1). If the fully converted sample is exposed to light again in the same way as before, no further reaction peak is observed (Curve 2). This DSC curve can be used as the blank.

When the sample is exposed to light, a small amount of the light is absorbed and converted to heat. This is the cause of the small DSC step at the beginning and end of the measurement (Curves 1 and 2).

Evaluation To determine the enthalpy of curing, the baseline must be known during the curing reaction. The reaction enthalpy can be determined in two ways:
1. Through integration of the peak using a horizontal baseline drawn through the equilibrium heat flow after the curing before switching off the light (horizontal baseline through the measured value at 20.8 min). Here it is important that curing is complete, i.e. that the curve at the end is horizontal (Curve 1).
2. Through subtraction of a blank curve: If the completely cured sample is exposed to UV light of the same intensity and duration, a blank curve is obtained (Curve 2, blank) that shows how much heat was absorbed during UV exposure. The difference between the two curves is the net heat power of the crosslinking reaction, which is then integrated in the normal way.

As shown in Figure 3.26, within the limits of measurement accuracy, calculation of the curing enthalpy using both methods result in the same value.

Conclusions An accessory to the DSC enables a sample to be exposed to UV or visible light during a measurement (photocalorimetry). This allows the influence of exposure time, light intensity and temperature on the curing process to be investigated. The optimum parameters needed for a sample to achieve an adequate degree of crosslinking can then be determined.

3.2 效应的 TGA 测试　Measurement effects with TGA

热重分析（TGA）用于定量测定混合物的主要成分：
- 水分和其他挥发性化合物
- 挥发性反应产物，例如来自缩合反应的
- 由热解或燃烧产生的聚合物含量
- 由于部分分解产生或作为残余物的填料和增强纤维
- 分解和燃烧残余物的灰分。

通常以 TGA 曲线的一阶微分（称作 DTG 曲线）来表示各个失重台阶，以提供更好的视觉分离。同步测试 DSC 或 SDTA 曲线扩展了解释的可能性，会显示没有失重伴随的效应。

Thermogravimetric analysis (TGA) is used to quantitatively determine the main constituents of a mixture：
- Moisture and other volatile compounds
- Volatile reaction products, for example from condensation reactions
- Polymer content through pyrolysis or combustion
- Fillers and reinforcing fibers through partial decomposition or as a residue
- Ash as decomposition and combustion residue.

The first derivative of the TGA curve, the so-called DTG curve, is usually displayed in order to provide a better visual separation the individual mass loss steps. The simultaneous measurement of the DSC or SDTA curves expands the possibilities for interpretation and shows effects that are not accompanied by a loss of mass.

3.2.1　热固性树脂升温时的质量变化　Mass changes on heating a thermoset

目的　描述和解释典型的 TGA 测试。
Purpose　To describe and interpret a typical TGA measurement.

样品　未固化的 KU600 环氧树脂粉末。
Sample　Uncured KU600 epoxy resin powder.

条件
Conditions

测试仪器：TGA/SDTA
坩埚：
30μL 铝坩埚，不带盖
样品制备：
坩埚装入 15.931 mg 松散粉末，即没有压实。
TGA 测试：
以 10K/min 从 30℃升温至 700℃
气氛：
吹扫气体，50ml/min：先氮气，然后在 600℃ 自动切换至空气，还有 20mL/min 氮气作为保护性气体。

Measuring cell：TGA/SDTA
Pan：
Alumina 30 μL, without lid
Sample preparation：
The crucible was filled with 15.931 mg loose powder, i. e. without compacting.
TGA measurement：
Heating from 30℃ to 700℃ at 10 K/min
Atmosphere：
Purge gas, 50 mL/min: first nitrogen, then switched automatically to air at 600℃, as well as 20 mL/min nitrogen as protective gas.

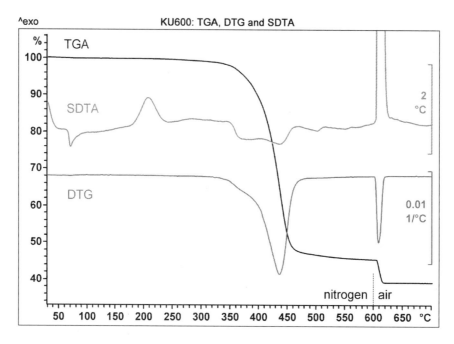

图 3.27　Fig 3.27

Interpretation　The TGA curve shows two large mass loss steps: first the decomposition of the polymer through pyrolysis between 320 and 480°C, and second, after switching to air at 600°C, the combustion of the carbon that had been mainly produced during pyrolysis. The slight loss of moisture up to 100°C is hardly noticeable. The two main steps can also be seen as large DTG peaks. The combustion of the carbon produces a large exothermic effect shown as an extremely large SDTA peak. In contrast, pyrolysis of the epoxy resin exhibits only a small endothermic effect. The TGA and DTG curves do not of course show the glass transition (seen as the small endothermic relaxation peak on the SDTA curve at 70°C) and the exothermic curing reaction between 180 and 240°C. The SDTA curve supplies similar information to DSC.

Evaluation　The content of polymer and filler can be directly estimated from the percent display of the TGA curve.

Conclusions　The TGA, DTG and SDTA curves show the pyrolysis and combustion processes exhibited by thermosets when they are heated in an inert gas atmosphere (pyrolysis) or air (combustion with oxygen). Effects that do not result in a change in mass, can only be observed in the SDTA curve.

解释　TGA 曲线显示两个大的失重台阶：第一个是聚合物在 320℃ 至 480℃ 之间的热分解，第二个是在 600℃ 切换到空气后，主要是热解生成的碳的燃烧。几乎观察不到 100℃ 前微量水分损失。这两个主要台阶也可表示为大的 DTG 峰。碳的燃烧产生大放热效应，呈现一个极大的 SDTA 峰。与之不同，环氧树脂的热解则只呈现一个小的吸热效应。TGA 和 DTG 曲线当然不显示玻璃化转变（在 SDTA 曲线上 70℃ 处显示为小的吸热松弛峰）和 180℃ 至 240℃ 之间的放热固化反应。SDTA 曲线提供与 DSC 相似的信息。

计算　能直接从 TGA 曲线估算聚合物和填料的百分比含量。

结论　TGA、DTG 和 SDTA 曲线显示热固性树脂在惰性气氛中（热解）或空气中（与氧燃烧）升温时呈现的热解和燃烧过程。只能在 SDTA 曲线上观察到不产生质量变化的效应。

3.2.2 含量测定:水分、填料和树脂含量
Content determination: moisture, filler and resin content

目的	说明含量的定量测定和炉体气氛(惰性和氧化)的影响。		
Purpose	To demonstrate the quantitative determination of content, and the influence of furnace atmosphere (inert and oxidative).		
样品	未固化的 KU600 环氧树脂粉末。		
Sample	Uncured KU600 epoxy resin powder.		
条件	测试仪器:TGA/SDTA	**Measuring cell**: TGA/SDTA	
Conditions	坩埚:	**Pan**:	
	70μL 铝坩埚,不带盖	Alumina 70μL, without lid	
	样品制备:	Sample preparation:	
	坩埚装入松散的粉末,即没有压实。	The crucible was filled with loose powder, i.e. without compacting.	
	TGA 测试:	TGA measurement:	
	以 10 K/min 从 40℃升温至 900℃	Heating from 40 to 900℃ at 10 K/min	
	气氛:	Atmosphere:	
	吹扫气体,70 mL/min:	Purge gas, 70 mL/min:	
	A:仅空气	A: only air	
	B:仅氮气	B: only nitrogen	
	C:先氮气,然后在 750℃自动切换至空气,另外以 20mL/min 氮气作为保护气体。	C: first nitrogen, then automatically switched to air at 750℃, and in addition 20 mL/min nitrogen as protective gas.	

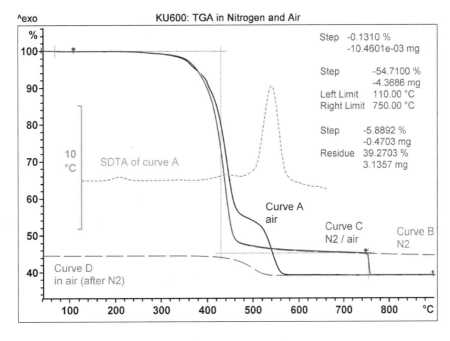

图 3.28 Fig 3.28

解释 将三个样品在不同气氛下升温(曲线 A、B 和 C)。第四条曲线

Interpretation Three samples were heated under different atmospheres (curves A, B and C). The fourth curve is that

是由样品 B 在空气中二次升温测得的(曲线 D)。

曲线 A:在高至约 480℃的第一个台阶观察到聚合物在空气中的分解。分解占了总失重的一半以上,没有发生燃烧,因为在 SDTA 曲线上观察不到大的放热效应。这个过程形成在高至 600℃时完全燃烧的残余物,SDTA 峰比 180℃处的固化反应峰(41.5 J/g)大约 50 倍,即根据初始样品质量的燃烧焓,因此约为 2500J/g(见前面的图和以后的 DSC 测试)。

曲线 B:聚合物分解发生在惰性气氛条件下,仍然形成残余物,这些在空气中第二次升温时被完全燃烧,因而灰分残余物与 A 一样。

曲线 C:高至 750℃,曲线与 B 是一样的。切换到空气时,碳化产物立即燃烧,因而剩余残留物的质量是一样的。对该曲线各个台阶进行数值计算:至 110℃的第一个台阶几乎观察不到,只有 0.13%(等于 10.5μg)的失重。在 110℃至 750℃之间的台阶显示 54.71%的失重,燃烧台阶为 5.89%。最后 900℃时的残余物仍有约 40%。各个台阶的范围用小旗标记。

计算 表 3.6 汇总了水分、聚合物和填料的含量。聚合物含量是热解台阶和氧化台阶之和。

obtained by heating the sample from B a second time but in air (curve D).

Curve A. The polymer decomposition in air is observed in the first step up to about 480℃. The decomposition represents more than half of the total mass loss and occurs without combustion because no large exothermic effect can be seen in the SDTA curve. During this process, residues are formed that burn completely up to 600℃. The SDTA peak is about fifty times greater than that of the curing reaction at 180℃ (41.5 J/g), i.e. the combustion enthalpy is therefore about 2500 J/g based on the original sample mass (see previous diagram and the DSC measurement later).

Curve B. Polymer decomposition occurs under inert conditions and residues are again formed. These are completely burned in the second heating run in air so that the ash residue is the same as in A.

Curve C. Up to 750℃, the curve is the same as B. On switching to air, the carbonized products immediately burn so that the mass of the remaining residue is the same. In this curve, the individual steps are numerically evaluated: The first step up to 110℃ is hardly visible and amounts to only 0.13% (corresponding to a mass loss of 10.5μg). The step between 110℃ and 750℃ shows a mass loss of 54.71%, and the combustion step 5.89%. The residue at the end at 900℃ is still about 40%. The individual step limits are marked by small flags.

Evaluation Table 3.6 summarizes the moisture, polymer and filler contents. The polymer content is the sum of the pyrolysis step and the oxidation step.

表 3.6 不同炉体气氛的 TGA 测试结果。
Table 3.6 Results of the TGA measurements performed under different furnace atmospheres.

炉体气氛 Furnace atmosphere	质量 Mass mg	至110℃水分 Moisture to 110℃ %	至750℃热解 Pyrolysis to 750℃ %	氧化 Oxidation %	900℃时填料(残余物) Filler(residue) at 900℃ %
A:空气 A: Air	9.871	0.16	—	60.77	39.07
B:氮气,然后第二次升温时空气 B: Nitrogen, then air on second heating run	8.808	0.14	54.88	5.42	(44.51) 39.03
C: N_2→空气 C: N_2→Air	7.985	0.13	54.71	5.89	39.27

| 结论 | 水分含量很低。聚合物含量为 60.8%，在 900℃时仍有 39.1% 残余物。这可以解释为填料。氮气下的升温表明，聚合物基体没有完全解聚，而是产生了含碳残余物。所以如果必须测定填料含量，样品必须在空气中燃烧。炉体气氛的选择也对含量测定有重要影响。在 600℃以上的温度，观察不到进一步的失重台阶，这意味着填料没有分解，似乎含有 $CaCO_3$。 | **Conclusions** The moisture content is very low. The polymer content is 60.8% and a residue of 39.1% remains at 900℃. This can be interpreted as filler. Heating under nitrogen shows that the polymer matrix does not completely depolymerize, but produces carbon-containing residues. So if only the filler content has to be determined, the sample must be burned in air. The choice of the furnace atmosphere also has an important influence on the content determination. At temperatures above 600℃, no further mass loss steps are visible, which means that the filler does not decompose, as would be the case with $CaCO_3$. |

3.2.3 苯酚—甲醛缩合反应的 TGA 分析
TGA analysis of a phenol-formaldehyde condensation reaction

目的	说明敞口坩埚中缩合反应的热重分析。由于干燥和反应形成的水分而发生蒸发失重。
Purpose	To demonstrate the thermogravimetric analysis of a condensation reaction in an open crucible. Mass losses occur due to drying and vaporization of the water formed in the reaction.
样品	含约 20% 苯酚-甲醛树脂的粉末酚醛模塑树脂。
Sample	Powdered PF molding resin containing about 20% phenol-formaldehyde resin.

| 条件
Conditions | 测试仪器：TGA
坩埚：
100μL 铝坩埚，不带盖
样品制备：
坩埚中装入 29.925 mg 松散粉末，即没有压实。
TGA 测试：
以 10K/min 从 30℃升温至 220℃
气氛：
吹扫气体：50mL/min 氮气；保护气体：20mL/min 氮气 | Measuring cell：TGA
Pan：
Aluminum 100μL, without lid
Sample preparation：
The crucible was filled with 29.925 mg loose powder, i.e. without compacting.
TGA measurement：
Heating from 30℃ to 220℃ at 10 K/min
Atmosphere：
Purge gas：50 mL/min nitrogen; protective gas：20 mL/min nitrogen |

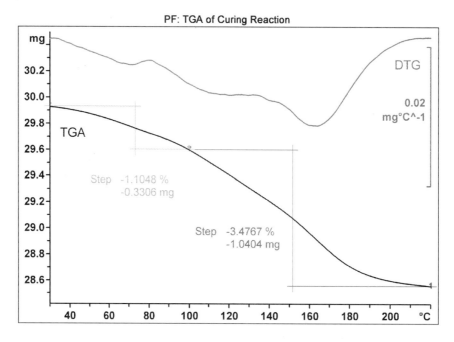

图 3.29　Fig 3.29

解释　TGA 升温曲线呈现连续失重,最大失重速率在 160℃（DTG 峰）。DTG 曲线上的小峰表示最大干燥速率在约 70℃。至 100℃时的失重为 1.1%。作为反应的结果,至 220℃有一个 3.5%的进一步失重。

Interpretation　The TGA heating curve shows a continuous loss of mass with the greatest rate of mass loss at 160℃ (DTG peak). The small peak in the DTG curve shows the maximum drying rate at about 70℃. The mass loss up to 100℃ is 1.1%. As a result of the reaction, there is a further loss of mass of 3.5% up to 220℃.

结论　在将 B 阶酚醛模塑树脂升温至 200℃以上的加工温度时,材料损失约 4.6%的水。由于缩合反应开始,干燥过程与失水重叠。在所用实验条件下无法将这两个效应相互完全分开。

Conclusions　On heating the PF molding resin in the B-stage to processing temperatures above 200℃, the material loses about 4.6% water. The drying process is overlapped by loss of water due to the beginning of the condensation reaction. The two effects cannot be properly separated from one another under the experimental conditions used.

3.3　效应的 TMA 测量　Measurement effects with TMA

热机械分析（TMA）通过测试样品尺寸与温度的关系来测定重要的性能和温度效应:
- 热膨胀（线性热膨胀系数 CTE）
- 玻璃化转变温度和软化点
- 蠕变和收缩行为,热学或力学松驰
- 固化凝胶点
- 弹性行为和弹性模量变化

Thermomechanical analysis (TMA) is used to determine important properties and temperature effects by measuring the dimensions of a sample as a function of temperature:
- Thermal expansion (coefficient of linear thermal expansion, CTE)
- Glass transition temperature and softening point
- Creep and shrinkage behavior, thermal or mechanical relaxation
- Gel point on curing
- Elastic behavior and change of the elastic modulus
- Dimensional stability, delamination

- 尺寸稳定性,分层
- 溶剂中溶胀

能测试宽广范围的样品尺寸(样品厚度),从非常薄的涂层(≤1μm)到20mm长的物体。SDTA曲线同步测试有助于解释和显示未涉及尺寸变化的效应。

- Swelling in solvents.

A wide range of sample dimensions (sample thickness) can be measured, from very thin coatings (≤1 μm) to objects that are 20 mm long. The simultaneous measurement of the SDTA curve facilitates interpretation and shows effects that do not involve a change of dimensions.

3.3.1 线膨胀系数的测定　Determination of the linear expansion coefficient

对于随温度而变化的材料的膨胀行为在建筑和机械工程中是非常重要的性能。对于复合材料,尤其如此。如果将膨胀系数迥然不同的材料连接在一起,当温度变化时,他们之间会产生大的应力。这甚至会导致层压板或复合材料的毁坏。因此,关于与温度有关的材料膨胀系数的详细信息对这些高性能材料的开发和使用是必不可少的。

The expansion behavior of a material on temperature change is a very important property in constructional and mechanical engineering. This is especially the case with composite materials. If materials with very different expansion coefficients are joined together, large stresses can occur between them when the temperature changes. This could even lead to the destruction of the laminate or composite. For this reason, detailed information about the temperature dependent expansion coefficients of materials is essential for the development and use of these high performance materials.

目的 **Purpose**	测定固化涂料粉末50℃至150℃间的热膨胀系数(CTE)。 To determine the coefficient of thermal expansion (CTE) of a cured coating powder between 50℃ and 150℃.	
样品 **Sample**	固化的KU600环氧树脂粉末形成的圆柱,直径6 mm,高度1.92 mm。 Cured KU600 epoxy resin powder formed as a cylinder, diameter 6 mm, height 1.92 mm.	
条件 **Conditions**	测试仪器:TMA 探头: 球点探头(石英玻璃);直径3 mm 样品制备: 圆柱样品的两平端用细砂纸打磨。石英玻璃片放置在样品与探头间和样品与样品支架间,以使探头力均匀分布在整个截面上,从而得到低压缩应力。 样品在测试前在TMA中升温一次以除去松弛效应。 TMA测试: 以5K/min从40℃各升温至160℃ 探头上的力:0.02 N 气氛:静态空气	**Measuring cell**:TMA probe: Ball-point probe (quartz glass); diameter 3 mm **Sample preparation**: The flat ends of the cylindrical sample were polished with fine sandpaper. Thin quartz glass disks were placed between the sample and probe and between the sample and sample support to distribute the force of the probe evenly over the entire cross-sectional area and so create a low compressive stress. The sample was heated in the TMA once before the measurement to remove relaxation effects. **TMA measurement**: Heating from 40℃ to 160℃ at 5 K/min **Force on the probe**:0.02 N **Atmosphere**:Static air

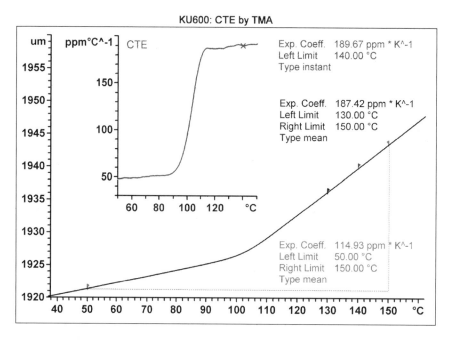

图 3.30　Fig 3.30

Interpretation The cylinder had an initial height of 1920 μm and underwent expansion on warming. The expansion up to about 100℃ is relatively small; afterward the slope of the TMA curve is noticeable steeper. The expansion rate, visible as the slope of the curve, is therefore a function of temperature. The change of slope indicates the glass transition, a temperature at which many physical properties change.

The smaller inserted diagram shows a plot of the instantaneous expansion coefficients. At 140℃, the expansion coefficient is 189.8 ppm/K. The mean expansion coefficient between 130℃ and 150℃ is calculated. Because the curve is a straight line in this temperature range, the two values are practically the same.

Evaluation The expansion coefficient is defined as follows:

$$\alpha = \frac{dL}{dT}\frac{1}{L_0}$$

dL: length change of the sample for a temperature change of dT

L_0: initial length at room temperature, 25℃

α: according to this definition is also called the instantaneous expansion coefficient.

The usual unit is $10^{-6} K^{-1}$.

Table 3.7 shows the expansion coefficients and the length of the sample at different temperatures as they can be read off in the inserted diagram.

表 3.7 样品长度和膨胀系数随温度的变化。
Table 3.7 Change of sample length and expansion coefficient with temperature.

温度 Temperature ℃	样品长度 Sample length μm	膨胀系数 α Expansion coefficient α $10^{-6} K^{-1}$
50.0	1921.34	47.19
70.0	1923.21	49.47
90.0	1925.17	54.39
110.0	1929.17	177.5
130.0	1936.21	188.76
150.0	1943.41	191.38

平均膨胀系数 $\bar{\alpha}$ 表示样品在温度 T_1 至 T_2 范围内的膨胀程度。

L_0:样品在 T_0（室温）的长度（参比长度）
L_1:样品在较低温度 T_1 的长度
L_2:样品在较高温度 T_2 的长度

上图中以虚线给出对温差 50℃ 至 150℃ 的 $\bar{\alpha}$ 测定的例子。对于该曲线形态，值为 114.9 $10^{-6} K^{-1}$。

膨胀系数由测试曲线减去空白曲线测定。因此需要两次测试，一次用样品，一次不用样品。

结论 固体材料的膨胀行为通过测试样品长度/高度与温度的关系来表示。线膨胀系数 α 可从 TMA 曲线的斜率计算，通常与温度有关。α 的台阶状变化表明材料中已经发生热效应。

小膨胀系数的测定需要相对高的样品（大 L_0）。为了避免大样品中的温度梯度，因而使用低升温速率（最大 5K/min）。应始终减去空白曲线以确保准确的结果。

数值计算必须区分瞬时系数和平均系数。平均膨胀系数表示一种材料在通常是大的温度区间内膨胀有多大，而瞬时系数表示特定温度点的膨胀。

The mean expansion coefficient $\bar{\alpha}$ indicates the extent to which a sample expands in the temperature range T_1 to T_2.

$$\bar{\alpha}_{T_1 T_2} = \frac{L_2 - L_1}{T_2 - T_1} \frac{1}{L} = \frac{\Delta L}{\Delta T} \frac{1}{L_0}$$

L_0: length of the sample at T_0 (room temperature) (reference length)
L_1: length of the sample at the lower temperature T_1
L_2: length of the sample at the upper temperature T_2

An example for the determination of $\bar{\alpha}$ is given in the above diagram through the dotted line for the temperature difference 50℃ to 150℃. With this curve shape, the value is 114.9 $10^{-6} K^{-1}$.

The expansion coefficient is determined by subtracting a blank curve from the measurement curve. Two measurements are therefore needed-one with, and one without the sample.

Conclusions The expansion behavior of a solid material is shown by measuring the sample length/height as a function of temperature. The linear expansion coefficient, α, can be calculated from the slope of the TMA curve. This is usually a function of temperature. A step-like change of α indicates that a thermal effect has occurred in the material.

The determination of low expansion coefficients requires relatively high samples (large L_0). Low heating rates (max. 5 K/min) are therefore used in order to avoid temperature gradients in the large samples. A blank curve should always be subtracted to ensure accurate results.

In the calculation of the numerical values, a distinction must be made between instantaneous and mean coefficients. The mean expansion coefficient shows how much a material expands over a usually large temperature interval, whereas the instantaneous coefficient shows the expansion at a particular temperature.

3.3.2 玻璃化转变的 TMA 测量 Measurement of the glass transition by TMA

因为协同重排需要额外的空间,所以液态的自由体积要比玻璃态的大。这意味着在玻璃化转变时膨胀系数 α 发生变化。用 TMA 能容易地测试该效应。此方法相当灵敏,因为对于未填充的热固性树脂,α 从约 50 增至 150×10^{-6}/K。如果玻璃化转变以上的黏度低,则材料软化,能够流动。这会引入测试误差。高探头力条件下的 TMA 测试对软化温度测定是理想的。因此该测试模式也能用来测试薄膜和涂料。推荐用 TMA 或 DLTMA 的弯曲测试来测定基体树脂含量低的材料的软化温度。

有关玻璃化转变性质的更多细节可参见 3.1.1 节的介绍。

已经有许多描述用 TMA 测定玻璃化转变温度的不同标准方法,例如,ASTM E1545、CEI-IEC 1006、IPC-TM 650、ISO 7884-8 和 ISO 11359-2。测试和计算参数定义清晰,因此可有意义地比较结果。

The free volume in the liquid state is greater than in the glassy state because the cooperative rearrangements need additional space. This means that at the glass transition, the coefficient of expansion, α, changes. This effect can easily be measured by TMA. The method is relatively sensitive because with unfilled thermosets, α increases from about 50 to 150 ppm/K. If the viscosity above the glass transition is low, the material softens and is able to flow. This introduces measurement errors. TMA measurements under conditions of high probe force are ideal for determining the softening temperature. This measuring mode can therefore also be used to measure thin films and coatings. Bending measurements using TMA or DLTMA are recommended for the determination of the softening temperature of materials with low contents of matrix resin.

Further details on the nature of the glass transition can be found in the introduction to Section 3.1.1.

A number of different standard procedures for the determination of the glass transition temperature by TMA have been described, for example, ASTM E1545, CEI-IEC 1006, IPC-TM 650, ISO 7884-8, and ISO 11359-2. The measurement and evaluation parameters are clearly defined so that the results can be meaningfully compared.

3.3.2.1 测定玻璃化转变的膨胀曲线

Determination of the glass transition by means of the expansion curve

目的	按照通常的标准方法测定玻璃化转变温度	
Purpose	To determine the glass transition temperature according to the usual standard procedures	
样品	固化的 KU600 环氧树脂粉末形成的圆柱,直径 6mm,高度 1.92mm。	
Sample	Cured KU600 epoxy resin powder formed as a cylinder, diameter 6 mm, height 1.92 mm.	
条件	测试仪器:TMA	**Measuring cell**:TMA
Conditions	探头:	**Probe**:
	球点探头(石英玻璃);直径 3 mm	Ball-point probe (quartz glass); diameter 3 mm
	样品制备:	**Sample preparation**:
	圆柱样品的两平端用细沙纸打磨。薄石英玻璃片放置在样品与探头间和样品与样品支架间,以使探头力均匀分布在整个截面上,从而得到低压缩应力。	The flat ends of the cylindrical sample were polished with fine sandpaper. Thin quartz glass disks were placed between the sample and probe and between the sample and sample support to distribute the force of the probe evenly over the entire cross-sectional area and so create a low compressive stress.

样品在 TMA 中升温两次以证实在第一次升温时的松弛效应。

TMA 测试：

以 5K/min 从 40℃加热至 160℃

探头上的力：0.02 N

The sample was heated in the TMA twice to demonstrate the relaxation effects on the first heating run.

TMA measurement：

Heating from 40℃ to 160℃ at 5 K/min

Force on the probe：0.02 N

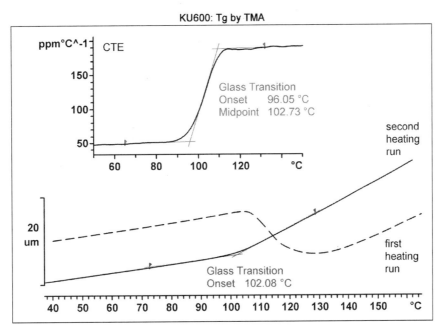

图 3.31　Fig 3.31

解释 图中所示为用上例中的相同曲线来计算的膨胀系数。此外，显示了第一次测试的曲线（虚线）。尽管小心地制备样品，它表明还是发生了轻微的软化和可能的热松弛。结果，石英圆片更均匀的压在样品上，因而用同样的探头力测试的第二次升温曲线就是正常的膨胀曲线。

计算 按所涉及的标准方法，在 TMA 玻璃化转变前后绘制切线。切线的交点就是玻璃化转变温度 T_g。

玻璃化转变的特征温度也可以依照通常的 DSC 方法从 CTE 曲线测定。这表示在上面的图中。

结论 TMA 是测定填充或增强热固性树脂和大样品的玻璃化转变温度 T_g 的一种很好的方法。通常，

Interpretation The figure shows the same curves used in the previous example to calculate the expansion coefficient. In addition, the curve of the first heating measurement is displayed (dotted line). This shows that slight softening and possibly thermal relaxation occur despite the careful sample preparation. As a result of this, the quartz disks are more uniformly pressed on the sample, so that the second heating curve measured with the same probe force is the normal expansion curve.

Evaluation In the standard method referred to, tangents are drawn on the TMA curve before and after the glass transition. The point of intersection of these tangents is the glass transition temperature, T_g.

A characteristic temperature of the glass transition can also be determined from the CTE curve according to the usual DSC methods. This is shown in the upper diagram.

Conclusions TMA is an excellent method for determining the glass transition temperature, T_g, of filled or reinforced thermosets, and of large samples. Usually, special sample

如果 T_g 能从第二次升温曲线中测定,就不需要特殊的样品制备。

preparation is not necessary if the T_g can be determined from the second heating curve.

3.3.2.2 薄涂层软化温度的测定
Determination of the softening temperature of thin coatings

目的 **Purpose**	测定软化温度,因而获得关于玻璃化转变温度范围的信息	To determine the softening temperature and thus obtain information about the glass transition temperature range.
样品 **Sample**	涂有约 27μm 厚的固化 KU600 环氧树脂粉末涂层的铝板,1.5mm 厚。	Aluminum plate, 1.5 mm thick, with a coating of cured KU600 epoxy resin powder about 27 μm thick.
条件 **Conditions**	测试仪器:TMA 探头: 球点探头(石英玻璃);直径 3 mm 样品制备: 从板上锯下约 5×5mm 的一片。探头直接放在涂层上,对样品施加大的压缩应力。 TMA 测试: 以 5K/min 从 40℃升温至 190℃ 负载:1.0N 气氛:静态空气	Measuring cell:TMA Probe: Ball-point probe (quartz glass); diameter 3 mm Sample preparation: A piece of about 5×5 mm was sawn out of the plate. The probe was placed directly on the coating so that it exerted a large compressive stress on the sample. TMA measurement: Heating from 40 to 190℃ at 5 K/min Load:1.0 N Atmosphere:Static air

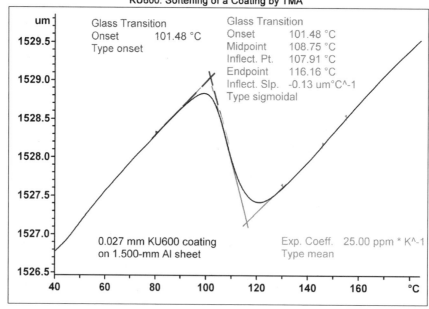

图 3.32　Fig. 3.32

解释 从球点探头的穿透测试(约 1.5μm)可清晰观察到薄涂层的软

Interpretation The softening of the thin coating is clearly apparent from the penetration measurement of the ball-point

化。探头的形状、所施加的力和填料阻止探头很深入地穿透涂层。效应前后的膨胀速率等于铝板的膨胀速率 $25 \cdot 10^{-6} K^{-1}$,因为薄涂层几乎不对膨胀作出贡献(100K 内约 $0.3 \mu m$)。因此,不可能进行类似于前例中的计算。

计算 玻璃化转变的测定是用软化温度进行的,类似于用穿透的外推起始点定义为 T_g 的 CEI-IEC 1006 方法。这里同样的,切线(红色虚线)位置可以有点变动。不过在膨胀曲线的情况下,这对起始点因而 T_g 的影响要小得多。

其他的穿透特征温度是拐点、终点和中点,它们在整个 S 形曲线的评估中给予了计算。

结论 TMA 是测定很薄的涂料或清漆涂层的软化温度因而间接获得玻璃化转变信息的很好方法。通常不需要特殊的样品制备,即涂层不必从支撑材料进行物理分离。软化温度直接从第一次升温曲线测定。

在引用结果时,始终必须说明所使用的测试和计算方法,因为探头力和探头末端的形状影响穿透的程度和位置。由于这个原因,所以穿透测试的软化温度与膨胀测试的玻璃化温度不是完全一致的。

probe (about $1.5 \mu m$). The shape of the probe, the applied force, and the fillers prevent the probe from penetrating very far into the coating. The rate of expansion before and after the effect corresponds to that of the aluminum plate of $25 \cdot 10^{-6} K^{-1}$ because the thin coating hardly contributes to the expansion (about $0.3 \mu m$ over 100 K). An evaluation similar that in the previous example is therefore not possible.

Evaluation The determination of the glass transition is performed using the softening temperature, analogous to the CEI-IEC 1006 method in which the extrapolated beginning of the penetration is defined as the T_g. Here again, the location of the tangents (red dashed lines) can vary somewhat. This however affects the onset and hence the T_g much less in the case of the expansion curve.

Further characteristic temperatures of the penetration are the point of inflection, the endpoint and the midpoint, which are calculated in the evaluation of the entire S-shaped curve.

Conclusions The TMA is an excellent method for determining the softening temperature of very thin paint or varnish coatings and thereby indirectly obtaining information about the glass transition. Usually, special sample preparation is not necessary, that is, the coating does not have to be physically separated from the support material. The softening temperature is directly determined from the first heating curve.

The measurement and evaluation method used must always be stated when quoting results because the probe force and shape of the probe tip influence the extent and location of the penetration. For this reason, the softening temperatures of penetration measurements are not exactly identical to the glass transition temperature of the expansion measurement.

3.3.2.3 由弯曲测试测定玻璃化转变
Determination of the glass transition from bending measurements

目的	用基于弹性行为变化的方法测定玻璃化转变温度。	
Purpose	To determine the glass transition temperature using a method based on the change of elastic behavior.	
样品	作为两块金属板之间的黏接层的固化 KU600 环氧树脂粉末。	
Sample	Cured KU600 epoxy resin powder as an adhesive layer between two metal plates.	
条件	测试仪器:TMA	**Measuring cell**:TMA
Conditions	探头:	**Probe**:

| 球点探头(石英玻璃);直径 3mm。 | Ball-point probe (quartz glass); diameter 3 mm, |
| 三点弯曲附件。 | 3-point bending accessory. |

样品制备:

KU600 环氧树脂粉末涂料在两片 0.1mm 厚的钢板间固化。它们是电动剃须刀片,16mm 长、5mm 宽。涂层厚度是 0.06mm。这产生了一个三明治结构,如同在黏合剂黏接中会发生的那样。已固化的样品放在弯曲附件上,因而可使用支座间的整个 14mm 长度。

Sample preparation:

The KU600 epoxy powder coating was cured between two 0.1mm thick steel plates. These were pieces of razor blades, 16mm long and 5mm wide. The thickness of the layer was 0.06mm. This produced a sandwich structure such as could occur in adhesive bonding. The cured sample was mounted on the bending accessory so that the full length of 14mm could be used between the supports.

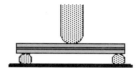

TMA 测试:

以 5K/min 从 50℃升温至 240℃
负载:每 6 秒(周期 12 秒)在 0.5N 与 1.0N 之间交变
气氛:静态空气

TMA measurement:

Heating from 50℃ to 240℃ at 5 K/min
Load: alternating between 0.5N and 1.0N every 6s (period 12s)
Atmosphere: Static air

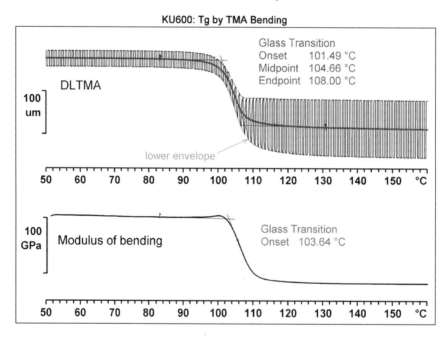

图 3.33　Fig 3.33

解释 由于对探头施加力而使三明治结构发生稍许弯曲。在低于玻璃化转变温度时,黏接层(即固化的环氧粉末)因 0.5N 力的变化而产生约 40μm 的弯曲变化。在玻璃化转变以上,样品弯曲更显著,在 1N 时

Interpretation The sandwich structure bends slightly due to the force applied to the probe. The change in force of 0.5N produces a change in bending of about 40μm as long as the adhesive layer (i.e. the cured epoxy powder) is below the glass transition temperature. Above the glass transition, sample bending is more pronounced, about 200μm at 1N. The change in

约 200μm。随变化力而变化的弯曲较大，约 130μm。这种变化记录在 DLTMA 曲线上，可由此测定软化即玻璃化转变温度。该温度可以计算为平均曲线(红色)的起始点或中点。

上下包络线(虚线)表示对于 0.5N 和 1.0N 探头力的弯曲曲线。包络线之差是样品弯曲模量弹性的量度。可以假定钢刀片的弹性模量几乎不变，因而变化就只由于热固性树脂而引起的。三明治结构的弯曲模量曲线也可以用来测定"黏合剂"的 T_g，如下面的图中起始点所示的。

计算　三明治结构的弯曲模量(E)可按照下式直接从 DLTMA 曲线计算：

式中 ΔF 是作用在探头上力的变化，ΔL 是相应的弯曲变化。由厚度(a)、宽度(b)和支座间的距离(d)定义样品的几何因子，因而 ΔL 是上下包络线间的距离。因此，DLTMA 振幅与模量成反比。弯曲模量的测定需要减去空白曲线。

软化温度从静态弯曲测定，例如从平均值曲线或一条包络线，就如在穿透测试中那样(3.3.2.2节)。因此软化温度不用变换负载就可以从一个测试计算得到，例如在 1N 的恒力下。在这种情况下，得到的曲线会近似等于下包络线。

玻璃化转变温度通常从弯曲模量曲线上测定为起始点，就象在 DMA 测试中那样。

表 3.8 汇总了几个计算的结果，表明用变化负载的 TMA 方法测定的玻璃化转变温度在一定程度上与计算方法有关。此外，温度也与频率有关(见 DMA 测试)。

bending with the changing force becomes larger, about 130 μm. This is recorded in the DLTMA curve and allows the softening or glass transition temperature to be determined. The temperature can be evaluated as the onset or midpoint of the mean curve (red).

The upper and lower envelopes (dashed lines) show the bending profiles for probe forces of 0.5N and 1.0N. The difference between these envelopes is a measure for the elasticity for the modulus of bending of the sample. It can be assumed that the elastic modulus of the steel blades hardly changes so that the changes are only due to the thermoset. The curve of the bending modulus of the sandwich structure can also be used to determine the T_g of the "adhesive" as shown by the onset in the lower diagram.

Evaluation　The bending modulus (E) of the sandwich structure can be directly calculated from the DLTMA curve according to the following formula:

$$E = \frac{\Delta F \cdot d^3}{4 \cdot \Delta L \cdot b \cdot a^3}$$

where ΔF is the change of the force acting on the probe and ΔL the corresponding bending change. The thickness (a), width (b) and distance between the supports (d) define the geometry factor of the sample. ΔL is therefore the distance between the upper and lower envelopes. It follows from this that the DLTMA amplitude is inversely proportional to the modulus. The determination of the bending modulus requires the subtraction of a blank curve.

The softening temperature is determined from the static bending, for example from the mean value curve or an envelope, just as in a penetration measurement (Section 3.3.2.2). The softening temperature can therefore be evaluated from a measurement without alternating load, for example at a constant force of 1N. In this case, the resulting curve would correspond approximately to the lower envelope.

The glass transition temperature is usually determined from the bending modulus curve as the onset, just like in DMA measurements.

Table 3.8 summarizes the results of several evaluations. It shows that glass transition temperatures determined by the TMA method with changing load to some extent depend on the evaluation. Furthermore, the temperatures are frequency dependent (see the DMA measurements).

表 3.8 由 DLTMA 测定的玻璃化转变温度
Table 3.8 Glass transition temperatures from DLTMA

计算曲线 Evaluated curve	起始点,Onset ℃	中点,Midpoint ℃
平均值曲线 Mean value curve	101.5	104.7
下包络线(1N) Lower envelope (1N)	101.1	104.9
上包络线(0.5N) Upper envelope (0.5N)	100.0	103.3
弯曲模量 Bending modulus	103.6	108.3

结论 带静态和变化探头力的弯曲测试是测定涂层、黏接层和填充热固性树脂的软化即玻璃化转变温度的灵敏方法。因为只关心弯曲行为变化的温度范围或模量，所以耗时的多次测试是不必要的，样品制备通常只是让样品的几何形状适合于弯曲测试附件。由于能够施加在TMA上的最大力是受限制的，所以弯曲样品应尽可能薄而可以测试最小弯曲和弯曲变化。

如同以前和其他方法提到的一样，当报告弯曲测试结果时，说明测试如何进行和结果如何计算是重要的。

除了 DMA 外，DLTMA 测试经常是测定高填充或高增强热固性树脂的软化点或玻璃化转变的唯一可能性。对于这样的材料，玻璃化转变时的热容和膨胀系数变化常常非常小。

3.3.3 固化反应的 TMA 测量

在固化反应中，树脂从液态变化至橡胶弹性态，然后达到高度交联的玻璃态。材料的强度大大提高。可以用 DLTMA 测试弹性行为来研究该过程。在一定反应转化率后，树脂形成凝胶体，黏度变得非常大。树脂的可塑性显著降低，这也可用 DLTMA 来研究。当 DLTMA 测

Conclusions Bending measurements with static and changing probe force are sensitive methods for determining the softening or glass transition temperature of coatings, adhesive bonding layers and filled thermosets. Since only the temperature region of the change of bending behavior or the modulus is of interest, no time-consuming multiple measurements are necessary, and sample preparation is usually simply a matter of adapting the geometry of the sample to the bending measurement accessory. Since the maximum force that can be applied in the TMA is limited, the bending samples should be as thin as possible so that a minimal bending and change in bending can be measured.

As previously mentioned with other methods, when reporting the results of bending measurements, it is important to state how the measurements were performed and how the results were evaluated.

Apart from DMA, a DLTMA measurement is often the only possibility to determine the softening point or glass transition of highly filled or highly reinforced thermosets. With such materials, the changes in heat capacity and expansion coefficient at the glass transition are often very small.

Measurement of the curing reaction by TMA

In a curing reaction, the resin changes from a liquid state to a rubbery elastic state and then to a highly crosslinked glassy state. The strength of the material increases tremendously. This process can be investigated by measuring elastic behavior using DLTMA. After a certain reaction conversion, the resin forms a gel and the viscosity becomes very large. The plasticity of the resin decreases very markedly, which can also be investigated by DLTMA. While DLTMA measures the elastic behavior, the

试弹性行为时,同步 SDTA 曲线就象 DSC 那样显示吸热或放热反应焓。这就能直接比较力学信息和焓变。

simultaneous SDTA curve shows endothermic or exothermic reaction enthalpies just like DSC. This enables a direct comparison of mechanical information and enthalpy changes to be made.

3.3.3.1 固化反应的弯曲测量研究
Investigation of the curing reaction using bending measurements

目的 **Purpose**	根据弹性行为变化测定固化反应和玻璃化转变温度。 To determine the curing reaction and the glass transition temperatures based on the change in elastic behavior.	
样品 **Sample**	金属刀片上涂层的未固化 KU600 环氧树脂粉末。 Uncured KU600 epoxy resin powder as a coating on a metal blade.	
条件 **Conditions**	测试仪器:TMA/SDTA 探头: 球点探头(石英玻璃);直径 3 mm 三点弯曲附件 **样品制备**: 将少量 KU600 环氧树脂粉末以薄层铺在 0.1mm 厚、16mm 长和 5mm 宽的钢刀片(一片电动剃须刀片)上,加热到 100℃,结果粉末熔融并结合,生成 0.5mm 厚度的薄膜。样品放在弯曲附件上,这样支座间有 14mm 的长度。探头直接放在涂层上。	**Measuring cell**:TMA/SDTA **Probe**: Ball-point probe (quartz glass); diameter 3 mm 3-point bending accessory **Sample preparation**: A small amount of KU600 epoxy resin powder was spread in thin layer over a steel blade 0.1 mm thick,16mm long and 5mm wide (a piece of a razor blade) and heated to 100℃ so that the powder melted and coalesced. This produced a film thickness of 0.5mm. This sample was mounted on the bending accessory so that there was a length of 14mm between the supports. The probe was placed directly onto the coating.

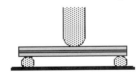

	TMA 测试: 以 5K/min 从 30℃ 至 260℃,对同一样品进行第一次和第二次升温测量。 力:每 6s 在 0.5N 与 1.0N 之间交变(周期 12s) 气氛:静态空气	**TMA measurement**: 1st and 2nd heating curves of the same sample, measured from 30℃ to 260℃ at 5K/min. Force: alternating between 0.5 and 1.0N every 6s (period 12s) **Atmosphere**:Static air

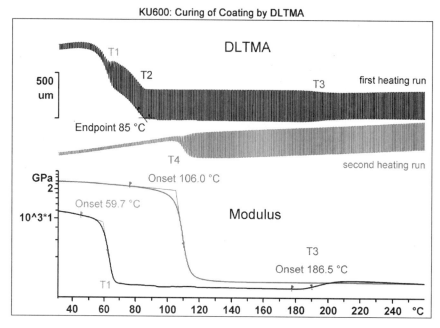

图 3.34　Fig. 3.34

解释　图 3.34 的上半部分,表示涂层在弯曲模式中第一次(黑色)和第二次(灰色)升温时的 DLTMA 曲线。图的下半部分,是纵坐标以对数刻度表示的从两条 DLTMA 曲线计算得到的两条弯曲模量曲线。

未固化的涂层在 T1 软化。由 DLTMA 振幅的加大明显观察到玻璃化转变,相应的模量下降。由于树脂黏度降低,探头进一步穿透涂层,直到它直接触及金属刀片(T2)。T2 至 T3 之间的弹性模量大约等于钢的弹性模量。

在 T3,模量重新增大,样品变得较硬,这是交联反应的结果。样品第二次升温的行为表明材料业已完全固化:低于 T4 时,DLTMA 振幅非常小,弯曲模量相应大。玻璃化转变温度显著高于 T1。

图 3.35 将反应放大,并一起展示了呈清晰放热峰的 SDTA 曲线。该曲线与热生成速率成正比,因而转化率曲线可以通过部分积分来计算。与模量曲线的比较表明,在观察到显著的力学行为的变化前,已有约 30% 的材料反应。

Interpretation　The upper part of Figure 3.34 shows the DLTMA curves of the coating in bending mode for the 1st (black) and 2nd (blue) heating runs. The lower part of the figure shows the two bending modulus curves on a logarithmic ordinate scale calculated from the two DLTMA curves.

At T1, the uncured coating softens. The glass transition is apparent as an increase in the DLTMA amplitude and the corresponding decrease of the modulus. Since the viscosity of the resin decreases, the probe penetrates further into the coating until it rests directly on the metal blade (at T2). The elastic modulus between T2 and T3 corresponds roughly to that of steel.

At T3, the modulus increases again, the sample becomes harder, a consequence of the crosslinking reaction. The fact that the material is fully cured is shown by the behavior of the sample in the 2nd heating run: the DLTMA amplitude below T4 is very small and the bending modulus correspondingly large. The glass transition temperature is significantly higher than T1.

In Figure 3.35, the reaction is zoomed and displayed together with the SDTA curve, which shows a clear exothermic peak. This curve is proportional to the rate of heat production, so that the conversion curve can be calculated by partial integration. A comparison with the modulus curve shows that about 30% of the material has already reacted before a significant change of mechanical behavior is observed.

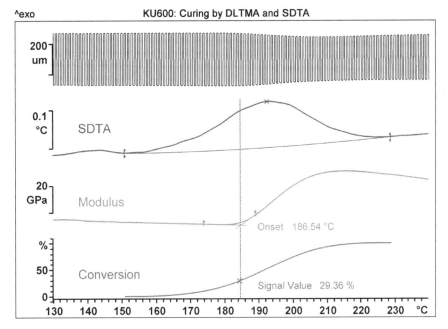

图 3.35　Fig. 3.35

计算　DLTMA 曲线用空白曲线作了修正。

表 3.9 汇总了几个计算的结果,表明由变化负载的 TMA 方法测定的玻璃化转变温度在一定程度上与计算方法有关。此外,温度也与频率有关(见 DMA 测试)。

Evaluation　The DLTMA curves were corrected using blank curves.

Table 3.9 summarizes the results of several evaluations. It shows that glass transition temperatures determined by the TMA method with changing load depend to some extent on the evaluation. Furthermore, the temperatures are frequency dependent (see the DMA measurements).

表 3.9　从弯曲模量曲线测定的特征温度
Table 3.9　Characteristic temperatures determined from the bending modulus curves.

效应 Effect	温度 Temperature	起始点 Onset in ℃
未固化涂层的玻璃化转变 Glass transition of the uncured coating	T1	59.7
穿透结束 End of the penetration	T2	85.0
固化 Curing	T3	186.5
固化涂层的玻璃化转变 Glass transition of the cured coating	T4	106.0

结论　探头力变化的弯曲测试(DLTMA 模式)不仅显示玻璃化转变时弯曲模量的变化,而且显示固化反应中发生的力学变化,即交联时刚度增大。同步测量的 DLTMA 和 SDTA 曲线可以比较力学和焓的变化。

Conclusions　The bending measurements with changing probe force (DLTMA mode) not only show the changes of the bending modulus at the glass transition, but also the mechanical changes that occur during the curing reaction, that is, the increase in stiffness on crosslinking. Simultaneously measured DLTMA and SDTA curves allow the mechanical and enthalpy changes to be compared.

3.3.3.2 凝胶时间的 DLTMA 测定　Determination of the gelation time by DLTMA

目的 **Purpose**	测定环氧树脂粉末的凝胶时间(凝胶点)。 To determine the gelation time (gel point) of an epoxy powder.	
样品 **Sample**	未固化的 KU600 环氧树脂粉末。 Uncured KU600 epoxy resin powder.	
条件 **Conditions**	测试仪器:TMA 探头: 球点探头(石英玻璃);直径 3mm 样品制备: 温度程序开始并达到温度 160℃后,打开炉体,将几毫克的涂料粉末直接撒在样品支架上,熔融的粉末黏合并在探头下形成一滴。然后立即关闭炉体。 测试后,将样品烧掉,伴随着一小缕火焰,在冷却后刷去残余物。 注意:残余物不可变为玻璃态而粘在探头或样品支架上。 TMA 测量: 在 160℃等温测试 40min 负载:每 6s 在 +0.05 N 与 -0.05 N 之间交变(周期 12s),即探头从样品支架上抬起 6s,然后放在样品上或陷入样品中 6s。 气氛:静态空气	Measuring cell:TMA Probe: Ball-point probe (quartz glass); diameter 3 mm Sample preparation: After the temperature program had been started and the temperature of 160℃ reached, the furnace was opened and several milligrams of coating powder spread directly on the sample support so that the molten powder coalesced and formed a drop under the probe. The furnace was then immediately closed. After the measurement, the sample was burned off with a small flame and the residue brushed away after cooling. Note: The residue must not become glassy and stick to the probe or sample holder. TMA measurement: Isothermal measurement at 160℃ for 40 min Load: every 6s alternating between +0.05N and -0.05N (period 12s), i.e. the probe is raised from the sample support for 6s, and rests on the sample or sinks into the sample for 6s. Atmosphere:Static air

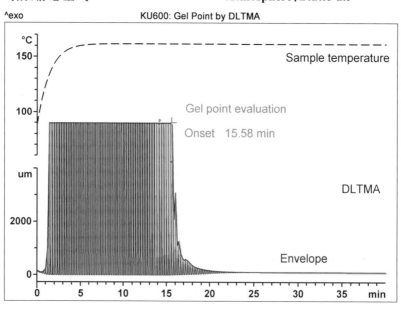

图 3.36　Fig. 3.36

解释 探头交替抬起放下产生大致符合 BS 3900：J3，DIN 55990 和 ISO 8130 标准的上下运动。如果样品是液态的，探头经过向上运动（拉伸力），因而到达上部机械停止点（在约 +5.6mm 处）。然后用 0.05N 的力（相当于约 5g）压入样品，然后停顿在样品支架上（在约 0.1mm 处）。探头能自由运动直至树脂凝胶化成为橡胶态。反应时间约 16min 后，可清楚观察到 DLTMA 曲线上包络线的急剧下降。包络线的起始点可定义为凝胶时间。通过进一步的交联，橡胶弹性运动还进一步下降，探头粘在样品中。

当炉体在 160℃ 打开一小段时间放入样品时，样品支架有点冷却，但是接着又升温到 160℃，如测试样品温度曲线所示。

计算 DLTMA 曲线通过上下运动呈现交替的最大值和最小值。"包络线"计算功能连接最小和最大点形成两条新曲线。然后上包络线用于测定起始点，如图所示。

结论 由于上下运动，探头重复接触样品并从样品撤回，这样能测定凝胶时间。当探头黏住反应样品时就达到了凝胶点。

为了估算特定温度下必需的测试时间，测试也可以非等温进行，即从室温以 5 K/min 升温。

3.4 效应的 DMA 测量

动态热机械分析 DMA 用于测定材料与温度和频率有关的弹性行为。能测量以下重要性能和效应：
- 剪切、弯曲和拉伸模量，损耗因子 tanδ
- 玻璃化转变温度
- 固化和凝胶点
- 填料和增强纤维对模量的影响

Interpretation The alternate raising and lowering of the probe creates an up-and-down movement that corresponds roughly to the BS 3900：J3，DIN 55990 and ISO 8130 standards. If the sample is liquid, the probe experiences an upward movement (tensile force), so that it reaches the upper mechanical stop (at about +5.6 mm). It is then pressed down into the sample with a force of 0.05N (corresponding to about 5g) and then rests on the sample support (at about 0.1mm). The probe is able to move freely until the resin gels and becomes rubbery-like. This is clear from the sharp decrease of the upper envelope of the DLTMA curve after a reaction time of about 16 min. The onset of the envelope can be defined as the gelation time. Through further crosslinking, the rubbery-elastic movements decrease still further and the probe becomes stuck in the sample.

When the furnace is opened for a short time at 160℃ to introduce the sample, the sample support cools somewhat but is then heated again to 160℃, as the curve of the measured sample temperature shows.

Evaluation The DLTMA curve exhibits alternate maxima and minima through the up-and-down movements. The "Envelope" evaluation function joins the minimum and maximum points to form two new curves. The upper envelope is then used to determine the onset as shown in the diagram.

Conclusions Due to the up-and-down movement, the probe repeatedly touches and recedes from the sample, which allows the gelation time to be determined. The gel point is reached when the probe sticks to the reacting sample.

To estimate the measurement time necessary for a particular temperature, the test can also be performed non-isothermally, i.e. by heating at 5 K/min from room temperature.

Measurement effects with DMA

Dynamic mechanical analysis, DMA, is used to determine the elastic behavior of a material as a function of temperature and frequency. The following important properties and effects can be measured:
- Shear, bending and tension moduli, loss factor, tanδ
- Glass transition temperature
- Curing and gel point
- Influence of fillers and reinforcing fibers on the modulus

- 力学松驰行为,形变和频率的影响
- Mechanical relaxation behavior, influence of deformation and frequency

剪切测试是最广泛使用的 DMA 操作模式。该模式产生重复性很好和非常准确的结果,因为温度测试是与样品接触进行的。三点弯曲测试是测定增强热固性树脂弹性模量的首选 DMA 操作模式。

The shear measurement is the most widely used DMA operating mode. This mode yields very reproducible and accurate results because temperature measurement is performed in contact with the sample. 3-point bending measurements are the preferred DMA operating mode for the determination of the elastic modulus of reinforced thermosets.

3.4.1 玻璃化转变的 DMA 测定　Determination of the glass transition by DMA

目的
Purpose

说明如何用 DMA 测量玻璃化转变和如何用不同的计算方法测定玻璃化转变温度。

To show how to measure the glass transition by DMA, and determine the glass transition temperature using different evaluation methods.

样品
Sample

固化的 KU600 环氧树脂粉末。

Cured KU600 epoxy resin powder.

条件
Conditions

测试仪器:DMA,剪切样品夹具

样品制备:

两个小圆柱体,直径 5mm、厚 0.56mm,通过压实细小的未固化 KU600 粉末制成。圆柱体装在剪切样品夹具中,以 2K/min 升温至 250℃固化。

DMA 测试:

测试在 1Hz 下以 2K/min 的升温和降温速率进行。样品先从 250℃ 冷却至 40℃,然后从 40℃ 升温至 160℃。

最大力振幅 5N;最大位移振幅 20μm;偏移控制为零。

气氛:静态空气

Measuring cell:DMA with the shear sample holder

Sample preparation:

Two small cylinders, 5mm in diameter and 0.56 mm thick were made by compressing fine uncured KU600 powder. The cylinders were mounted in the shear sample holder and heated to 250℃ at 2K/min for curing.

TMA measurement:

The measurements were performed at 1Hz at heating and cooling rates of 2K/min. The sample was first cooled from 250℃ to 40℃ and then heated from 40℃ to 160℃.

Maximum force amplitude 5N; maximum displacement amplitude 20μm; offset control zero.

Atmosphere:Static air

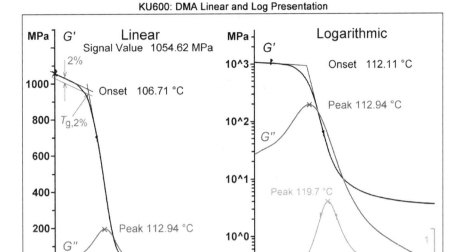

图 3.37　Fig. 3.37

解释　图 3.37 表示固化 KU600 的贮存模量(G')和损耗模量(G'')的升温曲线。在左图中,模量以线性纵坐标表示。在右图中,用对数纵坐标表示。

两图中,均在约 110℃ 观察到固化树脂的玻璃化转变。贮存模量台阶降低,损耗模量显示峰。损耗因子 tanδ 在玻璃化转变期间也呈现峰(右图)。

玻璃化转变温度的计算方法可以分成两组。一些标准例如 DIN 65583 是使用根据线性刻度的所谓 2% 法。如左边图所示,在该方法中,首先确定两条切线的交点为线性坐标图中 G' 的起始点。然后在曲线上定义比起始点温度低温度 50K 的点。然后以该点的贮存模量值降低 2% 即下降至 98% 来定义一个新点。然后绘制一条穿过该点与玻璃化转变前的切线相平行的直线,该直线与贮存模量曲线的交点温度就被作为玻璃化转变温度($T_{g,2\%}$)。在本示例中,G'' 的峰最大值相当于 G' 的拐点。

其他的标准例如 ASTM E1640,不是以对数刻度就是以线性刻度取 G' 的

Interpretation　Figure 3.37 displays the storage modulus (G') and loss modulus (G'') heating curves of cured KU600. In the left diagram, the modulus is displayed on a linear ordinate scale. In the right diagram, a logarithmic ordinate scale is used.

In both diagrams, the glass transition of the cured resin is observed at about 110℃. The storage modulus shows a step decrease and the loss modulus a peak. The loss factor, tanδ, also exhibits a peak during the glass transition. This is shown in the right diagram.

The evaluation procedures for the glass transition temperature can be divided into two groups. Some standards, e. g. DIN 65583, use the so-called 2% method based on the linear scale. As shown in the diagram on the left, in this method, the onset of G' in the linear scale diagram is first determined as the intersection point of the two tangents. A point on the curve is then defined at a temperature 50K lower than the onset temperature. The storage modulus value at this point is then reduced by 2%, i. e. to 98%, to define a new point. A line is then drawn through this point parallel to the tangent before the glass transition. The temperature at which this line intersects the curve of the storage modulus is taken as the glass transition temperature ($T_{g,2\%}$). The peak maximum of G'', in this presentation, corresponds to the point of inflection of G'.

Other standards, e. g. ASTM E1640, take the onset of G'' either in the logarithmic scale or the linear scale, the peak of

起始点、G''峰或 $\tan\delta$ 峰(见右面图)作为玻璃化转变的特征温度。这些性能的特征温度以不同的温度出现。用对数刻度计算的贮存模量 G' 的起始点也可以与用线性刻度计算的明显不同,如同在图中能看到的那样。因此,在报告玻璃化转变温度时详细说明计算的性能和刻度类型这两者非常重要。结果的比较只有在使用了相同的表达时才有意义。

在右图所示的对数表达中,$\log G''$ 峰的温度最大值与 $\log G'$ 台阶的起始点吻合良好。玻璃化转变温度后的贮存模量约在 4MPa 保持几乎不变(橡胶平台)。损耗模量在该区域下降。材料没有显示流动行为。对于固化的热固性树脂这是典型的。

从两图的比较可看到对数表达的明显优势。在低模量范围,贮存和损耗模量间的差异在对数表达中比在线性表达中要清晰得多,在线性表达中玻璃化转变后的所有值看起来几乎为零。

DMA 的实验条件例如升温或降温速率、所应用的频率等能影响玻璃化转变的测定。加热和冷却的效应显示在图 3.38 中。与以 2K/min 升温速率测定的相比,以 2K/min 降温速率测定的 $\tan\delta$ 的峰温向低温移动了约 2K。这个差异是真实的,是由于玻璃化转变中的松弛现象产生的。T_g 的速率依赖性证明了报告测试结果时详细说明实验条件的重要性。

G', or the peak of $\tan\delta$ (see diagram on the right) as the characteristic temperature of the glass transition. The characteristic temperatures of these properties occur at different temperatures. The onset of the storage modulus G' evaluated using in the logarithmic scale may also differ significantly from that using a linear scale, as can be seen in the diagram. It is therefore very important to specify both the evaluated properties and the scale type when reporting a glass transition temperature. Comparison of results only makes sense if the same presentation is used.

In the logarithmic presentation shown in the right diagram, the temperature maximum of the $\log G''$ peak agrees well with the onset of the step in $\log G'$. The storage modulus after the glass transition remains almost constant at about 4MPa (rubbery plateau). The loss modulus decreases in this region. The material shows no flow behavior. This is typical for a cured thermoset.

A clear advantage of the logarithmic presentation is apparent from a comparison of the two diagrams. The differences between the storage and loss modulus in the low modulus range are much clearer in the logarithmic presentation than in the linear presentation, where all the values after the glass transition appear to be practically zero.

The experimental conditions of a DMA measurement, for example the heating or cooling rate, applied frequency, etc., can influence the determination of the glass transition. The effect of heating and cooling is shown in Figure 3.38. The peak temperature of $\tan\delta$ determined at a cooling rate of 2K/min is shifted by about 2K to lower temperature compared to that determined at a heating rate of 2 K/min. This difference is real and is due to relaxation phenomena during the glass transition. The rate dependence of T_g demonstrates the importance of specifying the experimental conditions when reporting the test results.

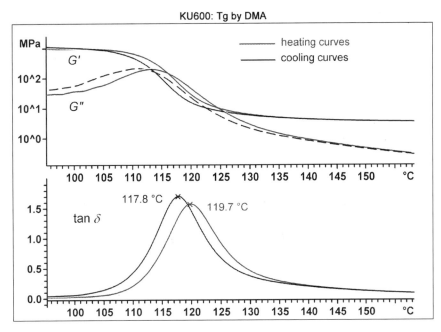

图 3.38　Fig. 3.38

结论　材料的玻璃化转变伴随着贮存模量的台阶下降和损耗模量与损耗因子的峰。玻璃化转变温度最好表征为对数刻度上 G' 的起始点或任一个刻度上 G'' 或 $\tan\delta$ 的峰温。

在报告转变温度时，必须详细说明影响结果的实验条件和计算方法。为了有意义地比较结果，必须使用相同的表达。

测试曲线的解释通常使用对数刻度而不是线性刻度来进行，因为模量在玻璃化转变处变化几个数量级，转变后在线性表达中模量的微小变化是不明显的。

Conclusions　The glass transition of a material is accompanied by a step decrease in the storage modulus and by peaks in the loss modulus and loss factor. The glass transition temperature is best characterized as the onset of G' on the logarithmic scale, or the peak temperatures of G'' or $\tan\delta$ on either scale.

Experimental conditions and evaluation procedures that influence the results must be specified when reporting transition temperatures. To compare results meaningfully, the same presentation must be used.

Interpretation of the measurement curves is normally done using a logarithmic scale rather than a linear scale because the modulus changes by several decades at the glass transition, and the slight change of the modulus after the transition is not apparent in a linear presentation.

3.4.2　玻璃化转变的频率依赖性　The frequency dependence of the glass transition

目的	说明在不同频率下测试时，力学模量和损耗因子在玻璃化转变处如何变化。
Purpose	To show how the mechanical modulus and loss factor change at the glass transition when measured at different frequencies.
样品	固化的 KU600 环氧树脂粉末
Sample	Cured KU600 epoxy resin powder
条件	测试仪器：DMA，剪切样品夹具 样品制备： 两个小圆柱体，直径 5mm、厚 1mm，通过压实细小的未固化 KU600
Conditions	Measuring cell：DMA with the shear sample holder Sample preparation： Two small cylinders, 5mm in diameter and 1 mm thick were made by compressing fine uncured

135

粉末制成。圆柱体装在剪切样品夹具中,以 2K/min 升温至 250℃ 固化树脂,然后降温到室温。

DMA 测试:
测试在 0.1、1、10、100 和 1000Hz 下进行,以 1K/min 升温速率从 70℃至180℃。最大力振幅5N;最大位移振幅 30μm;偏移控制为零。

气氛:静态空气

KU600 powder. The cylinders were mounted in the shear sample holder, heated to 250℃ at 2K/min to cure the resin, and then cooled to room temperature.

DMA measurement:
The measurement was performed at 0.1, 1, 10, 100, and 1000 Hz and a heating rate of 1K/min from 70℃ to 180℃. Maximum force amplitude 5 N; maximum displacement amplitude 30μm; offset control zero.

Atmosphere: Static air

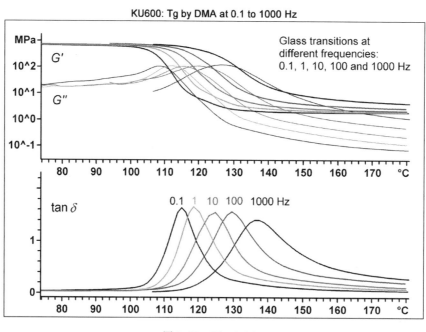

图 3.39　Fig. 3.39

解释　不同频率测试的与温度关系的曲线见图 3.39。在玻璃化转变区 G' 曲线呈台阶变化,而 G'' 和 tanδ 曲线呈现峰。可以观察到,在较高频率峰移向较高温度。同样清晰的是,在同一频率下测试的 G'' 和 tanδ 曲线的峰值并不在同一温度。由损耗模量测定的玻璃化转变低于从损耗因子测定的。

如以前所述(2.5.3.1节),玻璃化转变是一个松弛效应。它源自协同单元的分子活动性。随着温度升高,协同重排的频率增加。在低温下,这些重排的频率比所用的测试频率低得多。在这种情况下,样品

Interpretation　The curves measured at different frequencies are displayed as a function of temperature in the Figure 3.39. The G' curves show a step change in the glass transition region, whereas the G'' and tanδ curves show peaks. It can be seen that the peaks are shifted to higher temperatures at higher frequencies. It is also clear that the peak maxima of the G'' and tanδ curves measured at the same frequency are not at the same temper-ature. The glass transition temperature determined from the loss modulus is lower than that from the loss factor.

As mentioned previously (Section 2.5.3.1), the glass transition is a relaxation effect. It has its origins in the molecular mobility of cooperative units. With increasing temperature, the frequency of the cooperative rearran- gements increases. At low temperatures, the frequency of these rearrangements is much lower than the measurement frequencies used. In this case, the

显得坚硬,因而贮存模量大。在较高温度,协同重排的频率比测试频率高得多,因而材料显得柔软,贮存模量低。

在测量频率大致等于协同重排频率的温度范围内,在贮存模量上观察到一个台阶。这时,部分力学能量被消散,转化成热。这就是为什么贮存模量的台阶伴随着损耗模量的峰。

玻璃化转变和分子运动频率之间的关系再次反映在被测转变的频率依赖性中。

计算 玻璃化转变(松弛)的温度和频率依赖性可通过计算不同测试频率下 tanδ 峰最大值的温度来研究。通常情况下,用频率的对数对温度的倒数作图(以 K 为单位):

sample appears hard and the storage modulus is therefore large. At higher temperatures, the frequency of the cooperative rearrangements is much higher than the measurement frequency. The material then appears soft and has a low storage modulus.

In the temperature range in which the measurement frequency corresponds to about the frequency of the cooperative rearrangements, a step is observed in the storage modulus. In this case, part of the mechanical energy is dissipated. It is converted to heat. This is why the step in the storage modulus is accompanied by a peak in the loss modulus.

The relationship between the glass transition and the frequency of the molecular movement is again reflected in the frequency dependence of the measured transition.

Evaluation The temperature and frequency dependence of the glass transition (relaxation) can be investigated by evaluating the temperature of the tanδ peak maximum at different measurement frequencies. Normally, the logarithm of the frequency is plotted against the reciprocal temperature (in Kelvin):

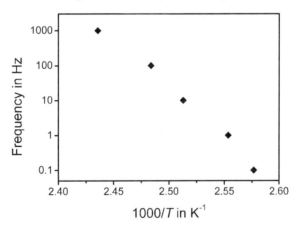

图 3.40 松弛过程(玻璃化转变)的活化图

Fig. 3.40 Activation diagram of the relaxation process (glass transition)

由于测试频率大致等于协同重排的平均频率,所以该曲线可被看作松弛过程的活化图。协同松弛过程如玻璃化转变总是显示本图所示类型的曲线,可由 Vogel-Fulcher 方程描述:

Since the measurement frequency corresponds to about the average frequency of the cooperative rearrangements, the curve can be looked on as the activation diagram of the relaxation process. Cooperative relaxation processes, such as the glass transition, always exhibit a curve of the type shown in this diagram, which is described by the Vogel-Fulcher equation:

$$\log\left(\frac{f}{f_0}\right) = \frac{B}{T - T_V}$$

式中 f_0 是高频极限值,T_v 是 Vogel 温度,B 是曲率参数。

Here f_0 is the upper frequency limit, T_v the Vogel temperature and B a curvature parameter.

Williams、Landel 和 Ferry 的 WLF 方程也常用来描述温度依赖性：

$$\log\left(\frac{f}{f_r}\right) = \frac{C_1(T-T_r)}{C_2+(T-T_r)}$$

式中 T_r 是任一参比温度，f_r 是 T_r 时的参比频率，C_1 和 C_2 是 WLF 常数。当然，这里的 WLF 常数取决于参比值的选择。

两个等式是相互等价的，但更推荐 Vogel-Fulcher 的描述，因为少使用一个参数。

结论 玻璃化转变是与频率有关的。在较高频率下，它移至较高温度。测试频率与协同重排的实际频率有相同的数量级。因为频率和转变温度之间的非线性关系，所以在较高频率下测得的玻璃化转变温度比较低频率下测得的更宽。当报告由 DMA 测定的玻璃化转变温度时，详细说明实验条件尤其是所用的频率是必不可少的。

The WLF equation of Williams, Landel and Ferry is also often used to describe the temperature dependence：

Here T_r is any reference temperature, f_r is the reference frequency assigned to T_r and C_1 and C_2 are the WLF constants. Of course, here the WLF constants depend on the choice of the reference values.

Both equations are equivalent to one another, but the Vogel-Fulcher description is preferred because one less parameter is used.

Conclusions The glass transition is frequency dependent. At higher frequencies it shifts to higher temperatures. The measuring frequency is of the same order as the actual frequency of the cooperative rearrangements. Because of the nonlinear relationship between the frequency and the transition temperature, a broader transition is measured at higher frequencies than at lower frequencies.

When reporting glass transition temperatures determined by DMA, it is essential to specify the experimental conditions and especially the frequency used.

3.4.3 动态玻璃化转变 The dynamic glass transition

目的	说明如何使用频率变化的等温测试（频率扫描）可得到关于松弛行为的信息。例如，本节中所讨论的玻璃化转变和模量的频率依赖性。
Purpose	To show how information can be obtained on relaxation behavior using isothermal measurements in which the frequency is varied (frequency sweep). For example, the glass transition and the frequency dependence of the modulus is discussed in this section.
样品	固化的 KU600 环氧树脂粉末。
Sample	Cured KU600 epoxy resin powder.
条件	测试仪器：DMA，剪切样品夹具 样品制备：两个圆柱体，直径 5mm、厚 0.56mm，通过压实细小的未固化 KU600 粉末制成。圆柱体装在剪切样品夹具中，以 2K/min 升温至 250℃ 固化树脂，然后降温到室温。 DMA 测试：测试在 115℃ 下在 1kHz 至 1mHz 频率范围内进行。最大力振幅 5N；最大位移振幅 20μm；偏移控
Conditions	Measuring cell：DMA with the shear sample holder Sample preparation：Two cylinders, 5mm in diameter and 0.56mm thick were made by compressing fine uncured KU600 powder. The cylinders were mounted in the shear sample holder, heated to 250℃ at 2K/min to cure the resin, and then cooled to room temperature. DMA measurement：The measurement was performed at 115℃ in the frequency range 1kHz to 1mHz. Maximum force amplitude 5N; maximum displacement amplitude

制为零。 20μm; offset control zero.
气氛：静态空气 Atmosphere: Static air

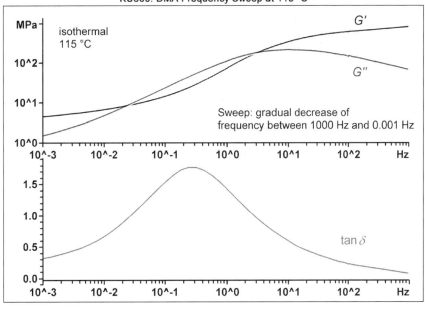

图 3.41　Fig. 3.41

解释　在等温测试中，模量随着频率而变化。在松弛范围内，贮存模量台阶式变化，高频下比在低频下大。这意味着在高频，与对应的协同重排频率相比，测试频率更高，与协同重排的速度相比，施加应力的速度更快。样品因此显得坚硬，贮存模量大。在低频，分子重排能与外应力发生作用，因而样品显得柔软，贮存模量低。

在松弛范围内，损耗模量 G'' 出现峰。在所测的样品中，峰最大值的频率是 10 Hz。损耗模量峰的形状相当于松弛时间的分布。宽峰意味着松弛时间的宽分布，因此在协同重排的可能性中有较大差异。这可能是由复杂的分子间或分子内的结构所引起的。

在松弛范围内，损耗因子 tanδ 也呈现一个最大值在较低频率的峰。在本测试实例中，在 2.75Hz。估算松弛行为的另一个数据来源是曲线形状。如果被测材料显示清晰的松弛

Interpretation　In an isothermal measurement, the modulus changes with the frequency. In the relaxation range, the storage modulus changes stepwise. At high frequencies, it is larger than at low frequencies. This means that at high frequencies, the measurement frequency is higher than the frequency of the corresponding cooperative rearrangements and that the stress is applied rapidly compared to the speed of the cooperative rearrangements. The sample therefore appears hard and has a large storage modulus. At low frequencies, the molecular rearrangements are able to react to the external stress. The sample appears soft and has a low storage modulus.

In the relaxation range, a peak occurs in the loss modulus, G''. In the sample measured, the frequency at the peak maximum was 10Hz. The shape of the peak in the loss modulus corresponds to a distribution of relaxation times. A broad peak means a broad distribution of the relaxation times and therefore a larger difference in the possibilities for cooperative rearrangements. This can be caused by a complex intermolecular or intramolecular structure.

In the relaxation range, the loss factor, tanδ, also exhibits a peak whose maximum is at lower frequencies. In the example measured, this was at 2.75Hz. Another source of data for the estimation of the relaxation behavior is the curve shape. If the material measured exhibits a distinct relaxation range, then the

范围,那么用对数－对数表示的 G'' 和 tanδ 峰的两侧应该是直线。如果预期行为发生偏差,如本例中低于 0.025Hz 和大于 2.1Hz 时,那么这非常灵敏地暗示存在其他与频率有关的过程。其可能的原因是由于内部表面(例如由于相分离)或复杂分子结构(例如形成分子网络结构)产生的松弛过程。

结论 等温条件下力学性能的频率依赖性测试是研究松弛行为的非常灵敏的方法。除了松弛强度和松弛的频率范围,宽度和曲线形状也是重要的:可以获得有关分子和超分子结构或它们的变化的信息。由于松弛范围宽,所以为了对实验结果求得解释,实验应该覆盖最大可能的频率范围。

由于许多工艺过程和材料应用涉及宽频率范围的动态应力,所以这种测试有助于对产品最优化或对故障进行分析。动态测试应尽可能在材料实际受到应力作用的频率范围内进行,因为在其他频率下的行为可能完全不同,外推并不能始终保证满意的结果。

sides of the peaks of G'' and tanδ in log-log presentation should be straight lines. If deviations from the expected behavior occur, as in the example considered at below 0.025Hz and above 2.1Hz, then this is a very sensitive indication of other frequency-dependent processes. Possible reasons for this are relaxation processes due to internal surfaces(e.g. through phase separation) or complex molecular structure (e.g. the formation of molecular networks).

Conclusions The measurement of the frequency dependence of the mechanical properties under isothermal conditions is a very sensitive method for the investigation of relaxation behavior. Besides the relaxation intensity and the frequency range of the relaxation, the width and curve shape are also important: information can be gained about molecular and supermolecular structures or their changes. Because the relaxation range is wide, for interpretation purposes experiments should cover the largest possible frequency range.

Since many technological processes and material applications involve dynamic stress over a wide frequency range, such measurements can also be useful for product optimization or for failure analysis. A dynamic measurement should as far as possible be performed in the frequency range in which the material is actually stressed because the behavior at other frequencies can be quite different and extrapolation does not always ensure satisfactory results.

3.4.4 等温频率扫描 Isothermal frequency sweeps

目的	进行不同温度下可用于绘制主曲线的等温频率扫描。如果所测试的频率范围足够宽,则可将力学松弛行为的等温频率依赖性测试看作是力学松弛频率谱。
Purpose	To perform isothermal frequency sweeps at different temperatures that can be used to construct a master curve. An isothermal frequency-dependent measurement of mechanical relaxation behavior can be considered as a mechanical relaxation spectrum if the frequency range investigated is sufficiently large.
样品	固化的 KU600 环氧树脂粉末。
Sample	Cured KU600 epoxy resin powder.
条件	测试仪器:DMA,剪切样品夹具 样品制备: 两个圆柱体,直径 5mm、厚 0.56mm,通过压实细小的未固化 KU600 粉
Conditions	Measuring cell:DMA with the shear sample holder Sample preparation: Two cylinders, 5mm in diameter and 0.56mm thick were made by compressing fine uncured

末制成。圆柱体装在剪切样品夹具中，以 2K/min 升温至 250℃，然后冷却到室温。

DMA 测试：

在 80℃ 至 190℃ 之间不同温度的等温条件下在 1mHz 至 1kHz 频率范围内进行测试。最大力振幅 5N；最大位移振幅 20μm；偏移控制为零。

气氛：静态空气

KU600 powder. The cylinders were mounted in the shear sample holder, heated to 250℃ at 2 K/min, and then cooled to room temperature.

DMA measurement：

The measurement was performed in the frequency range 1mHz to 1 kHz under isothermal conditions at different temperatures between 80 and 190℃. Maximum force amplitude 5 N; maximum displacement amplitude 20μm; offset control zero.

Atmosphere：Static air

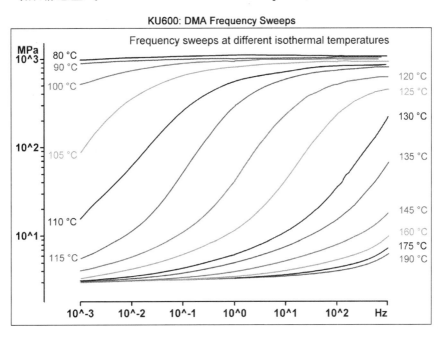

图 3.42　80℃ 至 190℃ 之间温度频率从 0.001 至 1000Hz 测试的剪切贮存模量。

Fig. 3.42　The shear storage modulus in the frequency window 0.001 to 1000Hz measured at temperatures between 80℃ and 190℃.

解释　温度足够低（80℃），则由不同频率测得的贮存模量相当大（1000MPa），几乎与频率无关。随着温度的升高，在低频下贮存模量下降，直至松弛范围台阶移出测试范围。在 120℃，在所测试的频率范围内整个台阶几乎都能测到。随着温度进一步升高，松弛范围移至更高频率。在 190℃ 低于 10Hz 的测试频率范围内，贮存模量约为 3MPa，只测到几乎与频率无关的橡胶平台。

Interpretation　In a frequency-dependent measurement at sufficiently low temperatures (80℃), the storage modulus is relatively large (1000MPa) and practically independent of the frequency. With increasing the temperature, the storage modulus decreases at low frequencies until the step of the relaxation range shifts across the measurement range. At 120℃, practically the entire step can be measured in the frequency range used. With further temperature increase, the relaxation range is shifted to higher frequencies. At 190℃ in the measured frequency range below 10Hz, only the almost frequency-independent rubbery plateau with a storage modulus of about 3 MPa is measured.

结论 在频率扫描实验中,样品在恒定温度下在实验可达到的频率范围测试。如果这样的实验在 T_1 和 T_2 两个温度进行(这里 $T_1 < T_2$),那么在 T_2 的频率范围内测得的行为与在 T_1 下以更低频率测得的一样。因此,通过改变温度,能将松弛区横向移出该频率范围。

换言之,在这样的测试中,较低温度下的曲线相当于参比温度下更高频率的曲线,较高温度下的曲线相当于参比温度下更低频率的曲线。在低于参比温度的温度下,就象在较低频下测试相同的松弛行为。

可认为接近玻璃化转变温度的非晶态聚合物的频率(时间)和温度行为之间存在一般的等值关系——在一定条件下具有橡胶态特性的聚合物,如果温度降低或观察的时间尺度减小,行为可以如玻璃。因为频率依赖性直接与相应的时间依赖性相关,所以该关系通常被称为时间-温度叠加原理。

温度与频率之间的关系由 Vogel-Fulcher 方程或 WLF 方程描述(见 3.4.2 节)。

Conclusions In a frequency sweep experiment, the sample is measured at constant temperature in an experimentally accessible frequency window. If such an experiment is performed at two temperatures, T_1 and T_2 (where $T_1 < T_2$), the behavior measured at T_2 in the frequency window is the same as that which would be measured at T_1 at lower frequencies. The relaxation range can therefore be shifted across the frequency window by varying the temperature.

In other words, in such measurements the curves at lower temperatures correspond to those at the reference temperature and higher frequencies, and the curves at higher temperatures to those at the reference temperature and lower frequencies. At temperatures below the reference temperature, the same relaxation behavior is measured as at lower frequencies.

It is thought that there is a general equivalence between the frequency (time) and temperature behavior of amorphous polymers close to the glass transition temperature — a polymer that has rubbery characteristics under certain conditions can behave as a glass if the temperature is reduced or the time scale of the observation is decreased. Since the frequency-dependence is directly related to a corresponding time-dependence, the relationship is usually referred to as the time-temperature superposition principle.

The relationship between temperature and frequency is described by the Vogel-Fulcher equation or the WLF equation (See 3.4.2).

3.4.5 主曲线绘制和力学松弛频率谱
Master curve construction and mechanical relaxation spectrum

目的 说明如何能对一个特定参比温度绘制主曲线。得到的频率谱包括在给定参比温度下在实际实验中无法达到的频率范围。主曲线能够获得无法直接测量的数据。

Purpose To show how a master curve can be constructed for a particular reference temperature. The resulting frequency spectrum includes frequency ranges that are not accessible in practical experiments at the given reference temperature. A master curve enables data to be obtained that cannot be directly measured.

样品 固化的 KU600 环氧树脂粉末。
Sample Cured KU600 epoxy resin powder.

测试:
条件和 DMA 曲线:见上节。

Measurements:
Conditions and DMA curves: see previous section.

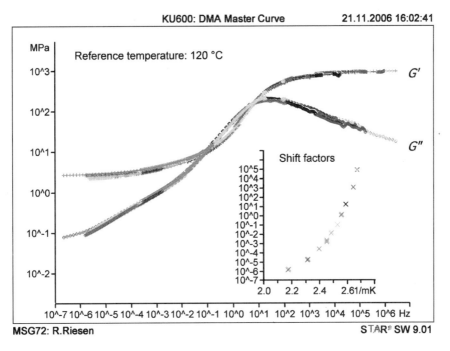

图 3.43　Fig. 3.43

计算　计算基于前面 3.4.4 节结论中所描述的时间-温度叠加（TTS）原理，上面的主曲线利用了该应用中图示的贮存模量测试曲线和相应的损耗模量曲线（未显示）。主曲线通过水平移动温度低于参比温度下测试的曲线至较高频率来绘制，用这种方法，贮存模量 G' 和损耗模量 G'' 的各条曲线与如此形成的相应复合曲线最大程度地叠加。同样地，在较高温度下测试的曲线移至较低频率。由于样品几何形状的变化产生的绝对模量的轻微差异可通过垂直向上或向下移动曲线加以考虑。

作为这些操作的结果，得到复合模量两个分量覆盖频率范围 10^{-9} 至 $10^9\,\mathrm{Hz}$ 的曲线。等温频率扫描的移动因子也予以绘制，如插图所示。本例绘制了对参比温度为 120℃ 的剪切模量和损耗模量的主曲线。

解释　在 10^{-9} 至 $10^{-4}\,\mathrm{Hz}$ 的低频，呈现贮存模量值约为 2.5 MPa 的橡

Evaluation　This evaluation is based on the principle of time-temperature superposition (TTS) described in the previous section 3.4.4 under Conclusions. The above master curve makes use of the storage modulus measurement curves shown in the figure in this application and the corresponding loss modulus curves (not shown). The master curve is constructed by shifting the curves measured at temperatures lower than the reference temperature horizontally to higher frequencies in such a way that the individual curves of the storage modulus, G', and the loss modulus, G'', overlap to the greatest possible extent with the corresponding composite curves so formed. Similarly, the curves measured at higher temperatures are shifted to lower frequencies. The slight difference of the absolute modulus due to the change in the sample geometry can be taken into account by moving the curves vertically up or down.

As a result of these operations, one obtains curves of both components of the complex modulus over a frequency range of 10^{-9} to $10^9\,\mathrm{Hz}$. The shift factors of the isothermal frequency sweeps are also plotted, as shown in the inserted diagram.

In this example, the master curves of the shear modulus and loss moduli have been constructed for a reference temperature of 120℃.

Interpretation　At low frequencies between 10^{-9} and $10^{-4}\,\mathrm{Hz}$, the storage modulus exhibits the rubbery plateau with a modulus

胶平台。橡胶平台的位置由交联度决定。低频下材料的橡胶弹性行为符合对交联环氧树脂体系的预期,事实表明其贮存模量总大于损耗模量。贮存模量接着显示一个约 2.5 个数量级的台阶,相伴着损耗模量中的峰。这是特征频率约为 20Hz (G'' 峰最大值处的频率)的玻璃化转变。在更高频率下,贮存模量几乎恒定在大约 1000MPa,对数—对数表示的损耗模量线性降低。

位移因子曲线可乘以对应于参比温度的参比频率,本例中为 20Hz。所得到的曲线可看作为玻璃化转变过程的活化能图。

value that is about 2.5 MPa. The position of the rubbery plateau is determined by the degree of crosslinking. The rubbery elastic behavior of the material at low frequencies, indicated by the fact that the storage modulus is always greater than the loss modulus, is as expected for a crosslinked epoxy resin system. The storage modulus then shows a step of about 2.5 decades associated with a peak in the loss modulus. This is the glass transition with a characteristic frequency of about 20Hz (frequency at the maximum of the G'' peak). At higher frequencies, the storage modulus is almost constant at about 1000 MPa. The loss modulus decreases linearly in the log-log presentation.

The shift factor diagram can be multiplied by a reference frequency corresponding to the reference temperature, 20Hz in this case. The diagram obtained can be looked on as an activation energy diagram of the glass transition process.

结论 可以根据时间—温度叠加原理绘制表示完整松弛频率谱的主曲线。对于得到的曲线的详细分析能够研究分子动力学和检测动态材料的性能是否适合于特定的应用。

如果各个模量曲线重叠得好,就如图示的情况那样,那么可认为该材料是流变简单的。这类材料包括那些分子间作用不受内部表面(例如由于混合或结晶的相分离)、结构形成、填料或化学反应影响,并且不同松弛范围不重叠的非晶态材料。时间—温度叠加原理实际上只对这样的材料有效。

Conclusions It is possible to construct a master curve that shows the entire relaxation spectrum based on the time-temperature superposition principle. A detailed analysis of the resulting curves allows one to study molecular dynamics and to check whether the dynamic material properties are suitable for a particular application.

If the individual modulus curves overlap well, as it is the case in the figure shown, then the material is said to be rheologically simple. Materials of this type include amorphous materials in which molecular interactions are not affected by internal surfaces (e.g. phase separation with blends or crystallization), structure formation, fillers or chemical reactions, and in which different relaxation ranges do not overlap. The time-temperature superposition principle is in fact only valid for such materials.

3.4.6 固化的 DMA 测量 Curing measured by DMA

目的 **Purpose**	本应用通过监测其力学行为来研究未固化环氧树脂体系的固化过程。 This application investigates the curing process of an uncured epoxy system by monitoring its mechanical behavior.	
样品 **Sample**	未固化的 KU600 环氧树脂粉末。 Uncured KU600 epoxy resin powder.	
条件 **Conditions**	测试仪器:DMA,剪切样品支架 样品制备: 两个小圆柱体,直径 5mm、厚 0.56mm,通过压实细小的未固化	Measuring cell:DMA with the shear sample holder Sample preparation: Two small cylinders, 5mm in diameter and 0.56 mm thick were made by compressing the fine

KU600 粉末制成。圆柱体装在剪切样品夹具中，在90℃预处理1h。

DMA 测试：

测试在1Hz进行，以速率2K/min 从40℃升温至250℃，然后冷却至90℃。

最大力振幅5N；最大位移振幅20μm；偏移控制为零。

气氛：静态空气

uncured KU600 powder. The cylinders were mounted in the shear sample holder and preconditioned at 90℃ for 1h.

DMA measurement：

The measurement was performed at 1 Hz and at a heating rate of 2 K/min from 40 to 250℃ and then cooling to 90℃.

Maximum force amplitude 5N; maximum displacement amplitude 20 um; offset control zero.

Atmosphere：Static air

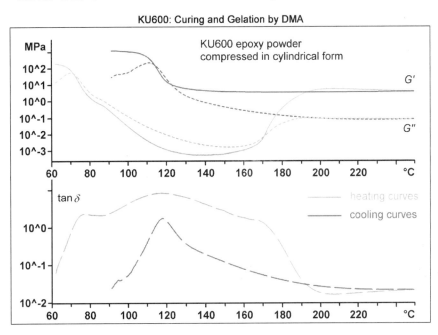

图3.44　Fig.3.44

解释　图表示未固化KU600环氧树脂升温时和随后完全固化KU600降温时贮存模量（G'）、损耗模量（G''）和 tanδ 的测试曲线。

在约70℃观察到玻璃化转变过程，分别在贮存模量上呈现大约2.5个数量级的台阶变化和在损耗模量与 tanδ 呈现一个峰。在约90℃，材料开始流动，伴随着贮存模量的加速下降和 tanδ 的增加。

树脂的固化过程开始于120℃左右，这时 tanδ 开始下降。开始时固化过程进行得非常慢，在160℃以上加速，这时贮存模量快速增加。在约200℃，材料的贮存模量最后达到了约5.3MPa的值，进一步升

Interpretation　The figure displays the measurement curves of the storage modulus（G'）, loss modulus（G''）and tanδ during heating of the uncured KU600 epoxy resin and subsequent cooling of the fully cured KU600.

The glass transition process is observed at about 70℃, shown as a step change of about 2.5 decades in the storage modulus and a peak in the loss modulus and tanδ respectively. At about 90℃, the material starts to flow, which is accompanied by an accelerated decrease in the storage modulus and an increase in tanδ.

The curing process of the resin starts at about 120℃, where the tanδ starts to decrease. The curing process proceeds very slowly at the beginning and accelerates above 160℃, where the storage modulus increases rapidly. The storage modulus of the material finally reaches a value of about 5.3 MPa at approximately 200℃ and remains practically constant on further heating. This

温时几乎保持不变。这表明固化过程在约200℃时已基本完成。

固化反应的凝胶点可确定为贮存和损耗模量的交点,这时tanδ等于1。在上图中约170℃处可观察到本测试体系的凝胶点。

在从250℃至90℃的降温期间,固化材料在约110℃显示从橡胶态至玻璃态的转变。贮存模量再次增加约2.5个数量级到1120MPa。由于固化玻璃化转变温度提高了40K。

结论 DMA能容易地检测热固性材料的固化反应和固化对其力学性能和物理转变的影响。DMA可非常灵敏地确定固化过程的起始点,这用热流测试来测定是相当困难的。此外,DMA也能用来测定固化反应的凝胶点,即贮存和损耗模量的交点。

indicates that the curing process is more or less complete at about 200℃.

The gel point of the curing reaction can be determined as the crossover point of the storage and loss modulus, where tanδ equals 1. The gel point of the measured system is observed at approximately 170℃ in the upper diagram.

The cured material shows a transition from the rubbery state to glassy state at about 110℃ during cooling from 250℃ to 90℃. The storage modulus again increases about 2.5 decades to 1120 MPa. The glass transition temperature increased by 40K due to curing.

Conclusions DMA can easily detect the curing reaction of a thermosetting material and the effects curing has on its mechanical properties and physical transitions. DMA can define the starting point of the curing process with great sensitivity. This is quite difficult to determine from heat flow measurements. In addition, DMA can also be used to determine the gel point of the curing reaction, which is the crossover point of the storage and loss moduli.

3.5 玻璃化转变的DSC、TMA和DMA测量比较
A comparison of the glass transition measured by DSC, TMA and DMA

测定玻璃化转变温度是热分析常见的应用。如同在前面的实例中已讨论的,物理性能诸如比热容、热膨胀系数和机械模量在玻璃化转变处发生变化。鉴于DSC、TMA和DMA测试原理不同,产生了关于应该使用哪种技术和测得的玻璃化转变温度在何种程度上具有可比性的问题。比较不同条件下测得的玻璃化转变温度会呈现几个开尔文度的差异。实际上,理解这种差异的原因是非常重要的,特别当比较材料时,例如在质量保证中。

务必牢记的是,玻璃态不是热力学平衡态,向橡胶(或液态)的转变是一个松弛过程,因此是受动力学控制的。由于这个原因,所以玻璃化转变并不象熔融那样出现在一个固

The determination of the glass transition temperature is a frequent application of thermal analysis. As has already been discussed in the previous examples, physical properties such as the specific heat capacity, the coefficient of thermal expansion, and the mechanical modulus change at the glass transition. In view of the fact that the measurement principles of DSC, TMA and DMA are different, the question arises as to which technique should one use and to what extent the measured glass transition temperatures are comparable. A comparison of glass transition temperatures measured under different conditions can show differences of several Kelvin. In practice, it is very important to understand the reason for such differences, in particular when comparing materials, for example in quality assurance.

A special point to bear in mind is that a glass is not in thermodynamic equilibrium and the transition to the rubbery (or liquid) state is a relaxation process and is therefore kinetically controlled. For this reason, the glass transition does not occur at a fixed temperature, as is the case with melting, but over a broad

定的温度,而是覆盖一个宽的温度范围。然而,为了测得在数字上可比较的温度,已经开发出不同的计算程序和相应的测试方法(见2.5.4节中的一览表)。这些只涉及一种技术,并不保证由 DSC、TMA 或 DMA 测定的玻璃化转变温度是相等的。

temperature range. To, nevertheless, determine numerically comparable temperatures, different evaluation routines and corresponding test methods have been developed (see the list in Section 2.5.4). These refer to just one technique and there is no guarantee that glass transition temperatures determined by DSC, TMA or DMA are identical.

目的　比较 DSC、TMA 和 DMA 三种不同测量技术测定的玻璃化转变,并讨论重要的因素。
Purpose　To compare the glass temperatures determined by the three different measuring techniques, DSC, TMA and DMA and to discuss important factors.

样品　固化的 KU600 环氧树脂粉末。
Sample　Cured KU600 epoxy resin powder.

条件　测试曲线和计算已在前面的实例中描述和讨论过。结果汇总于下。
Conditions　The measured curves and evaluations have been described and discussed in the previous examples. The results are summarized below.

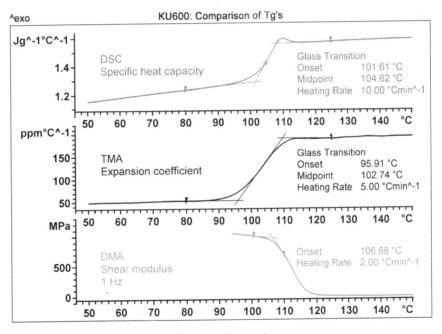

图 3.45　Fig. 3.45

解释　由比热容、膨胀系数和剪切模量三个物理性能的测试曲线表示 KU600 的玻璃化转变区。曲线比较表明,它们呈现不同的温度依赖性,并且转变区域处于不同的温度。由此可见,T_g 与测试技术(量热、体积测定、力学测量)、计算方法和样品的热历史有关。因此,从测试曲线得到的玻璃化转变温度是不能直

Interpretation　The measured curves of the three physical properties specific heat, expansion coefficient and shear modulus are displayed in the region of the glass transition of KU600. A comparison of the curves shows that they exhibit different temperature dependencies and that the transition regions are at different temperatures. From this, it is clear that the T_g depends on the measurement technique (calorimetric, volumetric, mechanical), the evaluation method and the sample's thermal history. The glass transition temperatures derived from the measured curves

接比较的。

图也显示了在玻璃化转变期间物理性能的变化多么强。这能对测试技术的灵敏度作出评估。在玻璃化转变时比热容变化约30%,膨胀系数可以增加多至300%,而模量变化高至3个数量级。因此DSC对TMA或对DMA的灵敏度比为1比10或1比1000,即对玻璃化转变,DMA测试比DSC测试要灵敏约1000倍。

计算 表3.10汇总了来自相应应用实例的各个玻璃化转变温度。热容台阶的中点几乎与膨胀系数台阶的中点相同。DMA和DLTMA线性刻度的模量变化起始点在DSC和TMA值的范围内。该行为是经常观察到的,但不是必然的情况。$\tan\delta$产生最高的温度。$\tan\delta$峰温的频率依赖性约为每数量级5K。这里没有显示升温和降温速率的影响,但不应该忽视。软化点(穿透和弯曲测试的起始值)与T_g也是可比较的,对此还必须考虑压缩应力的影响。

are therefore not directly comparable.

The diagram also shows how strongly the physical properties changes during the glass transition. This allows an estimate for the sensitivity of the measurement technique to be made. The specific heat capacity changes at the glass transition by about 30%, the expansion coefficient can increase by up to 300% and the modulus changes by up to 3 decades. The sensitivity ratio of the DSC to TMA, or to DMA is therefore about 1 to 10 or 1 to 1000, that is, the DMA measurement is about 1000 times more sensitive for the glass transition than the DSC measurement.

Evaluation Table 3.10 summarizes the individual glass transition temperatures from the corresponding application examples. The midpoint of the heat capacity step is practically the same as the midpoint in the step in the expansion coefficient. The onset of the modulus change on a linear scale in the DMA and DLTMA is in the range of the DSC and TMA values. This behavior is often observed but is not necessarily the case. $\tan\delta$ yields the highest temperatures. The frequency dependence of the peak temperature of $\tan\delta$ is about 5 K per decade. The influence of heating and cooling rates is not shown here, but should not be ignored. The softening points (onset values of penetration and bending measurements) are also comparable with the T_g, whereby the influence of the compressive stress must also be taken into account.

表3.10 用不同热分析技术在不同条件下测试的KU600环氧树脂的玻璃化转变温度(T_g)

Table 3.10 Glass transition temperatures (T_g) of KU600 epoxy resin, measured by different TA techniques under different conditions.

热分析技术 TA technique		T_g ℃	升温速率 Heating rate K/min	应用 Application
DSC	ASTM/IEC c_p 中点 Midpoint c_p	103.1 103.6	10	3.1.1.10
TMA	TMA起始点 Onset TMA 线膨胀系数中点 Midpoint CTE 穿透起始点 Onset penetration	102.1 102.7 101.5	5 5	3.3.2.1 3.3.2.2
DLTMA 0.08Hz	弯曲中点起始点 Onset bending midpoint 弯曲模量起始点(线性刻度) Onset bending modulus (linear scale)	101.5 103.5	5	3.3.2.3

热分析技术 TA technique		T_g ℃	加热速率 Heating rate K/min	应用 Application
DMA 1 Hz	起始点，G′(线性刻度) Onset, G′(linear scale)	106.7	2	3.4.1
	起始点，G′(对数刻度) Onset, G′(log scale)	12.1		
	峰，G″ Peak, G″	112.9		
	tan δ	119.7		
0.1Hz	峰，tan δ Peak, tan δ	115.0	2	3.4.2
10Hz	峰，tan δ Peak, tan δ	124.8		
1000Hz	峰，tan δ Peak, tan δ	137.4		

样品制备：在仪器中完全固化、不受控冷却。

结论 结果表明，可用所有不同的方法测试玻璃化转变。然而，每种方法测得不同的玻璃化转变温度值。因此，对于玻璃化转变温度不是只有一个"正确"值。如果升温速率和样品预处理是相同的，则 DSC 和 TMA 提供相似的结果。当用 DMA 的结果比较数值时，必须考虑到测试频率。根据参考文献，10K/min 的 DSC 或 TMA 的升温速率对应于约 2mHz。与 DSC 或 TMA 测试比较，1Hz(约 2 个数量级的更高频率)下的 DMA 测试测得高出约 10K 的 T_g。该差值是由于施加在样品上的不同类型的应力导致的。因此，所引用的每个玻璃化转变温度必须附以测试技术、升温速率、样品经热和机械预处理以及频率等详细信息。

2.5.3.2 节中的表可帮助评估对玻璃化转变温度的影响并解释差异。

参考文献 A. Hensel et. al，聚合物玻璃化转变区域中的温度调制量热法和介电光谱学，J. Thermal Analysis，Vol. 46 (1996) 935

Sample preparation: complete curing and uncontrolled cooling in the instrument.

Conclusions The results show that the glass transition can be measured by all the different methods. Each technique, however, yields different values for the glass transition temperature. There is therefore not just one "correct" value for the glass transition temperature. If the heating rate and sample pretreatment are the same, DSC and TMA provide similar results. When one compares values with DMA results, the measurement frequency must be taken into account. According to Reference, a DSC or TMA heating of 10 K/min corresponds to about 2 mHz. Compared with DSC or TMA measurements, the DMA measurement at 1 Hz (roughly 2 decades higher frequency) yields a T_g that is about 10 K higher. This difference is due to the different type of stress applied to the sample. Every glass transition temperature quoted must therefore be accompanied by details of the measuring technique, the heating rate, the thermal and mechanical sample pretreatment and, if appropriate, the frequency.

The tables in Section 2.5.3.2 are an aid for estimating influences on the glass transition temperature and for the interpretation of differences.

Reference A. Hensel et. al, Temperature-modulated calorimetry and dielectric spectroscopy in the glass transition region of polymers, J. Thermal Analysis, Vol. 46 (1996) 935

4 环氧树脂 Epoxy resins

4.1 影响固化反应的因素 Factors affecting curing reactions

4.1.1 固化条件(温度、时间)的影响
Influence of curing conditions (temperature, time)

目的	固化所用的条件影响材料固化的状态。可用 DSC 测量不充分的固化。本实验旨在检测后固化和研究固化不足对玻璃化转变温度的影响。
Purpose	The conditions used for curing influence the state of cure of a material. Inadequate curing can be measured by DSC. The purpose of the experiment was to check whether postcuring reactions could be detected and to investigate the influence of insufficient curing on the glass transition temperature.
样品	适合于制造建筑增强部件用板的由双组分环氧树脂体系制得的板材。 板材 1:在 21℃ 固化 72h 板材 2:在 21℃ 固化 72h 和在 40℃ 固化 24h 板材 3:在 21℃ 固化 72h、在 40℃ 固化 24h 和在 50℃ 固化 12h。 然后板材在室温后固化 7 个星期。
Sample	Plates from a two-component EP resin system used to make sheets for constructional reinforcement elements. Plate 1, cured at 21℃ for 72h Plate 2, cured at 21℃ for 72h and at 40℃ for 24h Plate 3, cured at 21℃ for 72h, at 40℃ for 24h, and at 50℃ for 12h. The plates were then postcured at room temperature for seven weeks.
条件	测试仪器:DSC 坩埚:40μl 标准铝坩埚,盖钻孔 样品制备: 用锋利刀片从每块板上切下重 1.0 至 1.5mg 的样品 DSC 测试: 第一次:以 10K/min 从 −20℃ 升温至 190℃ 降温:以 10K/min 从 190℃ 至 −20℃ 第二次:以 10K/min 从 −20℃ 升温至 300℃ 气氛:干燥空气,50mL/min
Conditions	Measuring cell:DSC Pan:Aluminum 40μL with pierced lid Sample preparation: A sample weighing 1.0 to 1.5mg was cut from each plate using a sharp knife. DSC measurement: 1st run:Heating from −20℃ to 190℃ at 10 K/min Cooling:from 190℃ to −20℃ at 10 K/min 2nd run:Heating from −20℃ to 300℃ at 10 K/min Atmosphere:Dry air,50mL/min

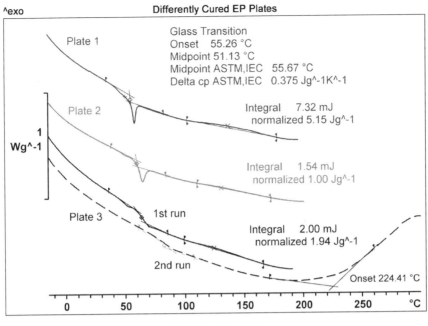

图 4.1　Fig. 4.1

解释　在前两个条件下固化的样品（板材 1 和 2）的曲线呈现玻璃化转变温度 T_g 区域的焓松弛峰。这影响了玻璃化转变温度的测定（起始点和中点，见板材 1 的曲线）。按照 ASTM/IEC，玻璃化转变温度确定为中点温度。较长的固化时间使玻璃化转变移向较高的温度。在所有三条曲线（第一次测试中的板材 1 到 3）上都可以看到高于 100℃ 的残余后固化反应。曲线难以计算，因为涉及的反应焓很小。

一般而言，玻璃化转变温度随着固化时间和温度的增加而增加。由完全固化的体系得到的 T_g 值（板材 3 的第二次测试）可清楚地证明这一点。T_g 值与反应焓的比较表明，进一步的固化即使已在很高的转化率（>98%）仍可导致 T_g 值的显著增加。

Interpretation　The curves of samples cured under the first two conditions (Plates 1 and 2) exhibit enthalpy relaxation peaks in the region of the glass transition temperature, T_g. This influences the determination of the glass transition temperatures (onset and midpoint, see the curve of Plate 1). The glass transition temperatures were determined as midpoint temperatures according to ASTM/IEC. Longer curing times cause the glass transition to shift to higher temperature. Residual postcuring reactions can be seen above 100℃ in all three curves (Plates 1 to 3 in the 1st runs). The curves are difficult to evaluate because of the very small reaction enthalpies involved.

In general, glass transition temperatures increase with increasing curing time or temperature. This is clearly demonstrated by the T_g value obtained for the fully cured system (2nd run of Plate 3). A comparison of the T_g values and reaction enthalpies shows that further curing can still result in a significant increase in T_g values even at very high conversions (>98%).

计算 Evaluation		第一次测试 1st run			第二次测试 2nd run
		板材 1 Plate 1	板材 2 Plate 2	板材 3 Plate 3	板材 3 Plate 3
玻璃化转变 T_g(ASTM/IEC) Glass transition T_g (ASTM/IEC)					
起始温度,℃ Onset temperature in ℃		55	59	60	75
中点温度,℃ Midpoint temperature in ℃		56	64	64	82
反应焓,J/g Reaction enthalpy in J/g		5.2	1.0	1.9	～0
转化率,%(100%:401.2J/g) Conversion in % (100%: 401.2 J/g)		98.7	99.8	99.5	100

结论 高转化率环氧树脂体系的固化度可用 T_g 值而不是反应焓值来清楚地表征。即使当后固化反应焓很小时,玻璃化转变温度仍然呈现显著的提高。这证明初始固化是不完全的。

Conclusions The degree of cure of EP resin systems at high conversions can be more clearly characterized by T_g values rather than by reaction enthalpy values. Even when the postcuring reaction enthalpy is very small, the glass transition temperature still shows a significant increase. This is clear evidence that the initial curing was incomplete.

4.1.2 组分混合比例的影响 Influence of the mixing ratio of the components

目的 树脂与硬化剂的不同比例会对模塑料的性能有重要的影响。本实验的目的是说明如何能用 DSC 测试来对此进行研究——特别是混合比例对反应性能和玻璃化转变的影响和热稳定性是否受到影响。

Purpose Different ratios of resin to hardener can have a significant effect on the properties of the molding compound. The purpose of the experiment is to show how this can be studied using DSC measurements-in particular the effects of the mixing ratio on reaction properties and the glass transition and whether thermal stability is affected.

样品 双组分环氧树脂黏合剂
产品:称为"UHU plus endfest"的商品化快速黏合剂
制备了树脂对硬化剂不同比例的混合物。

Sample Two-component EP resin adhesive
Product: commercial instant adhesive known as "UHU plus endfest"
Mixtures with different ratios of resin to hardener were prepared.

条件 测试仪器:DSC
Conditions 坩埚:40μl 标准铝坩埚,盖钻孔
样品制备:
由管子分配两个组分—挤出的长度用来计算混合比。用小铲混合组分,坩埚内加入 20 至 28mg 的黏合剂,快速放入仪器(一切在 4min 内

Measuring cell: DSC
Pan: Aluminum 40μL with pierced lid
Sample preparation:
The two components were dispensed from tubes-the lengths squeezed out were used to calculate the mixing ratio. The components were mixed using a spatula, the crucible filled with 20 to 28mg adhesive and quickly

完成）。

DSC 测试：

不同混合比的混合物：

第一次：以 10K/min 从 25℃升温至 175℃，175℃下恒温 1min

冷却：以 10K/min 从 175℃至 －50℃，在－50℃下恒温 1min

第二次：以 10K/min 从－50℃升温至 280℃

纯组分：

－50℃下恒温 1min，以 10 K/min 从－50℃加热至 360℃

气氛：干燥空气，50mL/min

inserted into the cell (this was all done within 4min).

DSC measurement：

Mixtures with different mixing ratios：

1st run：Heating from 25℃ to 175℃ at 10K/min, 1min at 175℃

Cooling：from 175℃ to －50℃ at 10K/min, 1min at －50℃

2nd run：Heating from －50℃ to 280℃ at 10K/min

Pure components：

1 min at －50℃, heating from －50℃ to 360℃ at 10K/min

Atmosphere：Dry air, 50mL/min

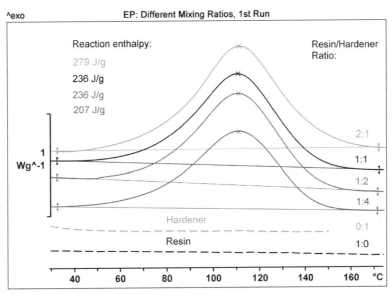

图 4.2　Fig. 4.2

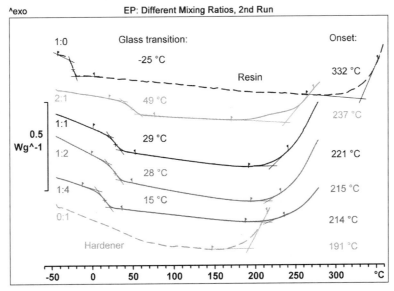

图 4.3　Fig. 4.3

解释 从图 4.2 所示的固化反应（第一次升温），不可能就决定是否采用制造商推荐的 1∶1 树脂/硬化剂混合比。特别是，反应焓的最大值不对应于 1∶1 v/v 混合比，而与树脂和硬化剂的活性基团的化学计量比有关。反应焓随着树脂含量的增加而增大。

第二次测试（图 4.3）清楚地表明玻璃化转变温度随着树脂体积含量的增加而提高。分解温度也随树脂含量的增加而稍有提高。

计算
Evaluation

Interpretation From the curing reactions shown in Figure 4.2 (1st heating run) it is not possible to decide whether the 1∶1 resin/hardener mixing ratio recommended by the manufacturer was employed. In particular, the maximum of the reaction enthalpy does not correspond to a 1∶1 v/v mixing ratio but depends on the stoichiometric proportions of the reactive groups of resin and hardener. The reaction enthalpy increases with increasing resin content.

The 2nd runs (Fig. 4.3) clearly show that the glass transition temperature increases with increasing resin volume content. The decomposition temperatures also increase a little with increasing resin content.

表 4.1 第一次测试——固化反应。
Table 4.1 1st run-curing reaction.

树脂/硬化剂比例 Resin/Hardener ratio	1∶0	2∶1	1∶1	1∶2	1∶4	0∶1
峰温,℃ Peak temperature in ℃	—	112	111	111	111	—
反应焓,J/g Reaction enthalpy in J/g	—	279	236	236	207	—

表 4.2 第二次测试——已固化的黏合剂。
Table 4.2 2nd run-cured adhesive.

树脂/硬化剂比例 Resin/Hardener ratio	1∶0	2∶1	1∶1	1∶2	1∶4	0∶1
玻璃化温度： 起始温度,℃ 中点温度,℃ Glass transition Onset temperature in ℃ Midpoint temperature in ℃	−29 −25	38 49	20 29	19 28	6 15	27
分解温度,℃ Decomposition temperature in ℃	332	237	221	215	214	191

图 4.4 表示测得的反应焓和相应的玻璃化转变温度之间的关系。树脂含量较大则反应焓较大表明形成了较大量的化学交联。这导致较大的网状结构密度而导致较高的玻璃化转变温度。

Figure 4.4 shows the relationship between the measured reaction enthalpy and the resulting glass transition temperature. Higher reaction enthalpy with larger resin content indicates the formation of a larger number of chemical crosslinks. This leads to a greater network density and hence to a higher glass transition temperature.

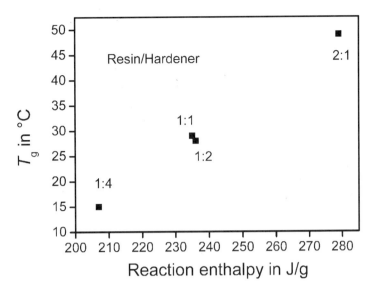

图 4.4　Fig. 4.4

结论　测试直至分解的树脂/硬化剂混合物的 DSC 曲线表明混合比对玻璃化转变和分解温度的影响。在上例中,较大的树脂含量导致较大的网状结构密度和较高的玻璃化转变温度。

测试表明,对于硬化剂含量较高的混合物,所推荐的黏合剂组分混合比(1∶1)的小偏离没有引起产物固化性能的显著变化。然而,过量的树脂或硬化剂导致玻璃化转变温度的显著改变。

Conclusions　DSC curves of resin/hardener mixtures measured up until decomposition show the effect that the mixing ratio has on the glass transition and decomposition temperatures. In the above case, a larger resin content leads to a greater network density and a higher glass transition temperature.

The measurements show that small deviations of the recommended mixing ratio of the adhesive components (1∶1) to mixtures with larger hardener content do not result in a significant change of the curing properties of the product. An excessive amount of resin or hardener however causes a significant shift of the glass transition temperature.

4.1.3　促进剂类型的影响 Influence of the type of accelerator

目的	通常将促进剂加到环氧树脂中以调节固化速率使之适应特殊的应用。本实验的目的是说明 DSC 如何能用来研究促进剂对等温固化反应的影响。
Purpose	Accelerators are normally added to epoxy resins to adapt the curing rate to the particular application. The purpose of the experiment is to show how DSC can be used to study the influence of the accelerator on the isothermal curing reaction.
样品	含两个不同促进剂(各为 GY260 的 1%)的环氧树脂(GY260/HY917): ZKXB5111 三氯化硼胺络合物 DY9577
Sample	EP resin (GY260/HY917) with two different accelerators (each 1% relative to GY260): ZKXB5111 DY9577, boron trichloride amine complex
条件	测试仪器:DSC 坩埚:40μl 标准铝坩埚,盖钻孔
Conditions	Measuring cell:DSC with Intralooler Pan:Aluminum 40μL with pierced lid

| 样品制备:称样品(8至12mg)放入坩埚。 | **Sample preparation**: The sample (8 to 12mg) was weighed into the crucible. |

样品制备:称样品(8至12mg)放入坩埚。
DSC 测试:
在140℃等温固化60min
固化后玻璃化转变的测定:以10 K/min 从50℃升温至160℃
气氛:氮气,50 mL/min

Sample preparation: The sample (8 to 12mg) was weighed into the crucible.
DSC measurement:
Isothermal curing at 140℃ for 60 min
Determination of the glass transition after curing: Heating from 50℃ to 160℃ at 10K/min
Atmosphere: Nitrogen, 50mL/min

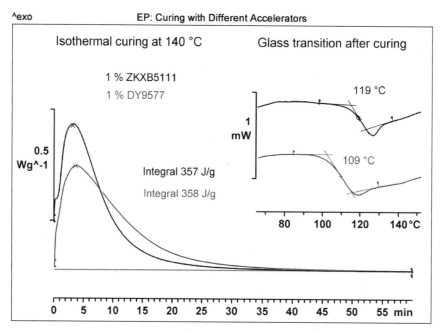

图 4.5　Fig. 4.5

解释　在起始强吸热热流已经达到 -20mW 值后,在坩埚放入 140℃ 的炉中后测试立即开始。固化反应迅即开始,大约一小时后逐渐停止。含 DY9577 的固化反应甚至在 60min 后仍没有彻底完成(DSC 曲线不是完全水平的)。

等温固化后的样品的升温曲线(右边插入图)表明,固化较快的树脂的玻璃化转变温度比另一体系的高。本例的促进剂有两个作用:第一是对反应速率,第二是对网状结构。即使反应焓几乎相同,但含 ZKXB5111 的比含另一种促进剂的反应更快,导致更大的交联度。

Interpretation　The measurement was started immediately after inserting the crucible into the furnace at 140℃ after the initial strong endothermic heat flow had reached a value of -20 mW. The curing reaction begins immediately and dies down after about an hour. The curing reaction with DY9577 is still not quite complete even after 60 min (the DSC curve is not perfectly horizontal).

The heating curves of the isothermally cured samples (inserted diagram, right) show that the glass transition temperature of the resin that cured more rapidly is higher than that of the other system. In this case, the accelerator has two effects: first on the reaction rate, and second on the network structure. The reaction is faster with ZKXB5111 than with the other accelerator and results in a greater degree of crosslinking, even though the reaction enthalpies are practically the same.

计算 Evaluation	计算结果汇总于下表 The evaluation results are summarized in the following table.		
		ZKXB5111	DY9577
	ΔH_{curing}, J/g ΔH_{curing} in J/g	357	358
	Tg, ℃ Tg in ℃	109	119
	$\alpha=90\%$ 的时间, min Time at $\alpha=90\%$ in min	14.7	20.0
	10min 后转化率, % Conversion after 10min, in %	78	63

结论 等温 DSC 测试非常快地表明不同促进剂对交联的影响。用这个方法能容易地对特定应用进行树脂体系的最优化。此外，后固化测试显示所达到的玻璃化转变温度。在开发中这也可用作最优化的标准。

在常规应用中，要检查合适的组分是否以合适的量加入到树脂中，短时间的等温测试就足够了。本例得出结果只需化费 10min。

Conclusions Isothermal DSC measurements very quickly show the influence that different accelerators have on crosslinking. Resin systems can easily be optimized for the particular application in this way. Postcuring measurements furthermore show the glass transition temperature that is reached. This can also be used as an optimization criterion in development.

In routine applications, a short isothermal measurement is sufficient to check whether the right components in the right quantities were added to the resin. In this case, it would only take 10 min to make this decision.

4.1.4 促进剂含量对固化反应的影响
Influence of accelerator content on the curing reaction

目的 **Purpose**	用 DSC 和 TGA 研究促进剂含量对环氧树脂－环脂肪酐体系的固化反应的影响。 To investigate the effect of the accelerator content on the curing reaction of an epoxy-cycloaliphatic anhydride system using DSC and TGA.
样品 **Sample**	双酚 A 的二缩水甘油醚类环氧树脂(Araldite F，Ciba Specialty Chemicals) 从甲基四氢苯酐制得的硬化剂(Araldite HY 905，Ciba Specialty Chemicals) 促进剂是起催化剂作用的叔胺(Araldite DY061，Ciba Specialty Chemicals) Epoxy resin based on diglycidyl ether of bisphenol A (Araldite F, Ciba Specialty Chemicals). Hardener derived from methyl-tetrahydrophthalic anhydride (Araldite HY 905, Ciba Specialty Chemicals). Accelerator: is a tertiary amine that acts as a catalyst (Araldite DY061, Ciba Specialty Chemicals).
条件 **Conditions**	测试仪器：DSC 和 TGA 坩埚：DSC：40μl 标准铝坩埚，盖钻孔 　　　TGA：70μl 氧化铝坩埚，带盖 样品制备： 环氧树脂和硬化剂以 100:100 w/w Measuring cell: DSC and TGA Pan: DSC: Aluminum 40μL with pierced lid 　　　TGA: Alumina 70μL with lid Sample preparation: The epoxy resin and the hardener were mixed in a

的比例混合。加入必要的催化剂量以得到下列促进剂浓度:0.25、0.5、1.0和2.2重量份(pbw)。样品重量8至9mg。

测试:

DSC:以10K/min从-90℃升温至410℃;

TGA:以10K/min从50℃至600℃。

气氛:氮气,50mL/min(DSC)和200mL/min(TGA)

100:100 w/w ratio. The quantity of catalyst was added necessary to give the following concentrations of accelerator: 0.25, 0.5, 1.0 and 2.2 parts by weight (pbw). The sample weight was 8 to 9 mg.

measurement:

DSC: Heating from -90℃ to 410℃ at 10K/min;

TGA: 50℃ to 600℃ at 10K/min.

Atmosphere: Nitrogen, 50mL/min (DSC) and 200mL/min (TGA)

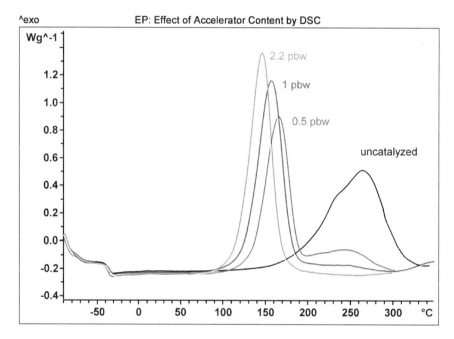

图 4.6 未催化体系和所示促进剂浓度(0.5、1.0和2.2 pbw)的催化体系的DSC曲线

Fig. 4.6 DSC curves of the uncatalyzed system and catalyzed systemsat the indicated accelerator concentrations (0.5, 1.0 and 2.2 pbw).

解释 未催化体系的DSC曲线在150℃至310℃呈现宽放热峰。促进剂浓度小于1pbw时,催化环氧-酐体系的动态DSC曲线呈现两个放热峰(图4.6)。在80℃至200℃的第一个峰尖锐、是可明确解释的,与促进剂含量有关。峰温随促进剂含量的增加而下降,即固化反应发生在更低温度且更快。

第二个峰出现在200℃至320℃之间,较宽且较平坦。它出现在未催化体系的放热峰的相同区域。对促进剂含量低于约1pbw的样品可观

Interpretation The DSC curve of the uncatalyzed system shows a broad exotherm between 150℃ and 310℃. At accelerator contents of less than 1 pbw, the dynamic DSC curves of the catalyzed epoxy-anhydride system shows two exothermic peaks (Fig. 4.6). The first peak between 80℃ and 200℃ is sharp and well defined, depending on the accelerator content. The peak temperatures decrease with increasing accelerator content, that is, the curing reaction occurs at lower temperature and more rapidly.

The second peak occurs between 200℃ and 320℃ and is broader and flatter. It appears in the same region as the exothermic peak of the uncatalyzed system. This peak is observed with samples in which the accelerator content is less than about 1 pbw. The

察到该峰。该效应表明,当促进剂含量低时,发生了无催化剂的固化反应。

未催化体系的 TGA 表明至约 250℃时有大约 25％的失重,最大失重速率在 195℃。与未催化体系相同的温度下,硬化剂也有失重。这意味着本体系中的该失重,就是可能由于硬化剂挥发产生的损失。不过,失重在 200℃左右变平,它大致为未催化体系固化的起始温度。

effect indicates that the curing reaction takes place without a catalyst when the accelerator content is low.

In TGA, the uncatalyzed system shows a mass loss of some 25% up to about 250℃, with a maximum rate of mass loss at 195℃. The hardener also loses mass at the same temperature as the uncatalyzed system. This suggests that the mass loss in this system corresponds to loss of hardener, possibly due to evaporation. Nevertheless, the loss of mass levels off at about 200℃, which is roughly the onset temperature of the curing in the uncatalyzed system.

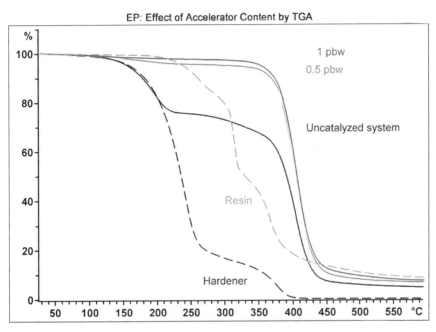

图 4.7　促进剂浓度 0.5 和 1pbw 的催化体系、未催化体系、Araldite F 树脂和 HY905 硬化剂的的 TGA 曲线

Fig. 4.7　TGA curves of catalyzed systems at accelerator concentrations of 0.5 and 1 pbw, the uncatalyzed system, Araldite F resin and HY905 hardener.

计算
Evaluation

本表汇总了不同促进剂浓度的催化体系的 DSC 结果
The table summarizes the DSC results for the catalyzed systems at different accelerator concentrations.

促进剂浓度,重量份 Accelerator content in pwb	第一个峰温,℃ 1st peak temperature in ℃	第二个峰温,℃ 2nd peak temperature in ℃	固化焓,J/g Enthalpy of curing in J/g
2.2	146	—	329
1.0	157	—	323
0.5	167	245	314
0.25	175	258	329
0（未催化的 uncatalyzed）	—	290	324

结论 含酸酐硬化剂的环氧树脂的固化反应需要催化剂（促进剂）使反应发生在合理的温度范围内，并防止二次反应发生。TGA 分析表明，未催化体系在固化期间有硬化剂损失。DSC 测试能测定促进剂的最佳浓度。

Conclusions The curing reaction of an epoxy with an anhydride hardener requires a catalyst (accelerator) for the reaction to take place over a reasonable temperature range and to prevent secondary reactions occurring. TGA analysis shows that the uncatalyzed system loses hardener during curing. DSC measurements allow the optimum concentration of accelerator to be determined.

参考文献 促进剂浓度对环氧树脂-酐体系的固化反应的影响；S. Montserrat, C. Flaqqué, M. Calafell, G. Andreu, J. Málek；Thermochimica Acta, Vol 269/270 (1995) 213-229.

Reference Influence of the accelerator concentration on the curing reaction of an epoxy-anhydride system; S. Montserrat, C. Flaqqué, M. Calafell, G. Andreu, J. Málek; Thermochimica Acta, Vol 269/270 (1995) 213-229.

4.1.5 环氧树脂：转化率行为的预测和验证
EP: Prediction of conversion behavior and verification

目的 基于仅几个 DSC 测试的结果，借助动力学分析来描述复杂树脂体系的反应行为从而预测在特定温度下反应体系的行为。
出于安全理由，预测必须通过转化率行为的实际测试独立验证。

Purpose To describe the reaction behavior of a complex resin system by means of kinetic analysis and so predict the behavior of a reactive system at particular temperatures based on the results of just a few DSC measurements.
For safety reasons, the predictions must be independently verified by practical measurements of conversion behavior.

样品 用于制造适合于建筑增强构件的薄板的双组分环氧树脂体系。

Sample Two-component EP resin system used to make sheets for constructional reinforcement elements.

条件 测试仪器：DSC
坩埚：40 μl 标准铝坩埚，盖钻孔
样品制备：
树脂－硬化剂体系在冰/水浴中混合，搅拌约 5min，混合物在冰/水浴中再放置 5min。然后在室温下加入铝坩埚。
用于测定反应动力学的坩埚在加入样品后约 2min 放入 DSC 测试池。为每个测试制备新的树脂－硬化剂体系。
对于要用非模型动力学来检查预测的样品，在 42℃ 的暖柜中贮存不同

Conditions Measuring cell: DSC
Pan: Aluminum 40 μL with pierced lid
Sample preparation: The resin-hardener system was mixed in an ice/water bath, stirred for about 5 min and the mixture left for a further 5 min in the ice/water bath. The aluminum crucibles were then filled at room temperature.
The crucible for the determination of the reaction kinetics was inserted in the DSC measuring cell about 2 min after it had been filled. A fresh resin-hardener system was prepared for each measurement.
Samples to check predictions using model free kinetics were stored for different periods in a warming cabinet

的时间。在树脂－硬化剂体系混合后,含样品的坩埚放在柜中约10min。同样的树脂-硬化剂制备用于所有的单独样品。

DSC测试:

非模型反应动力学:

以15K/min从25℃降温至0℃,在0℃恒温1min,以3、5和12.5K/min从0℃升温至240℃

检查和验证:

在暖柜中42℃贮存

以10K/min从25℃升温至240℃

气氛:干燥空气,50mL/min

at 42℃. The crucibles with the samples were placed in the cabinet about 10 min after mixing the resin-hardener system. The same resin-hardener preparation was used for all the individual samples.

DSC measurement:

Model free reaction kinetics:

Cooling from 25℃ to 0℃ at 15K/min, 1min at 0℃, Heating from 0℃ to 240℃ at 3, 5 and 12.5K/min

Checking and verification:

Storage at 42℃ in the warming cabinet

Heating from 25℃ to 240℃ at 10K/min

Atmosphere: Dry air, 50mL/min

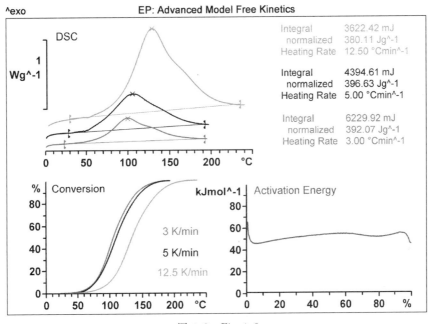

图4.8　Fig. 4.8

解释 根据Vyazovkin的高级非模型反应动力学方法(图4.8),进行至少三个不同升温速率的DSC测试。利用动态转化率曲线计算固化反应活化能与转化率的关系。为了检查动力学结果,样品先在42℃固化不同时间,然后进行DSC后固化测试以测定已达到的反应转化率(图4.9)。最上面的升温曲线表示完全的固化反应,其他曲线是已经在42℃固化了不同时间的样品的残余固化。

Interpretation The advanced model free reaction kinetics method of Vyazovkin (Fig. 4.8) is based on performing at least three DSC measurements at different heating rates. The dynamic conversion curves are used to calculate the activation energy of the curing reaction as a function of conversion. To check the kinetic results, samples were first cured for different periods at 42℃. DSC postcuring measurements were then performed to determine the reaction conversions that had been attained (Fig. 4.9). The uppermost heating curve shows the complete curing reaction, the other curves the residual curing of samples that had been cured for different periods at 42℃.

这能计算转化率（见表 4.3 和图 4.10，红色曲线）。树脂体系甚至在 42℃1.5 天后仍不是完全固化的。参见4.2.4节"玻璃化"。

This allows the conversion to be calculated (see Table 4.3 and Fig. 4.10, red curve). The resin system is still not completely cured even after 1.5 days at 42℃. See Section 4.2.4 Vitrification.

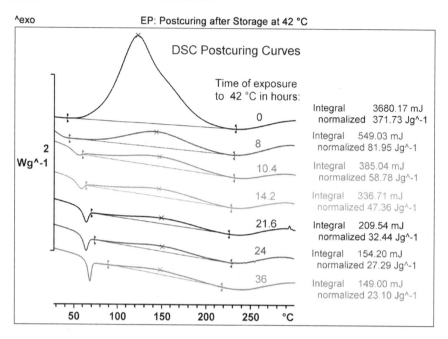

图 4.9　Fig. 4.9

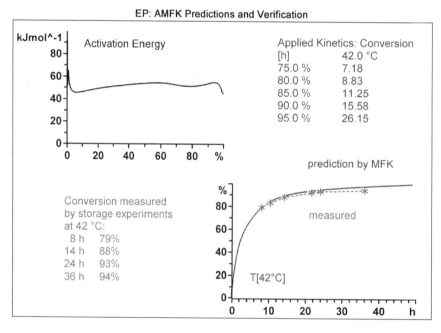

图 4.10　Fig. 4.10

计算　活化能曲线可用于计算对任何期望的温度和对任何转化率的反应时间的预测。图 4.10 是用非模型动力学求得的 42℃下转化率与

Evaluation　The activation energy curve can be used to calculate predictions for the reaction time at any desired temperature and for any conversion. In Figure 4.10, the prediction obtained using model free kinetics for conversion as a

时间关系曲线的预测(红线)与测试值作的比较。

function of time at 42℃ (red curve) is compared with the measured values.

表 4.3 Table 4.3

反应时间 h Reaction time in h	8	10.4	14.2	21.6	24	36
用高级非模型反应动力学得到的 42℃ 下的转化率,% Conversion using advanced model free reaction kinetics at 42℃ in %	78	84	89	93	94	97
实验测定的 42℃ 下的转化率,%（100%时等于 386.4 J/g） Experimentally determined conversion at 42℃ in % (100% corresponds to 386.4 J/g)	79	83	88	92	93	94

结论 用非模型反应动力学预测的转化率值与测试值吻合良好。

Vyazovkin 开发的非模型反应动力学方法(见 2.3.3 节)的主要优势是无需假定反应级数,还能应用于复杂反应。一般来说,对 10% 至 90% 范围内的转化率能作出非常可靠的预测。

在固化反应中,材料可能在低温下发生玻璃化。当比较由非模型动力学得到的预测时,必须考虑这一点。见 4.2.4 节"玻璃化"。

Conclusions The conversion values predicted using model free reaction kinetics agree well with the measured values.

The great advantage of the model free reaction kinetics approach developed by Vyazovkin (see Section 2.3.3) is that no assumptions have to be made about the order of the reaction and that it can also be applied to complex reactions. In general very reliable predictions can be made for conversions in the range 10% to 90%.

The material may possibly vitrify at low temperatures in curing reactions. This must be taken into account when predictions obtained from model free kinetics are compared. See Section 4.2.4 Vitrification.

4.1.6 环氧树脂固化的 DMA 测量
Curing of an EP resin measured by DMA

目的	固化反应使热固性树脂从低黏度液态转化成刚硬的固体。反应期间模量急剧变化。本应用说明在一次测试中如何研究这些变化、如何测定未反应的和固化的黏合剂的玻璃化转变。
Purpose	The curing reaction transforms a thermosetting resin from a low viscosity liquid to a rigid solid. The modulus changes dramatically during this reaction. The application shows how these changes can be investigated and how the glass transitions of the unreacted and cured adhesive can be determined in one measurement run.
样品	由 DGEBA(双酚 A 二缩水甘油醚)和 DDM(二胺基二苯基甲烷)组成的环氧树脂体系,未固化树脂
Sample	Epoxy system consisting of DGEBA (diglycidylether of bisphenol A) and DDM (diaminodiphenylmethane), uncured resin.
条件 **Conditions**	测试仪器:DMA　　　　　　　Measuring cell:DMA 预处理:如参考文献所述,活性混　Pretreatment:The reactive mixture was prepared

合物以 2∶1 的摩尔比制备为 DGEBA 和 DDM 的化学计量混合物。测试前混合物贮存在 −20℃。

样品制备：
用液体剪切夹具，保持夹具的三个部件在固定距离，以使液体树脂能够充满每个间距为 0.2mm 的两个缝隙中。这很重要，特别在 DMA 外室温下正确装填液体样品期间。

样品几何形状：直径 11.0mm，厚 0.2mm。

DMA 测试：
1) 以 3K/min 的速率从 −35℃ 升温至 210℃。用 40μm 的位移振幅和 35N 的最大力在 10Hz 频率下的剪切调制。
2) 接着以 3K/min 从 210℃ 降温至 70℃。剪切调制如上但用 2μm 的最大位移振幅。

气氛：静态空气

as a stoichiometric mixture of DGEBA and DDM in a molar ratio of 2∶1 as described in reference [1]. The mixture was stored at -20℃ before measurement.

Sample preparation： A shear clamp for liquids was used to keep the three parts of the clamp at a fixed distance so that the liquid resin could be filled into the two gaps of 0.2 mm each. This is important especially during proper loading of the liquid sample at room temperature outside of the DMA.

Sample geometry：11.0 mm diameter, 0.2 mm thick.

DMA measurement：
1) Heating from −35℃ to 210℃ at a rate of 3 K/min. Shear modulation at a frequency of 10 Hz using a displacement amplitude of 40μm and a maximum force of 35N.
2) Subsequent cooling from 210℃ to 70℃ at 3K/min. Shear modulation as above but using a maximum displacement amplitude of 2μm.

Atmosphere： Static air

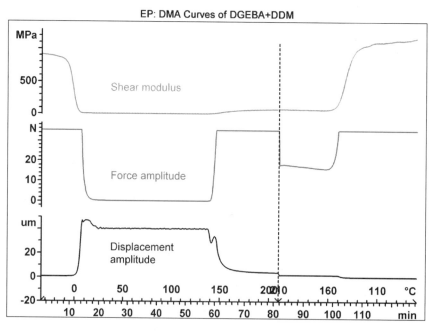

图 4.11　Fig. 4.11

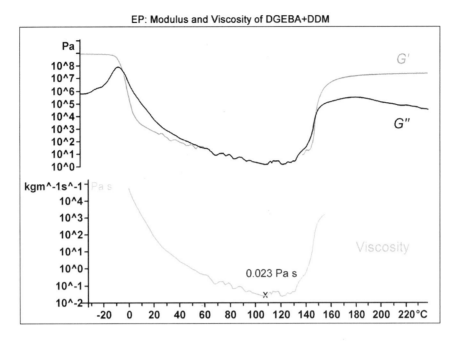

图 4.12　Fig. 4.12

解释　图 4.11 表示环氧树脂体系与时间关系的 DMA 升降温曲线。测试用 DMA/SDTA861^e 以剪切模式进行。图 4.12 表示与样品温度关系的 DMA 升温曲线。对液体范围计算了黏度(见下面的注释)。

新制备的样品在室温下是液态的。起始时在 -50℃，它处于玻璃态，升温时在约 -10℃ 变为液态。在 150℃ 交联反应时又成为固态。经过玻璃化转变区域时，模量 G' 下降超过 8 个数量级，从 900MPa 至 1Pa。在交联反应中，模量增加到 10MPa(橡胶平台)。

固化反应的凝胶点测定为贮存模量等于损耗模量(即 tanδ 为 1)的时间。在所给条件下，出现在 150℃。在随后的从 210℃ 至 70℃ 的降温期间(图 4.11)，材料呈现完全固化树脂的玻璃化转变，在约 155℃。在由橡胶态到玻璃态的转变期间，模量又增加了一个数量级。

Interpretation　Figure 4.11 shows the DMA heating and cooling curves of the epoxy system as a function of time. The measurements were performed in the shear mode using the DMA/SDTA861^e. Figure 4.12 shows the DMA heating curves as a function of sample temperature. The viscosity is calculated for the liquid range (see comments below).

A fresh sample is liquid at room temperature. Initially, at -50℃, it is in a glassy state and on heating becomes liquid at about -10℃. It becomes solid again during the crosslinking reaction at 150℃. The modulus, G', decreases by more than 8 orders of magnitude, from 900 MPa to 1Pa, on passing through the glass transition region. In the crosslinking reaction, the modulus increases to 10MPa (rubbery plateau).

The gel point of the curing reaction is determined as the time at which the storage modulus equals the loss modulus (i. e. tanδ is unity). Under the given conditions, this occurs at 150℃. During the subsequent cooling from 210℃ to 70℃ (Fig. 4.11), the material shows the glass transition of the fully cured resin at approximately 155℃. During this transition from the rubbery to the glassy state, the modulus is increased again by an order of magnitude.

计算 对黏流材料,黏度可用下式从剪切损耗模量 G'' 和频率计算:

Evaluation For viscous materials, the viscosity can be calculated from the shear loss modulus, G'', and the frequency using the following equation:

$$\eta' = \frac{G''}{\omega} = \frac{G''}{2\pi f}$$

对于其中树脂为液态的温度范围所得到的曲线示于图 4.12 中。

The resulting curve is shown in Figure 4.12 for the temperature range in which the resin is liquid.

结论 DMA/SDTA861^e 的剪切模式(只有这个模式)能够测试材料从玻璃态到液态和最终到刚硬固态时的模量(G')和 tanδ。在固化中,模量变化超过 6 个数量级。新制备的和固化的树脂的玻璃化转变和凝胶点只用单次实验就可测试。
用较低的升温速率和不同的测试频率检测交联反应期间的玻璃化也会是容易的。

Conclusions The DMA/SDTA861^e in the shear mode (only in this mode) allows the modulus (G') and tanδ to be measured when the material passes from the glassy state to the liquid state and finally to the rigid solid state. During curing, the modulus changes by more than 6 orders of magnitude. The glass transitions of the fresh and the cured resins, and the gel point have been measured in just one single experiment.
It would also be easily possible to detect vitrification during the crosslinking reaction by using lower heating rates and different measurement frequencies.

参考文献 J. E. K. Schawe,用 DSC 和温度调制 DSC 研究扩散控制对固化反应的影响,Journal of Thermal Analysis and Calorimetry,Vol. 64 (2001) 599—608.

Reference J. E. K. Schawe, Investigation of the influence of diffusion control on the curing reaction using DSC and temperature-modulated DSC, Journal of Thermal Analysis and Calorimetry, Vol. 64 (2001) 599—608.

4.1.7 预浸料固化的 DMA 测量 Curing of a prepreg measured by DMA

目的 航空工业中的许多测试方法是用 DMA 来表征树脂和预浸料的。本实验的目的是说明从这样的研究能得到的结果和解释怎样能够诠释测得的效应。

Purpose Many test methods in the aerospace industry use DMA to characterize resins and prepregs. The purpose of this experiment is to show the results that can be obtained from such an investtigation and explain how the measured effects can be interpreted.

样品 含碳纤维的环氧树脂预浸料

Sample Epoxy resin prepreg with carbon fibers.

条件
Conditions

测试仪器:
DMA,单悬臂样品支架

Measuring cell:
DMA with single cantilever sample holder

样品制备:
安装在样品支架中的一个预浸料有如下几何形状:长度 15.00mm、宽度 7.70mm、厚度 1.16mm。算得的几何因子为 280800m^{-1}。

Sample preparation: A piece of the prepreg mounted in the sample holder gave the following sample geometry: length 15.00mm, width 7.70 mm, thickness 1.16mm. The calculated geometry factor was 280800 m^{-1}.

DMA 测试:以 2K/min 从 0℃升温至 350℃,频率序列 0.1,1 和 10Hz

DMA measurement: Heating from 0℃ to 350℃ at 2K/min; frequency series of 0.1, 1 and 10Hz

气氛：静态空气　　　　　　　　　　　　　　Atmosphere: Static air

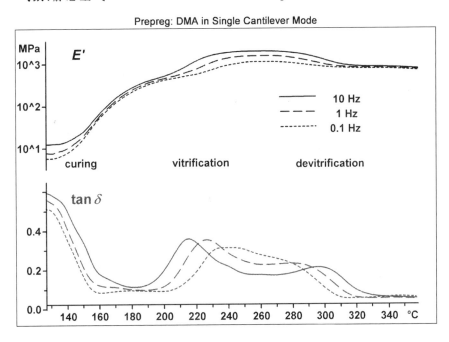

图 4.13　Fig. 4.13

解释　因为用碳纤维增强物作为支撑材料，所以能以弯曲模式测试树脂软状态。130℃时的弯曲模量（E'）只有 10MPa，这主要是由碳纤维织物产生的模量。在 140℃ 至 200℃ 的固化过程中，E' 几乎提高 2 个数量级，而 tanδ 呈显著下降。树脂由于升温速率低而玻璃化，其由 220℃ 处的 tanδ 峰和 E' 的增加显示。继续升温时，材料去玻璃化，于是处于橡胶弹性态。比较不同频率下测试的曲线清楚地表明玻璃化和去玻璃化是频率依赖的效应（见 tanδ 峰温的移动）。化学交联时模量曲线并不呈现随温度而变化，但随着频率增大而增大。

结论　可用 DMA 跟踪固化时的行为。即使当增强纤维含量高时，玻璃化和去玻璃化也可清晰观察到。使用不同调制频率可使测试效应得到可靠的解释。这用常规 DSC 测试不总是可能的。

Interpretation　The resin can be measured in the soft state in the bending mode because carbon-fiber reinforcement has been used as a support material. The bending modulus (E') at 130℃ is only 10 MPa. This is mainly due to the modulus of the carbon fiber fabric. During the curing process between 140℃ and 200℃, E' increases by almost two decades while tanδ shows a marked decrease. The resin vitrifies as a result of the low heating rate, which is shown by the peak in tanδ at 220℃ and the increase of E'. On further heating, the material devitrifies and is then in a rubbery-elastic state. A comparison of the curves measured at different frequencies clearly shows that vitrification and devitrification are frequency-dependent effects (see shift of peak temperature of tanδ). The modulus curves do not exhibit a temperature shift on chemical crosslinking, but increases with increasing frequency.

Conclusions　The behavior on curing can be followed by DMA. The vitrification and devitrification can be clearly seen even when the content of reinforcement fibers is high. The use of different modulation frequencies allows the measured effects to be reliably interpreted. This is not always possible with conventional DSC measurements.

4.1.8 粉末涂层的固化 Curing of a powder coating

粉末涂层是一项常用的表面修整技术。粉末状的涂料是带静电的，喷涂到基板的表面上，然后加热。粉末颗粒软化、接合和固化，形成连续的涂层。涂层的质量受粉末涂料的热性能的强烈影响。

Powder coating is a commonly used surface finishing technique. The powdered paint is electrostatically charged and sprayed onto the surface of a substrate. This is then heated. The powder particles soften, coalesce and cure to form a continuous coating. The quality of the coating is strongly influenced by the thermal behavior of the powder coating material.

目的	本实验的目的是用 TOPEM™ 测试来研究粉末涂层的热行为。用 DSC 显微镜方法同步观察粉末颗粒的软化和接合。	
Purpose	The purpose of this experiment is to investigate the thermal behavior of a powder coating using TOPEM™ measurements. The softening and coalescence of the powder particles is simultaneously observed using DSC microscopy.	
样品	未固化的 KU600 环氧树脂粉末涂料。	
Sample	Uncured KU600 epoxy resin powder coating.	
条件 Conditions	测试仪器： 带 DSC 显微镜附件的 HP DSC DSC 作 TOPEM™ 测试 坩埚：40μl 标准铝坩埚，无盖 样品制备：细粉末，常规 DSC 约 11mg，TOPEM™ 约 16mg，称量后放入坩埚。 DSC 测试： 以 1K/min 从 20℃ 升温至 260℃ TOPEM™：脉冲高度 0.5K、脉冲宽度 15 至 30s 气氛：静态空气	**Measuring cell**： HP DSC with DSC microscopy accessory DSC for TOPEM™ measurements **Pan**：Aluminum 40μL without lid **Sample preparation**：The fine powder, approx. 11mg for conventional DSC and 16mg for TOPEM™, was weighed into the crucibles. **DSC measurement**： Heating from 20℃ to 260℃ at 1 K/min TOPEM™：pulse height 0.5 K, pulse width 15 to 30s **Atmosphere**：Static air

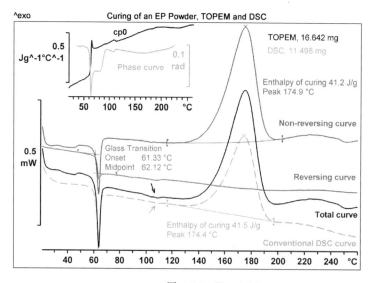

图 4.14　Fig. 4.14

解释 图示为由 TOPEM™ 计算得到的可逆、不可逆和总热流曲线,一并列出以 1K/min 测试的常规 DSC 曲线。

从 TOPEM™ 分析得到的总热流曲线与常规 DSC 曲线实际上是完全相同的。在约 60℃ 处观察到一个吸热峰。它可被解释为熔融峰,反应峰或两个重叠效应。分析可逆和不可逆曲线后,便可清楚解释该效应。可逆曲线呈现一个台阶变化,它是由于玻璃化转变。在相同的温度范围内从不可逆曲线上观察到一个大的吸热峰,是与玻璃化转变有关的焓松弛。

粉末的放热固化反应发生在 105℃ 至 205℃ 范围内,峰最大处在 175℃。从 TOPEM™ 测试的不可逆曲线得到的固化焓值 41.2J/g,实际上与从常规 DSC 曲线计算得到的值相同。

总 TOPEM™ 曲线与常规 DSC 曲线的仔细比较显示,在约 110℃ 处存在一个小效应。这在示于小插图中的相位曲线上更明显。反复的测试表明,该效应是完全可重复的。样品在该温度范围显然经历了一个变化。

为了找出关于这个效应的更多信息,用与 DSC 连接的显微镜和一系列拍摄到的图像视觉监测样品。下面显示了在 85℃、95℃、105℃ 和 115℃ 时的四张这样的图像。白色区域或白点是由坩埚底部的反光产生的。

在 85℃,即玻璃化转变以上约 20K,样品变软并开始收缩,各个颗粒之间形成间隙。在 95℃,粉末更像黏性液体但具有有限流动性能;颗粒开始接合。

当温度升高到 105℃ 时,液体黏度变小,分子的活动性增大。聚合物慢慢流动和相互扩散,越过颗粒边界发生缠结。按一般规则,相互扩散速率与样品温度和玻璃化转变温度之差($T - T_g$)直接有关。在

Interpretation The figure shows the reversing, non-reversing and total heat flow curves obtained from the TOPEM™ evaluation together with a conventional DSC curve measured at 1K/min.

The total curve from the TOPEM™ analysis and the conventional DSC curve are practically identical. An endothermic peak is observed around 60℃. This could be interpreted as a melting peak, a reaction peak or two overlapping effects. The interpretation of the effect becomes clear when the reversing and non-reversing curves are examined. The reversing curve shows a step change, which is due to a glass transition. A large endothermic peak is observed in the non-reversing curve in the same temperature range. This results from enthalpy relaxation associated with the glass transition.

The exothermic curing reaction of the powder occurs in the range 105℃ to 205℃ with a peak maximum at 175℃. The value of 41.2J/g obtained for the enthalpy of curing from the non-reversing curve of the TOPEM™ measurement is practically the same as that from the evaluation of the conventional DSC curve. A detailed comparison of the total TOPEM™ curve and conventional DSC curve shows the presence of a small effect at around 110℃. This is more apparent in the phase curve shown in the small inserted diagram. Repeated measurements showed that the effect was perfectly reproducible. The sample obviously undergoes a change in this temperature range.

To find out more about this effect, the sample was visually monitored using a microscope coupled to the DSC and a series of images captured. Four such images at 85℃, 95℃, 105℃ and 115℃ are displayed below. The white area or spots are caused by the reflection of light from the bottom of the crucible.

At 85℃, i.e. about 20 K above the glass transition, the sample becomes soft and begins to shrink, creating gaps between the individual particles. At 95℃, the powder is more like a viscous liquid but with limited flow properties; the particles begin to coalesce.

As the temperature increases to 105℃, the liquid becomes less viscous and the mobility of the molecules increases. The polymer flows slowly and interdiffusion and entanglement occur across particle boundaries. As a general rule, the rate of interdiffusion is directly related to the difference between the sample temperature and glass transition temperature ($T - T_g$). At

120℃,颗粒的接合几乎已经完成。接合过程引起样品中热流的变化。这反映在常规 DSC 和 TOPEM™ 测试曲线上,是在110℃处观察到的效应的由来。

120℃, coalescence of the particles is practically complete. The coalescence process causes a change in the heat flow in the sample. This is reflected in the conventional DSC and TOPEM™ measurement curves and is the origin of the event observed at 110℃.

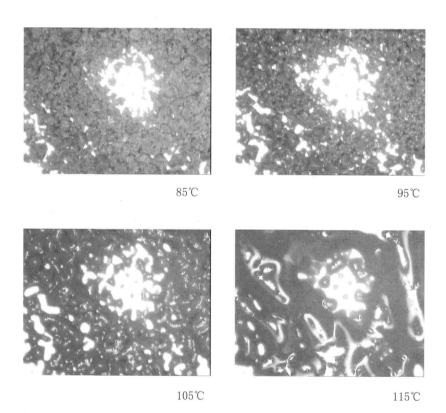

85℃   95℃

105℃   115℃

结论 用常规 DSC 和 TOPEM™ 测试能容易地表征粉末涂料的热行为,特别是玻璃化转变和固化反应。粉末颗粒的接合发生在玻璃化转变温度以上,在样品中产生热传递的变化。该效应能用 DSC 测试检测和用 DSC 显微镜法视觉监测。视觉观察对 DSC 曲线上意外效应的诠释常常是非常有用的。

Conclusion The thermal behavior of a powder coating and in particular the glass transition and curing reaction can be readily characterized using conventional DSC and TOPEM™ measurements. Coalescence of the powder particles occurs above the glass transition temperature and causes heat transfer in the sample to change. This event can be detected by DSC measurements and visually monitored using DSC-microscopy. Visual observation is often very useful for the interpretation of unexpected effects on DSC curves.

4.2 影响玻璃化转变的因素 Influences affecting the glass transition

4.2.1 重复后固化对玻璃化转变的影响
Effect of repeated postcuring on the glass transition

玻璃化转变温度的测定对含活性树脂的涂料的质量控制是很重要的。如果玻璃化转变温度高于一定值,则固化是足够的。这个测试也用于工艺控制,例如快速确定反应温度或反应时间的偏离。如果固化不足,在 DSC 测试曲线上就观察到放热的后固化峰。当再次测试样品时,玻璃化转变温度显示明显的提高。不过,即使不再呈现放热效应,玻璃化转变温度仍可能继续提高。

The determination of the glass transition temperature is important in the quality control of coatings that contain reactive resins. If the glass transition temperature is higher than a certain value, the curing is sufficient. This test is also used for process control, e. g. to quickly identify deviations from the reaction temperature or reaction time. If curing is insufficient, an exothermic postcuring peak is observed in the DSC measurement curve. The glass transition temperature shows a marked increase when the sample is measured again. The glass transition temperature can however continue to increase even when an exothermic effect is no longer visible.

目的	本应用旨在说明重复后固化(本例通过 DSC 重复的升温循环)如何引起玻璃化转变温度提高直至达到一个恒定值。
Purpose	The purpose of this application is to show how repeated postcuring (in this case through repeated heating cycles in the DSC) causes the glass transition temperature to increase until it reaches a constant value.
样品	已固化的 KU600 环氧树脂粉末。固化过程在相当低的温度 150℃进行,长达 4h 以上,以防涂层和基板可能产生热变化。
Sample	Cured KU600 epoxy resin powder: The curing process was performed at a relatively low temperature of 150℃ but over a long period of 4 hours to prevent possible thermal changes to the coating and the substrate.
条件	测试仪器:DSC
Conditions	Measuring cell:DSC
	坩埚:40μl 标准铝坩埚,盖钻孔
	Pan:Aluminum 40μL with pierced lid
	样品制备:分离涂层,称 15.807mg 样品放入坩埚。
	Sample preparation:The coating was detached and a sample of 15.807mg weighed into the crucible.
	DSC 测试:
	DSC measurement:
	以 10K/min 从 35℃升温至 260℃
	Heating from 35℃ to 260℃ at 10K/min
	气氛:氮气,50mL/min
	Atmosphere:Nitrogen,50mL/min

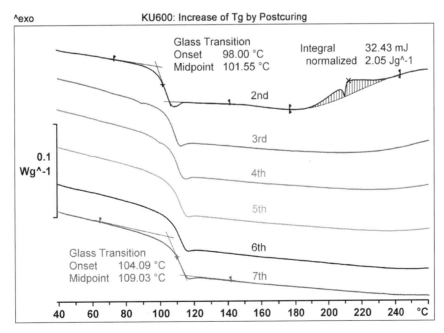

图 4.15　Fig. 4.15

计算　几乎完全固化的 KU600 的后固化反应在第二次 DSC 升温中(第一次升温是实际的固化过程)仍然可清晰地观察到小的放热峰。中点 101.7℃为几乎完全固化的材料的玻璃化转变。峰面积积分得到 2.05J/g 的后固化焓。

后固化反应期间 210℃处的小吸热峰是由双氰胺(促进剂成分)的熔融引起的。

后固化使材料的玻璃化转变提高 4.2K,从 101.7℃至 105.9℃,如在随后的第三次升温中可见的。对样品进行几次连续的升温,每次计算玻璃化转变。得到的数值汇总在表中,并标示在图 4.16 中。

Evalcuation　The postcuring reaction of the almost fully cured KU600 can still be clearly observed as a small exothermic peak in the 2nd DSC heating run (the first heating run is the actual curing process). The glass transition with a midpoint at 101.7℃ corresponds to the almost fully cured material. Integration of the area under the peak yields a postcuring enthalpy of 2.05 J/g.

The small endothermic peak at 210℃ during the postcuring reaction is caused by the melting of dicyandiamide (a constituent of the accelerator).

Postcuring causes the glass transition of the material to increase by 4.2K, from 101.7℃ to 105.9℃ as can be seen in the subsequent 3rd heating run. Several consecutive heating runs were performed on the sample and the glass transition evaluated each time. The values obtained are summarized in the table and plotted in Figure 4.16.

DSC 升温次数 No. of DSC heating run	玻璃化转变,℃ Glass transition in ℃
2	101.7
3	105.9
4	108.1
5	109.0
6	109.0
7	109.0

玻璃化转变温度从第三至第五次测试稍有提高,表明少量的后固化还在发生,尽管在 DSC 曲线上没有观察到明显的放热效应。进一步的多次升温对玻璃化转变不再产生影响,保持恒定在 109.0℃。这表明已经形成稳定的聚合物网状结构。

The glass transition temperature increases slightly from the 3rd to the 5th run, indicating that a small amount of postcuring is still occurring, although no obvious exothermic effects were observed in the DSC curve. Additional heating runs have no further effect on the glass transition temperature, which remains constant at 109.0℃. This indicates that a stable polymer network has been formed.

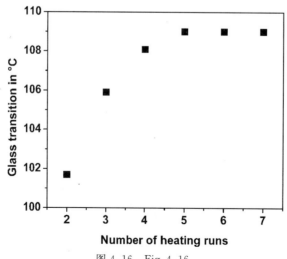

图 4.16　Fig. 4.16

结论 DSC 测试能够快速检查固化状态。玻璃化转变温度常常用作质量保证或工艺控制的标准。不完全的固化通常在更高温度下观察到放热后固化反应,它引起玻璃化转变温度提高。用高温可使玻璃化转变温度提高,即使观察不到明显的放热效应,可一直继续直到交联完全。对同一样品反复的升温(后固化)常常使玻璃化转变温度升高。

Conclusions DSC measurements allow the state of cure to be rapidly checked. The glass transition temperature is often used as a criterion for quality assurance or process control. Incomplete curing is usually observed as an exothermic postcuring reaction at higher temperatures, which causes the glass transition temperature to increase. The application of high temperatures can result in an increase in the glass transition temperature even though no apparent exothermic effect is visible. This continues until crosslinking is complete. Repeated heating runs (postcuring) of the same sample often result in an increase in the glass transition temperature.

4.2.2 化学计量对固化和最终玻璃化转变温度的影响
The effect of stoichiometry on curing and the resulting glass transition temperature

目的	研究化学计量对含过量胺或环氧的环氧-胺体系的玻璃化转变温度 T_g 的影响。 胺对环氧的化学计量比可定义为 $r=|A|/|E|$,式中 $|A|$ 和 $|E|$ 是胺和环氧单体的克当量数。本实验中分析的组成是化学当量组成 $r=1$,非化学当量组成 $r<1$,指环氧过量,而 $r>1$,指胺过量。
Purpose	To investigate the effect of the stoichiometry on the glass transition temperature, T_g, of an epoxy-amine system with excess of the amine or the epoxy. The stoichiometric ratio of amine to epoxy may be defined as $r=|A|/|E|$, where $|A|$ and $|E|$ are the number of gram-equivalents of amine and epoxy monomer. The compositions analyzed in this experiment are the stoichiometric one with $r=1$ and non-stoichiometric ones with $r<1$, which implies an excess of epoxy, and $r>1$, which indicates an excess of amine.
样品	双酚 A 二缩水甘油醚类环氧树脂(Araldite F,Ciba Specialty Chemicals)。 硬化剂是 $x+y+z≈5.3$ 的聚氧丙稀三胺(Jeffamine T403,Huntsman) $$\begin{array}{c} CH_2-[-OCH_2CH(CH_3)-]_X-NH_2 \\ | \\ CH_3CH_2CCH_2-[-OCH_2CH(CH_3)-]_Z-NH_2 \\ | \\ CH_2-[-OCH_2CH(CH_3)-]_Y-NH_2 \end{array}$$
Sample	The epoxy resin is based on diglycidyl ether of bisphenol A (Araldite F, Ciba Specialty Chemicals). The hardener is polyoxypropylenetriamine(Jeffamine T403, Huntsman) with $x+y+z≈5.3$ $$\begin{array}{c} CH_2-[-OCH_2CH(CH_3)-]_X-NH_2 \\ | \\ CH_3CH_2CCH_2-[-OCH_2CH(CH_3)-]_Z-NH_2 \\ | \\ CH_2-[-OCH_2CH(CH_3)-]_Y-NH_2 \end{array}$$
条件	测试仪器:DSC 坩埚:40 μl 标准铝坩埚,盖钻孔
Conditions	Measuring cell:DSC Pan:Aluminum 40 μL with pierced lid

样品制备:将环氧和硬化剂按不同的组成混合,得到化学当量的($r=1$)和非化学当量的($r<1$ 和 $r>1$)的样品。在60℃固化3h,在180℃后固化2h,然后以20K/min降温至常温。样品重8至9mg。

DSC测试:

以10K/min从0℃升温至120℃。

气氛:氮气,50mL/min

Sample preparation:The epoxy and the hardener were mixed to give different compositions: stoichiometric ($r=1$) and non-stoichiometric ($r<1$ and $r>1$). The samples were cured at 60℃ for 3h, postcured at 180℃ for 2h, and then cooled to ambient temperature at 20K/min. The sample weight was 8 to 9mg.

DSC measurement:

Heating from 0℃ to 120℃ at 10K/min

Atmosphere:Nitrogen, 50mL/min

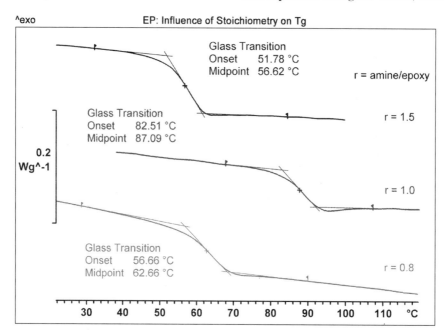

图4.17 化学当量体系($r=1$)和含过量环氧树脂($r=0.8$)或过量胺($r=1.5$)的非化学当量体系的环氧-胺体系的DSC曲线

Fig. 4.17 DSC curves of the epoxy-amine system for the stoichiometric system ($r=1$), and the non-stoichiometric systems with an excess of epoxy resin ($r=0.8$) or an excess of amine ($r=1.5$).

解释 化学当量体系的玻璃化转变温度 T_g 为87.1℃(中点法测量)。非化学当量体系显示较低的玻璃化转变温度:$r=0.8$ 为62.6℃,$r=1.5$ 为56.6℃。

在 $r=1$ 的树脂体系中,所有的胺氢原子与所有的环氧基团反应,产生最大交联密度的网状结构。

与之不同,在含过量环氧树脂($r=0.8$)的体系中,所有的 NH_2 和 NH 基团仅与 80% 的环氧基团反应。这意味着 20% 的环氧基团没有反

Interpretation The glass transition temperature, T_g, of the stoichiometric system is 87.1℃, measured as the midpoint. The non-stoichiometric systems exhibit lower glass transition temperatures: 62.6℃ for $r=0.8$ and 56.6℃ for $r=1.5$).

In the resin system with $r=1$, all the amine hydrogen atoms react with all the epoxy groups, resulting in a network with the maximum crosslinking density.

In contrast, in the system with excess epoxy resin ($r=0.8$), all the NH_2 and NH groups react with 80% of the epoxy groups. This means that 20% of the epoxy groups do not react, which produces a network with some bulky groups situated at the end

应,形成含有一些大基团位于环氧链末端的网状结构。这些大基团造成自由体积的增加,导致树脂的玻璃化转变温度的下降。

在含有过量胺($r=1.5$)的树脂中,所有的环氧基团与所有的 NH_2 基团反应,但只与一部分 NH 基团反应。存在的未反应的 NH 基团生成含支化结构的网状结构。这也促成了自由体积的增加,因而 T_g 降低。

计算 图 4.18 和下表表示图 4.17 中所示的三个体系($r=0.8$、1 和 1.5)和其他非化学当量比($r=0.6$、1.2 和 1.4)的不同化学计量混合物的玻璃化转变温度。

of the epoxy chains. These bulky groups give rise to an increase of the free volume, which results in a decrease of the glass transition temperature of the resin.

In the resin with excess amine ($r=1.5$), all the epoxy groups react with all the NH_2 groups but only some of the NH groups. The presence of unreacted NH groups produces a network with a branched structure. This also contributes to an increase the free volume and hence to a decrease of T_g.

Evaluation The Figure 4.18 and the following table show the glass transition temperatures for different stoichiometric mixtures for the three systems shown in Figure 4.17 ($r=0.8$, 1 and 1.5) and other non-stoichiometric ratios ($r=0.6$, 1.2 and 1.4).

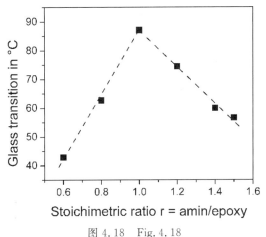

图 4.18 Fig. 4.18

$r=$胺/环氧 $r=$amine/epoxy	T_g,℃ T_g in ℃
0.6	42.9
0.8	62.7
1.0	87.1
1.2	74.5
1.4	60.0
1.5	56.6

如上所述,化学当量体系显示最高的 T_g,而非化学当量体系的 T_g 值较低。其原因是非化学当量体系由于大基团(过量环氧)或支化链(过量胺)的存在产生了不够完美的网状结构,两者均增加了体系的自由体积。

结论 化学当量的环氧—胺树脂体系的玻璃化转变温度高于非化学当量体系的 T_g 值。这在其他热固化体系中也可观察到,能用作校验热固性树脂体系化学计量的方法。

As discussed above, the stoichiometric system exhibits the maximum T_g, whereas the T_g values of non-stoichiometric systems are lower. The reason for this is that non-stoichiometric systems produce a network that is less perfect due to the presence of bulky groups (excess epoxy) or branched chains (excess amine), both of which increase the free volume of the system.

Conclusions The glass transition temperature of the stoichiometric epoxy-amine resin system is higher than the T_g values of the non-stoichiometric systems. This is observed in other thermosetting systems and can be used as a method to verify the stoichiometry of a thermosetting resin system.

参考文献 Y. Calventus, S. Montserrat, J. M. Hutchinson；非化学当量环氧－胺体系的焓松弛；Polymer Vol. 42, 7081 (2001)

Reference Y. Calventus, S. Montserrat, J. M. Hutchinson; Enthalpy relaxation of non-stoichiometric epoxy-amine systems; Polymer Vol. 42, 7081 (2001)

4.2.3 活性稀释剂对最终玻璃化转变温度的影响
Influence of reactive diluents on the resulting glass transition temperature

目的 研究不同环氧－酸酐树脂的固化度与玻璃化转变温度 T_g 之间的关系。

研究了两个环氧－酸酐体系,分别含有和不含活性稀释剂。活性稀释剂在环氧树脂配方中用作降低起始混合物的黏度和改善树脂的"可加工性"的添加剂。

Purpose To study the relationship between the degree of cure and the glass transition temperature, T_g, for different epoxy-anhydride resins.

Two epoxy-anhydride systems were investigated, one with a reactive diluent and another without. The reactive diluent is used as an additive in epoxy resin formulations to reduce the viscosity of the initial mixture and to improve the "processability" of the resin.

样品 双酚A二缩水甘油醚类环氧树脂(Araldite F, Ciba Specialty Chemicals)(DGEBA)。

从甲基四氢苯酐制得的硬化剂(Araldite HY 905, Ciba Specialty Chemicals)(MTHPA)。

促进剂是一种起催化剂作用的叔胺(Araldite DY061, Ciba Specialty Chemicals)(Accl)。

活性稀释剂是低黏度的脂族二环氧甘油醚类化学物(Araldite DY026)(RD)。

研究了两种环氧－酸酐体系。一种无活性稀释剂,称为FRD0,另一种含活性稀释剂,称为FRD30。重量份组成为：

FRD0：DGEBA/MTHPA/RD/Accl 比例为 100∶100∶0∶1

FRD30：DGEBA/MTHPA/RD/Accl 比例为 100∶100∶30∶1

Sample An epoxy resin based on diglycidyl ether of bisphenol A (Araldite F, Ciba Specialty Chemicals) (DGEBA).

A hardener derived from methyltetrahydrophthalic anhydride (Araldite HY 905, Ciba Specialty Chemicals) (MTHPA).

The accelerator is a tertiary amine that acts as a catalyst (Araldite DY061, Ciba Specialty Chemicals) (Accl).

A reactive diluent based on a low-viscosity aliphatic diglycidyl ether (Araldite DY026) (RD)

Two epoxy-anhydride systems were investigated. One without the reactive diluent, named FRD0, and another containing the reactive diluent, named FRD30. The compositions in parts by weight were

FRD0：DGEBA/MTHPA/RD/Accl in the ratio 100∶100∶0∶1

FRD30：DGEBA/MTHPA/RD/Accl in the ratio 100∶100∶30∶1

条件 测试仪器：DSC
Conditions 坩埚：40μl 标准铝坩埚,盖钻孔

样品制备：环氧和硬化剂以100∶100 w/w 的比例混合。加入1.0重量份催化剂。配制的混合物为

Measuring cell：DSC

Pan：Aluminum 40μL with pierced lid

Sample preparation：The epoxy and the hardener were mixed in a ratio of 100∶100 w/w. The catalyst was added at 1.0 part by weight. The

无活性稀释剂的树脂（**FRD**0）。对FRD30，加入30重量份的活性稀释剂。

样品在测试前于60℃、70℃和80℃固化1至120h，如图中所示。样品重量为8至9mg。

DSC 测试：

以10K/min从-80℃升温至300℃

气氛：氮气，50mL/min

resulting mixture corresponds to the resin without the reactive diluent, (FRD0). For FRD30, 30 parts by weight of reactive diluent was added.

The samples were cured before measurement at 60℃, 70℃ and 80℃ for 1 to 120h as shown. The sample weight was 8 to 9mg.

DSC measurement：

Heating from -80℃ to 300℃ at 10K/min

Atmosphere：Nitrogen, 50mL/min

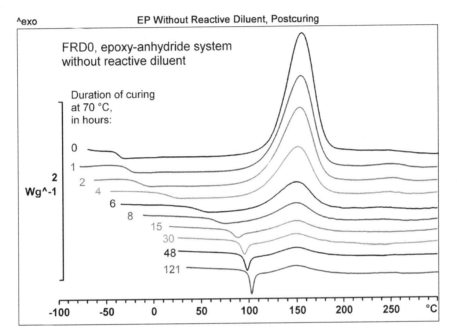

图 4.19 无活性稀释剂的环氧—酸酐催化体系 FRD0 的 DSC 曲线。
（样品已经在 70℃ 预固化了 1 至 121h。也显示了新制备的 FRD0
（固化时间 0h，即未固化样品）的曲线。）

Fig. 4.19 DSC curves of FRD0, the epoxy-anhydride catalyzed system without the reactive diluent.
(The samples had previously been cured at 70℃ for 1 to 121h.
The curve of fresh FRD0 (curing time 0h, i.e. uncured) is also shown.)

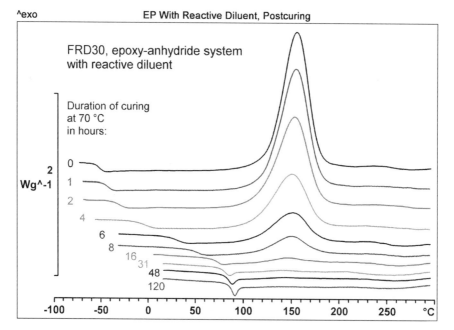

图 4.20 含活性稀释剂的环氧-酸酐催化体系 FRD30 的 DSC 曲线。
（样品已经在 70℃预固化了 1 至 120h。也显示了新制备的 FRD30
（固化时间 0h，即未固化样品）的曲线。）

Fig. 4.20 DSC curves of FRD30, the epoxy-anhydride catalyzed system containing the reactive diluent. The samples had been previously cured at 70℃ for 1 to 120 h. The curve of fresh FRD30 (curing time 0 h, i. e. uncured) is also shown.

解释　DSC 曲线呈现两个热效应：由热流的吸热台阶表示的玻璃化转变和等于部分固化树脂后固化反应的放热峰。

定性地看，随着固化时间的增加，由于固化度提高，后固化反应面积减小。同时，由于体系网状结构中更多的交联玻璃化转变温度提高。

固化 15h 或以上的样品的玻璃化转变伴随着一个松弛峰。这个效应是由于玻璃化，出现在当 T_g 等于固化温度 $T_c=70℃$ 时。在这些条件下，体系成为类玻璃，而实际上固化反应停止了。

当体系的 T_g 低于 T_c 时，反应动力学是受化学控制的，但是当 T_g 等于或高于 T_c 时，动力学是受扩散控制的。

为了研究不同转化率的环氧树脂体系，样品先在一个设定的温度 T_c 固化不同时间。样品的转化率 由

Interpretation　The DSC curves show two thermal effects: the glass transition, shown by an endothermic step in the heat flow, and an exothermic peak that corresponds to the postcuring reaction of the partially cured resins.

Qualitatively it can be seen that the area of the postcuring reaction decreases with increasing curing time due to the increase in the degree of cure. At the same time, the glass transition temperature increases due to more crosslinking in the network of the system.

The glass transitions of samples cured for 15 h or longer are accompanied by a relaxation peak. This effect is due to vitrification, which occurs when the T_g equals the curing temperature $T_c=70℃$. Under these conditions, the system becomes glass-like and the curing reaction practically stops.

When T_g of the system is less than T_c, the kinetics of the reaction is chemically controlled, but when T_g is equal to or higher than T_c, the kinetics is diffusion controlled.

In order to investigate the epoxy systems at different degrees of conversion, the samples were first cured at a defined temperature, T_c, for different times. The degree of conversion, α, of a sample was determined from the postcuring enthalpy of the partially cured sample and the total enthalpy of curing of a fresh sample. The total enthalpies of curing determined for fresh FRD0 and FRD30 samples were 306 and 335 J/g respectively. At the same time, these scans allow the glass transition temperature to be determined. Figures 4.19 and 4.20 show these thermal effects for the FRD0 and FRD30 epoxy systems partially cured at 70℃. Additional measurements were performed at curing temperatures of 60℃ and 80℃. The results obtained are summarized in the following figure.

部分固化样品的后固化焓和新制备样品的总固化焓来测定。对新制备的 FRD0 和 FRD30 的总固化焓分别测定为 306 和 335J/g。同时,这些扫描能测定玻璃化转变温度。图 4.19 和图 4.20 表示在 70℃ 部分固化的 FRD0 和 FRD30 体系的这些热效应。另外的测试在固化温度 60℃ 和 80℃ 进行。得到的结果汇总结在下图。

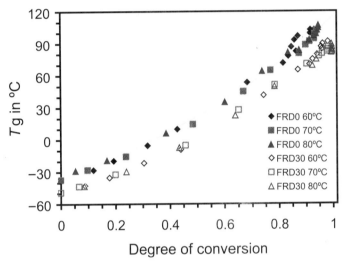

图 4.21 两个环氧树脂体系玻璃化转变温度与转化率的关系。
(图示包括 70℃(正方形)、60℃(菱形)和 80℃(三角形)得到的数值对(α, T_g)。
实心和空心符号分别对应着 FRD0 和 FRD30 样品。)

Fig. 4.21 The glass transition temperatures are shown as a function of the degree of conversion for the two epoxy systems. The plot includes the pairs of values (α, T_g) obtained at a curing temperature of 70℃ (squares), at 60℃ (diamonds) and 80℃ (triangles). The filled and open symbols correspond to the FRD0 and FRD30 samples, respectively.

计算 强调下列要点是重要的:在转化率与 T_g 之间存在着与固化温度无关的直接关系。这个关系由 DiBenedetto 方程描述:

Evaluation It is important to highlight the following points: There is a direct relationship between the degree of conversion and T_g, independent of the curing temperature. This relation is described by the DiBenedetto equation:

$$\frac{T_g - T_{g0}}{T_{g\infty} - T_{g0}} = \frac{\lambda \alpha}{1-(1-\lambda)\alpha}$$

式中 T_{g0} 和 $T_{g\infty}$ 分别等于新制备(未固化)和完全固化体系的玻璃化转变,λ 是 0 和 1 之间的可调整参数。图 4.21 中的连续线为 DiBenedetto 方程的拟合线,对 FRD0 和 FRD30 的参数列于下表。

where T_{g0} and $T_{g\infty}$ correspond to the glass transition of the fresh (uncured) and fully cured systems, respectively, and λ is an adjustable parameter between 0 and 1. The continuous lines in Figure 4.21 correspond to the fit of the DiBenedetto equation; the parameters for the FRD0 and FRD30 systems are shown in the following table.

	T_{g0}, ℃ T_{g0} in ℃	$T_{g\infty}$, ℃ $T_{g\infty}$ in ℃	λ
FRD0	−37	115	0.63
FRD30	−49	90	0.61

每个体系有一条 T_g 对转化率的特征曲线。对于给定的转化率，活性稀释剂的加入降低了体系的 T_g。

结论 DSC 曲线对含和不含活性稀释剂的两个体系显示相同的热效应。然而，活性稀释剂的加入增加了链段的活动性。对于相同的转化率，未固化 FRD30 体系的 T_g 比未固化 FRD0 体系的低约 12K。固化体系的 T_g 差值约为 20K。不过，与 FRD0（无活性稀释剂）相比，FRD30 固化反应的峰温没有显著改变。

对于一个给定的环氧树脂体系，在 T_g 与转化率之间存在特有的关系，它可由 DiBenedetto 方程来描述。这个关系使从单纯测量 T_g 来确定转化程度成为可能，如果已知未反应的和完全固化的树脂的玻璃化转变温度以及 λ 参数。

参考文献 1. S. Montserrat, G. Andreu, P. Cortés, Y. Calventus, P. Colomer, J. M. Hutchinson and J. Málek;活性稀释剂加入到催化环氧－酸酐体系 I. 对固化动力学的影响; Journal of Applied Polymer Science, Vol. 61, 1663 (1996)

2. S. Montserrat;环氧树脂等温固化中的玻璃化和进一步结构松弛; Journal of Applied Polymer Science, Vol. 44, 545 (1992)

Each system has a characteristic curve for T_g versus the degree of conversion. The addition of a reactive diluent reduces the T_g of the system for a given degree of conversion.

Conclusions The DSC curves show the same thermal effects for both systems, with and without the reactive diluent. However, the addition of the reactive diluent increases the mobility of the segmental chains. For the same degree of conversion, T_g of the uncured FRD30 system is about 12K lower than that of the uncured FRD0 system. The difference in T_g for the cured systems is about 20K. Nevertheless, the temperature of the curing reaction peak does not significantly change for FRD30 compared to FRD0 (without the reactive diluent).

For a given epoxy system, there is a characteristic relationship between T_g and the degree of conversion, which can be described by the DiBenedetto equation. This relationship allows the degree of conversion to be determined from a single measurement of T_g if the glass transition temperatures of the unreacted and the fully cured resin, and the λ-parameter are known.

References 1. S. Montserrat, G. Andreu, P. Cortés, Y. Calventus, P. Colomer, J. M. Hutchinson and J. Málek; Addition of a reactive diluent to a catalyzed epoxy-anhydride system. I. Influence on the cure kinetics; Journal of Applied Polymer Science, Vol. 61, 1663 (1996)

2. S. Montserrat; Vitrification and further structural relaxation in the isothermal curing of an epoxy resin; Journal of Applied Polymer Science, Vol. 44, 545 (1992)

4.2.4 玻璃化 Vitrification

4.2.4.1 玻璃化转变温度与转化率关系的测定
Determination of the dependence of the glass transition temperature on conversion

当固化产物的玻璃化转变温度高时,了解玻璃化转变温度对转化率的依赖性对环氧-胺体系是特别重要的。其原因是,对这样的反应体系,反应动力学的最佳固化条件,取决于反应混合物的瞬时玻璃化转变温度。测定玻璃化转变温度和转化率两者的一个有效方法是测试样品的后固化反应。

Knowledge of the dependence of the glass transition temperature on the degree of conversion is particularly important for epoxy-amine systems when the glass transition temperature of the cured product is high. The reason for this is that with such reaction systems the reaction kinetics and therefore the optimum curing conditions depend on the momentary glass transition temperature of the reacting mixture. An efficient method for determining both the glass transition temperature and conversion is to measure the postcuring reaction of samples.

目的	通过测试后固化来测定玻璃化转变温度与转化率的关系。		
Purpose	To determine the dependence of the glass transition temperature on the conversion by measuring postcuring.		
样品	DGEBA 和 DDM 的化学当量混合物		
Sample	Stoichiometric mixture of DGEBA and DDM.		
条件 **Conditions**	测试仪器:DSC	**Measuring cell**:DSC	
	坩埚:40μl 标准铝坩埚	**Pan**:Aluminum 40μL	
	样品制备:3 至 5mg 混合物加入坩埚,密封。测试前一直贮存在 −35℃。	**Sample preparation**:The crucibles were filled with 3 to 5mg of the mixture and then sealed. They were then stored at −35℃ until they were measured.	
	DSC 测试: 测试前,样品在 100℃保持不同时间固化至不同程度。然后由自动进样器从测试池内取出样品。DSC 测试温度范围为−35℃~220℃,升温速度为 10K/min。	**DSC measurement**:Before the measurement, the samples were held at 100℃ for different periods to be cured to different extents. The samples were then removed from the measuring cell by the automatic sample robot. The DSC measurement was performed from −35℃ to 220℃ at 10K/min.	
	气氛:氮气,50mL/min	**Atmosphere**:Nitrogen,50mL/min	

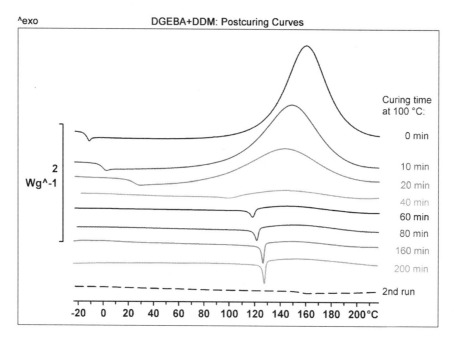

图 4.22 样品在 100℃ 等温固化了不同时间后的后固化 DSC 曲线。
Fig. 4.22 DSC postcuring curves of samples after curing isothermally at 100℃ for different times.

解释 最上面是新制备的未固化的样品的曲线（0min）（图 4.22）。在约 10℃ 呈现玻璃化转变和在 170℃ 由于固化反应产生的大放热峰。后固化曲线（10 至 200min）描述的是在 100℃ 保持了不同时间的样品，在该时间内发生了部分固化。这些曲线表明，部分固化样品的玻璃化转变温度随固化时间增加而增高。后固化开始于玻璃化转变之上。对于固化时间在 40min 以上的样品，玻璃化转变温度较高，甚至后固化反应与玻璃化转变同时出现。对较长的反应时间，还观察到由于焓松弛导致的吸热峰（参见 4.2.3 节）。为了说明后固化后材料的性能，显示了已完全固化的同一样品典型的二次测试 DSC 曲线（图 4.22，第二次测试）。这时只观察到约 160℃ 的玻璃化转变。

计算 对新制备样品测试曲线反应峰的积分，得比反应焓 $\Delta h_{c,1}$ 为

Interpretation The uppermost curve (0 min) is that of a fresh, uncured sample (Fig. 4.22). This exhibits a glass transition at about $-10℃$ and a large exothermic peak at 170℃ due to the curing reaction. The postcuring curves (10 to 200 min) relate to samples that had been held at 100℃ for different times during which partial curing occurred. These curves show that the glass transition temperatures of the partially cured samples increase with increasing curing time. Postcuring begins above the glass transition. For samples with a curing time of 40 min or more, the glass transition temperature is so high that the postcuring reaction begins as soon as the glass transition occurs. At longer reaction times, an endothermic peak due to enthalpy relaxation is also observed (see Section 4.2.3).

To demonstrate the behavior of the material after postcuring, the DSC curve of a typical second measurement (Fig. 4.22, 2nd run) of the same (fully cured) sample is shown. In this case, only a glass transition at about 160℃ is observed.

Evaluation Integration of the reaction peak in the measurement curve of the fresh sample yields a specific reaction enthalpy Δh_{c1},

407J/g。转化率 α 是由部分固化和完全固化材料的峰面积依照等式 $\alpha(t)=1-\Delta h_p/\Delta h_{c,1}(t)$ 计算得到的,式中 $\Delta h_p(t)$ 是对在100℃先前已经固化了时间 t 的样品测量的后固化比反应焓。此外,在100℃已经固化了不同时间的样品的玻璃化转变温度可直接由后固化曲线测定。这就是说,在100℃反应中发生的玻璃化转变温度和转化率的变化均可得到。相应的曲线示于图4.23。

of 407 J/g. The conversion, α, is calculated from the peak areas of the partially cured and fully cured material according to the equation $\alpha(t)=1-\Delta h_p/\Delta h_{c,1}(t)$. Here, $\Delta h_p(t)$ is the specific postcuring reaction enthalpy measured for a sample that had been previously cured at 100℃ for a time t. Furthermore, the glass transition temperatures of the samples that had been cured for different times at 100℃ can be determined directly from the postcuring curves. This means that both the change in the glass transition temperature and the conversion that occurred during the reaction at 100℃ are obtained. The corresponding curves are displayed in the Figure 4.23.

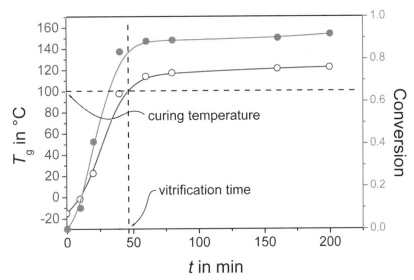

图4.23 在100℃温度玻璃化转变温度(红色)和转化率(灰色)与时间 t 的关系。
Fig. 4.23 Glass transition temperature (red) and conversion (grey) as a function of the reaction time, t, at a temperature of 100℃.

由图可见,玻璃化转变温度和转化率 α 增加较快。只要玻璃化转变温度显著低于反应温度,情况就是这样。反应发生在液体状态,是化学可控的。

随着反应进行,玻璃化转变温度接近反应温度即会发生。样品经历化学引发的玻璃化转变,样品玻璃化,此时处于固体(类玻璃)状态,分子运动被限制,随着玻璃化增强,固化反应越来越变为受扩散控制。结果,反应速率下降甚至反应停止。转化率曲线显示,在100℃的反应温度达到约90%的最大转化率。即使材料在该反应温度下贮存很长时间,转

It can be seen that the glass transition temperature and the conversion, α, show a rapid increase. This is the case as long as the glass transition temperature is significantly lower than the reaction temperature. The reaction takes place in the liquid state and is chemically controlled.

As the reaction proceeds, the glass transition temperature approaches the reaction temperature. The sample undergoes a chemically induced glass transition. The sample vitrifies and is then in the solid (glassy) state. Molecular mobility is restricted and the curing reaction becomes more and more diffusion controlled as vitrification increases. As a result of this, the reaction rate decreases to such an extent that the reaction practically stops. The conversion curve shows that at a reaction temperature of 100℃, a maximum conversion of about 90% is reached. Even if the material is stored for a long time at the reaction temperature, there is no further increase in the degree of

化程度也没有进一步增加。只有提高反应温度才能使转化率提高。

玻璃化转变温度等于反应温度时的时间称为玻璃化时间 t_v,是反应变为扩散控制的时间。在 100℃下,玻璃化时间为 45min。

玻璃化转变温度与转化率的关系可通过测量值作图来表示。也包含了完全固化样品(二次测试)的 T_g 以求图示完整,该点转化率 α 等于 1。

conversion. This can only be achieved by increasing the reaction temperature.

The time at which the glass transition temperature equals the reaction temperature is called the vitrification time, t_v, and is the time at which the reaction becomes diffusion controlled. At 100℃, the vitrification time is 45min.

The dependence of the glass transition temperature on the conversion can be obtained by displaying the measured values in a suitable diagram. To complete the picture, T_g of the fully cured sample (second measurement), at which the degree of conversion, α, equals 1, is also included.

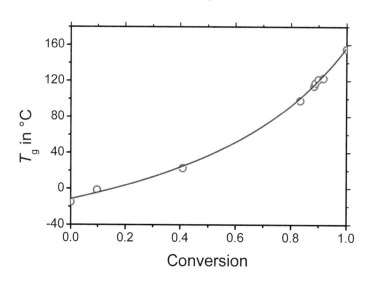

图 4.24 玻璃化转变温度与转化率的关系

Fig. 4.24 Glass transition temperature as a function of conversion.

可以看到玻璃化转变温度随着转化率的升高而增大。在转化率相当低(低于 60%)时,T_g 的变化相当小,而在较高转化程度时玻璃化转变温度的变化较大。描述这个关系的函数是 DiBenedetto 方程:

It can be seen that the glass transition temperature increases with increasing conversion. Whereas at relatively low conversion (less than 60%) the change in T_g is relatively small, the increase of the glass transition temperature at higher degrees of conversion is much larger. A function that describes this relationship is the DiBenedetto equation:

$$T_g(\alpha) = \frac{\lambda\alpha(T_{g1} - T_{g0})}{1-(1-\lambda)\alpha} + T_{g0}$$

式中 λ 是曲线参数,T_{g0} 和 T_{g1} 为未固化和完全固化材料的玻璃化转变温度。从测试曲线可确定 $T_{g0} = -18.3℃$ 和 $T_{g1} = 157℃$ 的值,曲线拟合得到 λ=0.32。

Here λ is a curve parameter, and T_{g0} and T_{g1} are the glass transition temperatures of the uncured and completely cured material. Values of $T_{g0} = -18.3℃$ and $T_{g1} = 157℃$ were determined from the measurement curve. Curve fitting yields λ = 0.32.

结论 后固化实验提供了关于热固性树脂行为的许多重要信息。例

Conclusions Postcuring experiments provide much important information about the behavior of thermosets. For example, it is

如，能同步测定转化率及其对应的玻璃化转变温度。能得到玻璃化时间 t_v 和在该反应温度下能达到的最大转化率 $α_{max}$。这两个量在固化体系的工艺中是重要的，因为 t_v 是合理反应时间的量度，$α_{max}$ 决定材料的力学性能和稳定性。

possible to simultaneously determine the conversion and the corresponding glass transition temperature. This allows one to obtain the vitrification time, t_v, and the maximum conversion $α_{max}$ that can be attained at the reaction temperature. These two quantities are important in the processing of curing systems because t_v is a measure of a reasonable reaction time and $α_{max}$ determines the mechanical properties and the stability of the material.

4.2.4.2 等温固化反应中化学引发玻璃化转变的温度调制 DSC 测量
Chemically induced glass transition in an isothermal curing reaction measured by temperature-modulated DSC

目的	说明用温度调制 DSC 同步研究等温反应和热容。 用常规 DSC 研究在等温固化反应中发生玻璃化时由化学引发的玻璃化转变，需要一系列的后固化测试（4.2.4.1 节）。温度调制 DSC 能同步测试反应热流和比热容，兹以 DGEBA-DDM 环氧树脂体系在 100℃ 的等温测试为例予以说明。
Purpose	To demonstrate the simultaneous investigation of an isothermal reaction and heat capacity using temperature-modulated DSC. The investigation of the chemically induced glass transition when vitrification occurs during an isothermal curing reaction using conventional DSC requires a series of postcuring measurements (Section 4.2.4.1). Temperature-modulated DSC allows simultaneous measurement of the reaction heat flow and the specific heat capacity. This is shown using an isothermal measurement of the DGEBA-DDM epoxy resin system at 100℃ as an example.
样品	DGEBA 和 DDM 的化学当量混合物。
Sample	Stoichiometric mixture of DGEBA and DDM.
条件 **Conditions**	测试仪器：DSC　　　　　　　　　**Measuring cell**：DSC 坩埚：40μl 铝坩埚　　　　　　　**Pan**：Aluminum 40μL 样品制备：坩埚中加入 4.91mg 的起始混合物，立即测试。　　**Sample preparation**：A crucible was filled with 4.91 mg of the starting mixture and immediately measured. DSC 测试：在平均温度 100℃ 进行准等温 ADSC 测试，调制振幅为 0.5K，周期为 60s。　　**DSC measurement**：Quasi-isothermal ADSC measurement at a mean temperature of 100℃. The modulation amplitude was 0.5K and the period 60s. 气氛：氮气，50mL/min　　　　**Atmosphere**：Nitrogen，50mL/min

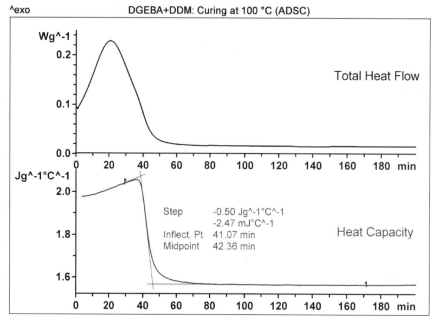

图 4.25　Fig. 4.25

解释　由 ADSC 测试计算总热流和复合比热。总热流等于常规 DSC 测试得到的热流。曲线表示在 20min 为最大的放热反应峰。热容曲线表示在反应中的比热容变化。在本体系中,热容随着反应时间的增加而增加。反应产物的热容大于起始混合物的热容。

在大约 40min 时,热容突然下降。达到台阶高度一半的时间 $t_{1/2}$ 为 42.4min。热容曲线台阶是由于化学引发的玻璃化转变而产生的。在反应时间短时,样品是液态的,具有比玻璃化后玻璃(固)态更大的热容。可以从约 40min 处的热流曲线观察到玻璃化对反应过程的影响。热流迅速下降。这是由于在玻璃态下的反应受扩散控制所导致的反应速率下降引起的。

将 $t_{1/2}$ 与从后固化实验(4.2.4.1 节)测定的玻璃化时间作比较,两者吻合良好。$t_{1/2}$ 稍小于 t_v,原因是 ADSC 是与频率有关的实验,在较低频率热容台阶向较长反应时间移动。对于实际用途,$t_{1/2}$ 近似良好地等于 t_v。

Interpretation　The total heat flow and the complex specific heat capacity were calculated from the ADSC measurements. The total heat flow corresponds to that obtained in a conventional DSC measurement. The curve shows the exothermic reaction peak with a maximum at about 20min. The heat capacity curve shows the change of the specific heat capacity during the reaction. In this system, the heat capacity increases with increasing reaction time. The heat capacity of the reaction product is greater than that of the starting mixture.

At about 40min, the heat capacity suddenly decreases. The time, $t_{1/2}$, at which half the step height is reached is 42.4min. The step in the heat capacity curve is due to a chemically induced glass transition. At short reaction times, the sample is liquid and has a larger heat capacity than in the glassy (solid) state after vitrification. The influence of vitrification on the course of the reaction can be seen in the heat flow curve at about 40 min. The heat flow decreases rapidly. This is caused by the decrease in the reaction rate due to the fact that the reaction is diffusion controlled in the glassy state.

A comparison of $t_{1/2}$ with the vitrification time determined from the postcuring experiments (Section 4.2.4.1) shows good agreement. $t_{1/2}$ is slightly less than t_v. The reason for it is that ADSC is a frequency-dependent experiment and that at lower frequencies the heat capacity step is shifted to longer reaction times. For practical purposes, to a good approximation $t_{1/2}$ equals t_v.

| 结论 | 准等温 ADSC 能同步测量反应进程(总热流曲线)和热容变化。由热容曲线可快速测定玻璃化时间。 | **Conclusions** | Quasi-isothermal ADSC allows the course of the reaction (in the total heat flow curve) and the change in the heat capacity to be simultaneously measured. The vitrification time can be quickly determined from the heat capacity curve. |

4.2.4.3 非模型动力学和固化过程中的玻璃化
Model free kinetics and vitrification during curing

目的	研究玻璃化对升温曲线非模型动力学分析结果的影响。 如 4.1.5 节中所示,固化反应动力学可用非模型动力学(MFK)来描述。因此可用 MFK 预测反应进程。如果材料在测试过程中发生玻璃化,则反应速率会显著的下降,从而引起动力学的变化。
Purpose	To investigate the influence of vitrification on the results of the MFK analysis of heating curves. As shown in Section 4.1.5, the kinetics of a curing reaction can be described using model free kinetics (MFK). MFK can then be used to make predictions about the course of the reaction. If the material vitrifies during the measurement, there is a significant decrease in the reaction rate and hence a change in the kinetics.
样品	DGEBA 环氧树脂与 DDM 的化学当量混合物。
Sample	Stoichiometric mixture of DGEBA epoxy resin with DDM.
条件 **Conditions**	测试仪器:DSC **Measuring cell**:DSC 坩埚:40μl 铝坩埚 **Pan**:Aluminum 40μL 样品制备:坩埚中加入 3 至 5mg 混合物,然后密封。在 −35℃ 贮存备测。 **Sample preparation**:The crucibles were filled with 3 to 5 mg of the mixture and then sealed. They were then stored at −35℃ until they were measured. DSC 测试:以 1、2、5、10K/min 从 −40℃升温测试至 220℃ **DSC measurement**:Heating measurements from −40℃ to 220℃ at 1, 2, 5, 10 K/min 气氛:氮气,50mL/min **Atmosphere**:Nitrogen,50 mL/min

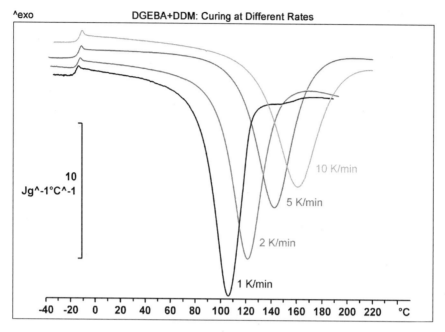

图 4.26 固化反应的比热容 c_p 曲线。（以对样品质量和升温速率的归一化表示）

Fig. 4.26 c_p curves of the curing reaction displayed as specific heat capacity. This represents normalization with respect to sample mass and heating rate.

计算 图 4.26 所示为以不同升温速率测试的 DSC 曲线。为了更好地比较，曲线用热容的归一化单位表示。可观察到在约 −15℃ 处玻璃化转变以后的反应峰。加热速率升高则峰移向较高温度。为了进行 MFK 计算，从测试曲线计算了转化率曲线。这里，重要的是保证转化率曲线相互不相交，特别是在反应开始和结束时。通过仔细选择计算界限就能做到这一点。下图（图 4.27）表示从四条 DSC 曲线计算得到的转化率曲线。

如果有至少三条在不同加热速率下测试的转化率曲线，用 MFK 就可计算与转化率关系的表观活化能。然后用它作为预测计算的基础。插图所示为 8% 至 98% 转化率之间的活化能曲线。曲线 A 从加热速率 2、5 和 10K/min 的转化率曲线计算得到，而曲线 B 从 1、2 和 5K/min 的转化率曲线计算得到。

Evaluation Figure 4.26 shows the DSC curves measured at different heating rates. For better comparison, the curves are displayed normalized in heat capacity units. The reaction peak can be seen after the glass transition at about −15℃. The peak is shifted to higher temperatures at higher heating rates. Conversion curves were calculated from the measurement curves for the MFK evaluation. Here it is important to ensure that the conversion curves do not cross over one another, especially at the beginning or end of the reaction. This can be achieved by carefully choosing the calculation limits. The following figure (Fig. 4.27) shows the conversion curves calculated from the four DSC curves.

The apparent activation energy can be calculated as a function of conversion using MFK if at least three conversion curves measured at different heating rates are available. This is then used as the basis for the calculation of predictions. The inserted diagram shows the activation energy curves for conversions between 8% and 98%. Curve A was calculated from the conversion curves for heating rates of 2, 5 and 10 K/min and Curve B from those for 1, 2 and 5 K/min.

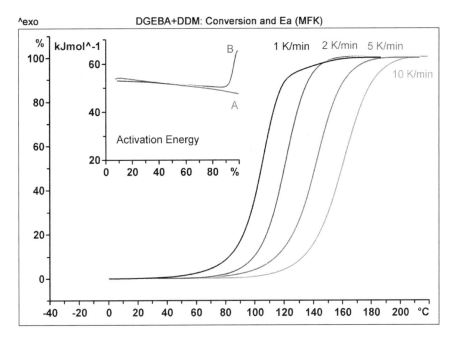

图 4.27 主图：从 DSC 曲线计算的转化率曲线。
插图：由 MFK 计算的表观活化能与转化率的关系
（曲线 A 从 2、5 和 10K/min 曲线得到，曲线 B 从 1、2 和 5K/min 曲线得到）

Fig. 4.27 Main diagram: Conversion curves calculated from the DSC curves.
Inserted diagram: Apparent activation energy as a function of the conversion calculated by MFK,
(for Curve A from the 2, 5 and 10 K/min curves and for Curve B from the 1, 2 and 5 K/min curves.)

解释 活化能曲线 A 表明，活化能与转化率几乎无关，活化能约 50kJ/mol。这类曲线是一步反应所特有的。如果很低的升温速率也用来计算活化能曲线，则在高于 80% 转化率时曲线有一个明显变化。在曲线 B 上观察到这一行为，表明在低速率下达到高转化率值时，固化反应动力学发生变化。反应动力学的变化是由于升温期间样品的玻璃化。如果已知玻璃化转变温度与转化率的关系，就能描述玻璃化。为了说明这一点，采用在 4.2.4.1 节描述的后固化实验的结果（玻璃化转变温度与转化率关系的测定）。用 DiBenedetto 方程从转化率曲线计算玻璃化转变温度与时间的关系，数据示于图 4.28，虚线是样品温度。

Interpretation The activation energy curve A shows an activation energy of about 50kJ/mol that is almost independent of conversion. This type of curve is typical for a one-step reaction. If very low heating rates are also used to calculate activation energy curves, there is a significant change of the curve at conversions above 80%. This behavior is observed in Curve B and indicates that the kinetics of the curing reaction changes when large conversion values are reached at low heating rates. This change in the reaction kinetics is due to vitrification of the sample during heating. The vitrification can be demonstrated if the dependence of the glass transition temperature on conversion is known. To show this, the results of postcuring experiments described in Section 4.2.4.1 (Determination of the dependence of the glass transition temperature on conversion) are used. The glass transition temperatures are calculated as a function of time from the conversion curves using the DiBenedetto equation. This data is shown in Figure 4.28. The dashed curves are the sample temperatures.

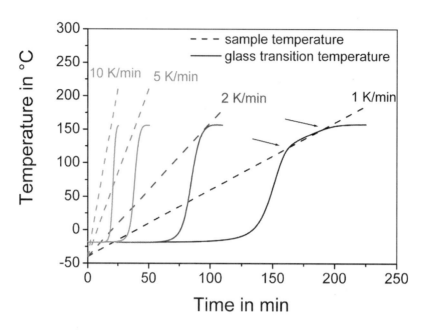

图 4.28 玻璃化转变温度和样品温度与测试时间的关系
Fig. 4.28 Glass transition temperature and sample temperature as a function of measurement time.

在测试开始时,样品温度低于玻璃化转变温度。材料处于玻璃态,因此是玻璃化的。升温时,样品温度升高,达到未反应样品的玻璃化转变温度。材料反玻璃化,变为液体。进一步升温,固化反应开始,引起玻璃化温度提高。然后样品继续反应,达到玻璃化转变温度等于样品温度的程度。如果加热速率足够高,则瞬时玻璃化转变温度始终低于瞬时样品温度。唯一例外是 1K/min,其中可看到大于 150min 时玻璃化转变温度高于样品温度,材料玻璃化。当样品温度达到固化材料的玻璃化转变温度时,样品又反玻璃化。曲线上玻璃化和反玻璃化发生的点用箭头表示。

玻璃化的样品至少部分处于像玻璃的固态。降低的分子运动影响反应动力学。这就是为什么在 MFK 计算中,与曲线 A 比,曲线 B 发生活化能变化的原因。曲线 B 是用包括 1K/min 在内的低升温速率计算的,而较高的升温速率用于计算曲线 A。

At the beginning of the measurement, the sample temperature is lower than the glass transition temperature. The material is in the glassy state and is hence vitrified. On heating, the sample temperature increases and reaches the glass transition temperature of the unreacted sample. The material devitrifies and becomes liquid. On further heating the curing reaction begins causing the glass temperature to increase. If the heating rate is high enough, the momentary glass transition temperature is always below the momentary sample temperature. The only exception is at 1K/min where it can be seen that above 150 min the glass transition temperature is higher than the sample temperature. The material vitrifies. The sample then continues to react to the extent that the glass transition temperature corresponds to the sample temperature. When the sample temperature reaches the glass transition temperature of the cured material, the sample again devitrifies. The points on the curve at which vitrification and devitrification occur are shown by arrows. The vitrified sample is at least partially in a glassy solid state. The reduced molecular mobility affects the reaction kinetics. This is the reason why in the MFK evaluation, a change in the activation energy occurs in Curve B compared to Curve A. Curve B was calculated using the lower heating rates, which include 1 K/min, whereas the higher heating rates were used to calculate Curve A.

结论 在低升温速率下,玻璃化可发生在固化反应过程中,因为玻璃化转变温度比样品温度提高得更快。如果发生玻璃化,由于分子运动性降低,反应变为主要受扩散控制。如果用 MFK 计算这样的曲线,则能从表观活化能曲线观察到扩散控制的影响。如果只有无玻璃化的曲线用于 MFK 分析,则 MFK 描述化学控制的反应动力学。MFK 能对化学控制的反应进程作出预测。

Conclusions At low heating rates, vitrification can occur during the curing reaction because the glass transition temperature increases more rapidly than the sample temperature. If this occurs, the reaction becomes largely diffusion controlled due to the reduced molecular mobility. If such curves are evaluated with MFK, the influence of diffusion control can be seen in the apparent activation energy curve. If only curves without vitrification are used for the MFK analysis, then MFK describes chemically controlled reaction kinetics. MFK allows predictions to be made for the course of chemically controlled reactions.

4.2.4.4 固化过程中玻璃化的测试
Measurement of vitrification during curing

上节描述的固化反应中的玻璃化过程是化学引发的玻璃化转变,所以预料热容会下降。然而在常规 DSC 中,该相当小的变化被大的反应峰所掩盖。温度调制技术能将热容变化与反应热流分开,这就使得玻璃化和反玻璃化过程能够被直接鉴别和测量。

Since the vitrification process in the curing reaction described in the previous section is a chemically induced glass transition, the heat capacity is expected to decrease. In conventional DSC, however, the relatively small change is masked by the large reaction peak. Temperature-modulated techniques can separate the heat capacity change from the heat flow of the reaction. This allows vitrification and devitrification processes to be directly identified and measured.

目的 Purpose	本应用的目的是说明如何用温度调制 DSC(TMDSC)能容易地测试玻璃化。 The aim of the application is to show how vitrification can be easily measured using temperature-modulated DSC (TMDSC).	
样品 Sample	DGEBA 和 DDM 的化学当量混合物。 Stoichiometric mixture of DGEBA and DDM.	
条件 Conditions	测试仪器:DSC 坩埚:$40\mu l$ 铝坩埚 样品制备:坩埚中加入 3 至 5mg 混合物,然后密封。在 $-35°C$ 贮存备测。 DSC 测试: 以基础升温速率 0.5、1 和 2K/min、温度振幅 0.5K 和周期 48s 作 ADSC 测试。 气氛:氮气,50mL/min	Measuring cell:DSC Pan:Aluminum $40\mu L$ Sample preparation:The crucibles were filled with 3 to 5 mg of the mixture and then sealed. They were then stored at $-35°C$ until they were measured. DSC measurement: ADSC measurements at underlying heating rates of 0.5, 1 and 2K/min, a temperature amplitude of 0.5K and a period of 48s. Atmosphere:Nitrogen,50mL/min

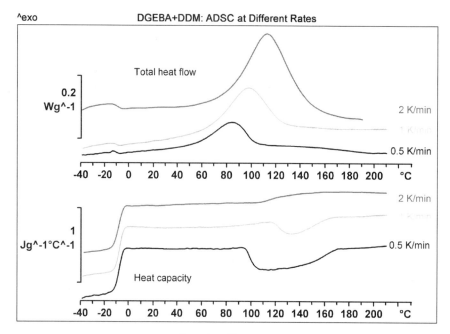

图 4.29 Fig. 4.29

解释 ADSC 曲线用来计算总热流和复合热容。总热流相当于常规 DSC 曲线，观察到的主要效应是反应峰。热容曲线表明，起始混合物的玻璃化转变在约 −10℃。就如从 4.2.4.3 节测试所知（非模型动力学和固化过程中的玻璃化），在 2K/min 的升温速率不发生玻璃化。热容在固化中增大，因为反应产物的热容比起始混合物的大。

在更低升温速率下，则首先观察到反应过程中热容的下降。原因是化学引发的玻璃化转变。在玻璃化转变时反应速率剧烈下降。继续升温，样品只反应到分子运动允许的程度。在约 160℃，在完全固化的材料的玻璃化转变区域，可观察到热容的增大。比较 1K/min 和 0.5K/min 测试的热容曲线表明，升温速率较高，则玻璃化的台阶高度稍小一些。这表明材料在较高速率下只部分玻璃化。升温速率较低则玻璃化程度较高。

结论 温度调制 DSC 是能用于跟踪固化反应过程中玻璃化和反玻璃

Interpretation The ADSC curves were used to calculate the total heat flow and the complex heat capacity. The total heat flow corresponds to the conventional DSC curve. The main effect observed is the reaction peak. The heat capacity curves show the glass transition temperature of the starting mixture at about −10℃. As already known from the measurements in Section 4.2.4.3 (Model free kinetics and vitrification during curing), vitrification does not occur at a heating rate of 2K/min. The heat capacity increases during the reaction because the reaction products have a somewhat larger heat capacity than the starting mixture.

At the lower heating rates a decrease in heat capacity is first of all observed during the reaction. The reason for this is the chemically induced glass transition. The reaction rate decreases drastically at the glass transition. On further heating, the sample reacts only to the extent that molecular mobility permits it to. At about 160℃, in the region of the glass transition of the fully cured material, an increase in heat capacity can be seen. The sample devitrifies. A comparison of the heat capacity curves measured at 1 K/min and 0.5K/min shows that the step height at vitrification is slightly smaller at the larger heating rate. This indicates that the material only partially vitrifies at the higher rate. The degree of vitrification is greater at the lower heating rate.

Conclusions Temperature-modulated DSC is a simple and relatively rapid method that can be used to follow vitrification

化过程的简单和相当快速的方法。从热容曲线可得到最重要的信息。

and devitrification processes during a curing reaction. The most important information is derived from the heat capacity curves.

4.2.5　TTT 图的测定 Determination of a TTT diagram

如前面的章节所示,当固化材料的玻璃化转变温度高时,可能在环氧树脂的固化反应中发生玻璃化。如果发生玻璃化,反应动力学从化学控制彻底地改变为扩散控制。事实上玻璃化是化学诱导的玻璃化转变。玻璃化后,材料处于玻璃态,迁移过程被"冻结"。反应几乎停止,转化率几乎没有进一步增加。
在等温反应中,如果反应温度低于固化材料的玻璃化转变温度,样品就会玻璃化。
了解玻璃化条件对于工业固化过程、稳定性预测和材料开发的优化是重要的。对于玻璃化的材料,弹性模量类性能经历久时会发生显著变化。材料是不稳定的,因为该过程是一个缓慢松弛过程从而使反应会慢慢继续下去。
表征反应中玻璃化的一个方法是时间－温度－转换图(TTT 图)。本节描述测定该图的各种方法。

As shown in the previous sections, vitrification can occur in the curing reaction of epoxy resin systems when the glass transition temperature of the cured material is high. If vitrification takes place, the kinetics of the reaction change drastically from being chemically controlled to diffusion controlled. Vitrification is in fact a chemically induced glass transition. After vitrification the material is in a glassy state and the transport processes are "frozen in". The reaction almost stops and there is hardly any further increase in conversion.
In isothermal reactions, the sample vitrifies if the reaction temperature is below that of the glass transition temperature of the cured material.
Knowledge of vitrification conditions is important for the optimization of technical curing processes, stability predictions and material development. With vitrified materials, properties such as the elastic modulus can change significantly over a long period. The material is not stable because the reaction continues on slowly as a result of the slow relaxation processes.
One way to characterize vitrification during the reaction is the Time-Temperature-Transformation diagram (TTT diagram). This section describes different possibilities for determining the diagram.

4.2.5.1　TTT 图:由后固化实验测定
TTT diagram: Determination from postcuring experiments

目的	测定 TTT 图的经典方法是根据 4.2.4.1 节所述的后固化行为的测试。本应用的目的是说明如何从这些测试能获得 TTT 图。
Purpose	The classical method to determine a TTT diagram is based on measurements of postcuring behavior as described in Section 4.2.4.1. The aim of this application is to show how a TTT diagram can be obtained from these measurements.
样品	DGEBA 和 DDM 的化学当量混合物。
Sample	Stoichiometric mixture of DGEBA and DDM.
条件 **Conditions**	测试仪器:DSC　　　　　　　　　**Measuring cell**: DSC 坩埚:40μl 铝坩埚　　　　　　　**Pan**: Aluminum 40μL 样品制备:坩埚中加入 3 至 5 mg 的混合物,密封,贮存在 －35℃ 备测。　**Sample preparation**: The crucibles were filled with 3 to 5 mg of the mixture, sealed and stored at －35℃ until they were measured.

DSC measurement: As described in Section 4.2.4.1, the samples were cured in the DSC for different times at the isothermal reaction temperature, then rapidly cooled to a temperature below the glass temperature. The postcuring reaction was then measured at 10K/min. Besides the measurements at 100℃ already described, isothermal reaction temperatures of 120, 80 and 60℃ were also used.

Atmosphere: Nitrogen, 50mL/min

Evaluation As shown in Section 4.2.4.1, the reaction at the reaction temperature T_c was stopped after a certain time by removing the sample and cooling it rapidly. It was then reinserted into the DSC cell and a heating run recorded at 10K/min. The glass transition temperatures were determined from the measurement curves and plotted as a function of curing time. The curing time at which the glass transition temperature equals the reaction temperature is the vitrification time. This time is recorded in the TTT diagram (see figure 4.30) on a logarithmic abscissa scale. The ordinate is the reaction temperature.

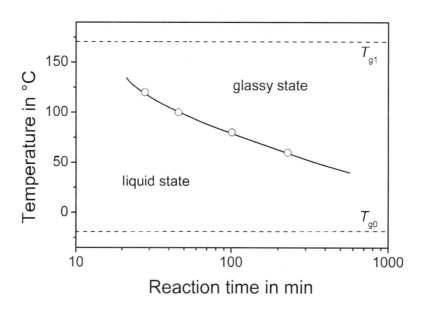

Fig. 4.30 Schematic TTT diagram, determined from postcuring measurements performed at temperatures from 60℃ to 120℃.

Interpretation The TTT diagram in Figure 4.30 also shows the glass transition temperatures of the uncured resin, T_{g0}, and the completely cured material, T_{g1}, (dashed lines). The glass transition of the partially cured mixture is always between these

介于这两个值之间。在高于反应混合物的玻璃化转变温度时且在短暂反应时间内,样品处于液体状态。随着反应进行,材料玻璃化,经历从液态到玻璃态的转变,这由 TTT 图中的曲线所表示。该曲线表征这样一种时间,在该时间之后玻璃化转变温度就等于反应温度。如果样品处于玻璃态,反应是受扩散控制的,因此实际上反应已经停止。

在较高的反应温度,玻璃化时间较短,但玻璃化发生时的转化率较大。如果反应温度在固化材料的玻璃化转变温度之下约 20K,玻璃化时间又增加了。如在下一节中将表明的,材料仅部分玻璃化,对反应动力学的影响较小。

结论 TTT 图对确定进行固化的体系的最佳反应条件是有用的。TTT 图的玻璃化曲线表明了反应实际上停止的条件。然而相应的材料是不稳定的,在使用中会继续固化。

two values. At a temperature higher than the glass transition temperature of the reacting mixture and at short reaction times the sample is in the liquid state. As the reaction proceeds, the material vitrifies and undergoes a transition from the liquid to the glassy state, which is shown by a curve in the TTT diagram. This curve characterizes the time after which the glass transition temperature corresponds to the reaction temperature. If the sample is in the glassy state, the reaction is diffusion controlled and has therefore practically stopped.

At higher reaction temperatures, the vitrification time is shorter but the conversion at which vitrification occurs is larger. If the reaction temperature is about 20K below the glass transition temperature of the cured material, the vitrification time increases again. As will be shown in the next section, the material only partially vitrifies and the influence on the reaction kinetics is smaller.

Conclusions A TTT diagram is useful for determining the optimum reaction conditions for a system that undergoes curing. The vitrification curve in the TTT diagram indicates the conditions under which the reaction practically stops. The corresponding material is however not stable and can continue to cure in use.

4.2.5.2 TTT 图:温度调制 DSC 的应用
TTT diagram: Application of temperature-modulated DSC

目的 本应用的目的是说明温度调制 DSC 如何能用来测试反应中的热容。在常规 DSC 测试中热容会被反应峰覆盖。

因为玻璃化是化学引发的玻璃化转变,所以预期反应期间热容有变化。玻璃化时间能用温度调制 DSC 直接测量,且数据能用于绘制 TTT 图。ADSC 和 TOPEM™ 测试适合于此用途。在所示实例中,使用了 ADSC 技术(也可参见 4.2.4.2)。

Purpose The aim of the application is to show how temperature-modulated DSC can be used to measure the heat capacity during a reaction. In a conventional DSC measurement the heat capacity would be overlapped by the reaction peak.

Since vitrification is a chemically induced glass transition, a change in heat capacity is to be expected during the reaction. Vitrification times can be directly measured by temperature-modulated DSC and the data used to construct a TTT diagram. ADSC and TOPEM™ measurements are suitable for this purpose. In the example shown, the ADSC technique was used (see also Section 4.2.4.2).

样品 DGEBA 和 DDM 的化学当量混合物。
Sample Stoichiometric mixture of DGEBA and DDM.

| 条件 Conditions | 测试仪器:DSC
坩埚:40μl 铝坩埚
样品制备:坩埚中加入 3 至 5mg 的混合物,密封,贮存在 −35℃ 备测。

DSC 测试:
在 40℃ 至 160℃ 之间以 10K 为间隔进行等温测试。新制备的样品放入预加热的 DSC 炉中,然后立即开始测试。ADSC 测试参数:
温度振幅:0.5K
周期:24s。
气氛:氮气,50mL/min | **Measuring cell**:DSC
Pan:Aluminum 40μL
Sample preparation:The crucibles were filled with 3 to 5 mg of the mixture, sealed and stored at −35℃ until they were measured.

DSC measurement:The isothermal measurements were performed at 10 K intervals between 40℃ and 160℃. Fresh samples were inserted in the preheated DSC furnace. The measurement began immediately afterward. ADSC measurement parameters:
Temperature amplitude:0.5K
Period:24s.
Atmosphere:Nitrogen,50mL/min |

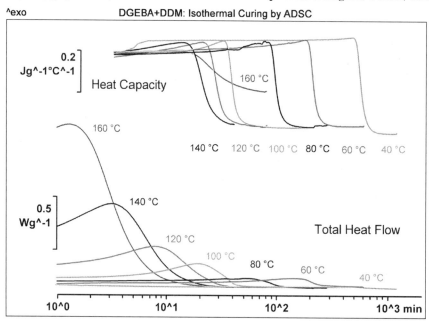

图 4.31　不同反应温度的总热流和热容与反应时间的关系曲线
（时间轴以对数刻度表示）

Fig. 4.31　Curves of the total heat flow and the heat capacity for different reaction temperatures as a function of reaction time. The time axis is shown on a logarithmic scale.

结果　如图 4.31 所示,可从测试曲线测定热容和总热流。可见热容先随反应时间而增加,原因是在这种情况下反应产物的热容大于反应混合物的热容。
然后热容因为样品玻璃化而突然下降了约 0.4J/gK。玻璃化时间可定义为台阶一半高度时的时间。
160℃ 的热容台阶要小得多,因为在该温度下样品只部分玻璃化。

Results　As shown in Figure 4.31, the heat capacity and the total heat flow were determined from the measurement curves. It can be seen that the heat capacity first increases with reaction time. The reason is that in this case the heat capacity of the reaction product is greater than that of the reaction mixture.
The heat capacity then suddenly decreases by about 0.4J/gK because the sample vitrifies. The vitrification time can be defined as the time at the half height of the step.
The heat capacity step at 160℃ is much smaller because at this temperature the sample only partially vitrifies. This results in

这造成扩散控制对反应动力学的影响较小。

diffusion control having less influence on the reaction kinetics.

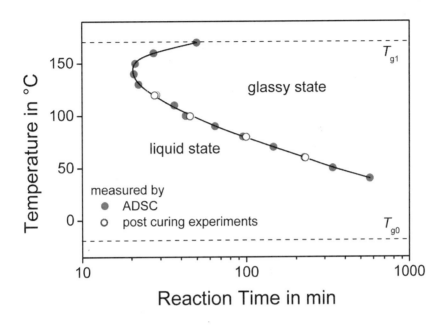

图 4.32　DGEBA 和 DDM 化学当量混合物的 TTT 图

Fig. 4.32　TTT diagram of a stoichiometric mixture of DGEBA and DDM.

解释　在图 4.32 中,从所有反应温度的 ADSC 实验得到的玻璃化时间画作灰点。为了进行比较,也显示了从 4.2.5.1 节的后固化实验得到的结果(红圈)。两套数据表明良好的一致性,因而玻璃化时间可从单个温度调制 DSC 测试测定。用常规 DSC,对每一个点需要一完整系列的测试。

Interpretation　In Figure 4.32, the vitrification times from all the ADSC experiments for all the reaction temperatures are plotted as grey points. For comparison, the results from the postcuring experiments from Section 4.2.5.1 are also shown (red circles). The two sets of data show good agreement. The vitrification time can thus be determined from one single temperature-modulated DSC measurement. With conventional DSC, a whole series of measurements is needed for each point.

结论　温度调制 DSC 是一种用一小部分常规 DSC 需要的实验时间就能高精确地测定 TTT 图的方法。

Conclusions　Temperature-modulated DSC is a method that allows the TTT diagram to be determined with high accuracy in a fraction of the experimental time needed by conventional DSC.

4.2.5.3　玻璃化和非模型动力学 Vitrification and model free kinetics

目的　如 4.2.4.3 节所示,如果选择适当的测试条件,化学控制的固化反应可用非模型动力学(MFK)来描述。本应用的目的是说明如何从这些数据可测定 TTT 图。

Purpose　As shown in Section 4.2.4.3, chemically controlled curing reactions can be described by model free kinetics (MFK), if suitable measurement conditions are chosen. The purpose of the application is to show how a TTT diagram can be determined from this data.

样品　DGEBA 和 DDM 的化学当量混合物。

purpose	Stoichiometric mixture of DGEBA and DDM.	
条件	测试仪器:DSC	**Measuring cell**:DSC
Conditions	坩埚:40μl 铝坩埚	**Pan**:Aluminum 40μL
	样品制备:坩埚中加入3至5 mg的混合物,密封,贮存在-35℃备测。	**Sample preparation**: The crucibles were filled with 3 to 5 mg of the mixture, sealed and stored at -35℃ until they were measured.
	DSC 测试:	**DSC measurement**:
	使用 4.2.4.3 节不发生玻璃化的测试(升温速率 2K/min 和以上)。	The measurements from Section 4.2.4.3 were used in which vitrification did not occur (heating rates of 2 K/min and higher).
	为检查 MFK 预测,在 40℃ 至 160℃ 温度之间以 10K 为间隔进行另外的等温测试。	To check the MFK predictions, additional isothermal measurements were performed at temperatures between 40℃ and 160℃ at intervals of 10 K.
	气氛:氮气,50mL/min	**Atmosphere**:Nitrogen,50mL/min

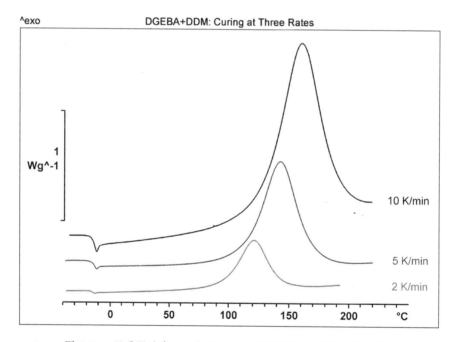

图 4.33 以升温速率 2、5 和 10K/min 测试的固化反应 DSC 曲线。

Fig. 4.33 DSC curves of the curing reaction measured at heating rates of 2, 5 and 10 K/min.

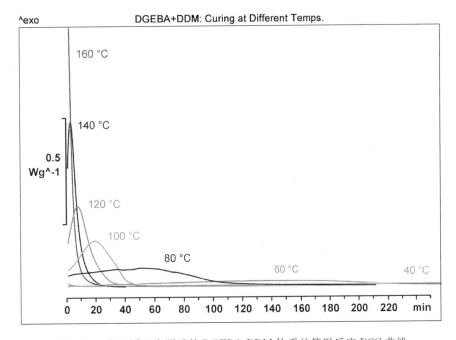

图 4.34 在不同温度测试的 DGEBA-DDM 体系的等温反应 DSC 曲线

Fig. 4.34 DSC curves of isotherma reactions of the DGEBA-DDM system measured at different temperatures.

计算 反应焓由三条升温测试曲线（图 4.33）的积分测定。得到的平均值为 $\Delta H_{react}=406 J/g$。转化率曲线也从这些测试测定，用 4.2.4.3 节所述的 MFK 计算。这样得到表观活化能与转化率的关系。该曲线用于计算对等温反应转化率曲线的预测。预测的准确性通过计算从等温测试得到的转化率曲线来评估。同时，必须考虑等温测试由于玻璃化而没有达到完全的转化率，所达到的最大转化率 α_{max} 是从反应焓 ΔH_{react} 和相应的等温测试的峰面积 ΔH_{iso} 计算的：

Evaluation The reaction enthalpy is determined by integration of the three heating measurement curves (Fig. 4.33). The mean value obtained is $\Delta H_{react}=406 J/g$. The conversion curves were also determined from these measurements and evaluated by MFK as described in Section 4.2.4.3. This yields the apparent activation energy as a function of conversion. This curve is used to calculate predictions for conversion curves of isothermal reactions. The accuracy of the predictions was estimated by calculating conversion curves from the isothermal measurements. At the same time, one has to take into account that the isothermal measurements do not reach complete conversion due to vitrification. The maximum conversion reached, α_{max}, is calculated from the reaction enthalpy, ΔH_{react}, and the peak area of the corresponding isothermal measurement, ΔH_{iso}：

$$\alpha_{max}=1-\frac{\Delta H_{iso}}{\Delta H_{react}}$$

一些转化率曲线示于下图中。虚线曲线是从 MFK 预测的，实线曲线是从等温测试测定的。

Some of the conversion curves are shown in the following figure. The curves with dashed lines are predictions from MFK; the curves with continuous lines were determined from the isothermal measurements.

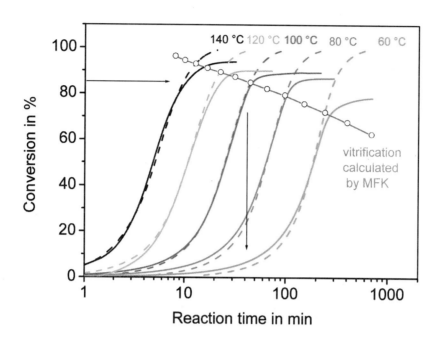

图 4.35 等温测试(—)和 MFK 预测(- - -)的 DGEBA+DDM 固化反应转化率曲线。MFK 预测计算的玻璃化线(红色)包含从其他温度测试得到的点。(这些温度的曲线未在图中显示)

Fig. 4.35 Conversion curves of the curing reaction of DGEBA+DDM, measured isothermally (—) and predicted by MFK (- - -). The vitrification line (red) calculated from the MFK predictions contains points from measurements at other temperatures that are not shown here.

解释 如所预期的,反应由于玻璃化而慢下来,所以测得的转化率曲线没有达到100%转化率。在较低反应温度下,所达到的最大转化率较小。

MFK 的预测始终达到 100% 转化率。值得注意的是,测试曲线与直至玻璃化发生点前的预测是重合的(在测试准确性范围内)。在这个范围内,反应主要是受化学控制的。这部分的反应可用 MFK 极好地描述。

现在提出的问题是能否用 MFK 得到 TTT 图的数据。如果认为玻璃化发生在反应温度正好与瞬时玻璃化转变温度相等($T_g(t_v) = T_{react}$)的时候,则玻璃化时间 t_v 的测定需要关于玻璃化转变温度转化率依赖性的信息。如 4.2.4.1 节所描述的,这能从后固化实验得到。

为了减少必需的实验时间,对已在 100℃经过了不同反应时间的样品

Interpretation As expected, the measured conversion curves do not reach 100% conversion because the reaction slows down due to vitrification. The maximum conversion reached is smaller at lower reaction temperatures.

The predictions of MFK always reach 100% conversion. It is worth noting that the measurement curves coincide with the predictions right up to point at which vitrification occurs (within the limits of measurement accuracy). In this range, the reaction is mainly chemically controlled. This part of the reaction can be described extremely well by MFK.

The question now arises whether MFK can be used to obtain data for the TTT diagram. If one assumes that vitrification occurs when the reaction temperature is the same as the momentary glass transition temperature ($T_g(t_v)$ = Treact), the determination of the vitrification time t_v requires information about the conversion dependence of the glass transition temperature. As described in Section 4.2.4.1, this can be obtained from postcuring experiments.

To reduce the necessary experimental time, it is sufficient to perform four postcuring measurements of samples that have already undergone reaction at 100℃ for different times. The reaction times were 10, 20, 40 and 80 min. Furthermore, data from the pure resin and the fully cured sample (2nd measurement) were used. The data are shown in Figure 4.36. The measurement curves were fitted using the DiBenedetto equation

$$T_g(\alpha) = \frac{\lambda\alpha(T_{g1}-T_{g0})}{1-(1-\lambda)\alpha} + T_{g0}$$

where T_{g0} and T_{g1} are the glass transition temperatures of the unreacted and completely reacted samples and λ is a fit parameter. $T_{g0} = -18.3℃$ and $T_{g1} = 157℃$ were determined from the measurement curves, and $\lambda = 0.32$ by curve fitting. The parameters of the DiBenedetto equation are therefore practically identical to those given in Section 4.2.4.1.

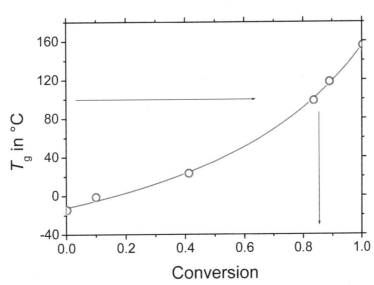

图 4.36 DGEBA+DDM 体系玻璃化转变温度与转化率的关系

Fig. 4.36 Glass transition temperature as a function of conversion for the system DGEBA+DDM

If the glass transition temperature is replaced by the reaction temperature in the DiBenedetto equation, rearrangement of the equation gives the conversion at the vitrification time t_v:

$$\alpha(t_v) = \frac{T_{react}-T_{g0}}{\lambda(T_{g1}-T_{react})+(T_{react}-T_{g0})}$$

All that now has to be determined is the time MFK evaluations predict for a particular conversion at a particular reaction temperature. The relationship between conversion and reaction time is displayed above in Figure 4.35 by the conversion curves. For example, if $T_{react}=100℃$, then $\alpha(t_v)$ is 0.866 according to the above equation. From the MFK prediction (Fig. 4.35), a reaction time of 50 min is obtained at 100℃ and 86.6%. The

86.6%处得到反应时间为50min。这样,数据对100℃、50min便可确定TTT图上的一个点。

如果用对应的反应温度对这些玻璃化时间作图,就得到TTT图的预测。为了校验预测,4.2.5.2节的ADSC数据绘入了图4.37以作比较。

data pair 100℃, 50 min then defines a point in the TTT diagram.

If one plots the corresponding reaction temperatures for these vitrification times, one obtains a prediction for the TTT diagram. To verify the predictions, the ADSC data from Section 4.2.5.2 was entered in the Figure 4.37 for comparison.

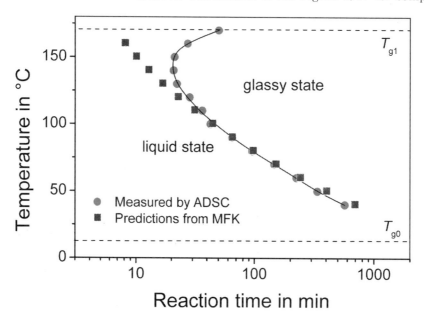

图 4.37　DGEBA-DDM 体系测试值(ADSC ●)和预测值(MFK ■)的 TTT 图。
Fig. 4.37　TTT diagram of the DGEBA-DDM system showing measured values (ADSC ●) and predicted values (MFK ■).

显然,MFK 能很好地预测高至130℃的玻璃化时间。在更高的温度,反应很快,能达到的最大转化率至少是90%。

在这些高温下,在实际应用中常可忽略玻璃化对反应的影响。TTT图对实际应用是重要的,特别在低温下。在这范围内,预测与测试吻合很好。

结论　TTT 图对具有高玻璃化转变温度的固化体系的应用很重要。如果用常规 DSC 测定此图,必须进行大量的后固化实验,这需要测试许多样品。用温度调制 DSC 能减少 DSC 实验的数量。如果应用非

It is evident that MFK allows a very good prediction of the vitrification time up to a temperature of 130℃. At higher temperatures, the reaction is very fast and the maximum conversion that can be reached is at least 90%.

At these high temperatures, the influence of vitrification on the reaction can often be neglected in practical applications. The TTT diagram becomes important for practical applications especially at low temperatures. In this range the predictions agree very well with the measurement.

Conclusions　The TTT diagram can be very important for the application of curing systems with high glass transition temperatures. If conventional DSC is used to determine the diagram, a large number of postcuring experiments have to be performed that require the measurement of many samples. The number of DSC experiments can be reduced by using

模型动力学,用七个样品进行八次测试就能测定完整的TTT图。

- 三个不同升温速率的升温测试,但要足够快以致不发生玻璃化。
- 一个固化样品的测试。
- 四个在一定温度下固化不同时间的材料测试。

这样在约一个工作日内就能测定一个完整的TTT图。

temperature-modulated DSC. If model free kinetics is applied, a complete TTT diagram can be determined with seven samples and eight measurements:

- Three heating measurements at different heating rates, but fast enough so that vitrification does not occur.
- One measurement of a cured sample
- Four measurements of material cured for different times at a certain temperature

This enables a complete TTT diagram to be determined in about one working day.

参考文献 J. Schawe, Thermochimica Acta, 288 (2002) 299—312.

References J. Schawe, Thermochimica Acta, 288 (2002) 299—312.

4.2.6 等温固化的凝胶点和力学玻璃化转变
Gel point and mechanical glass transition during isothermal curing

在固化反应中,材料的力学性能变化几个数量级。通常反应从低黏度液体开始。随着分子尺寸的稳定提高,黏度增加,最终生成聚合物固体。力学性能的变化可用DMA测试。DGEBA-DDM环氧—胺体系的等温固化反应是一个良好示例。

In a curing reaction, the mechanical properties of a material change by several orders of magnitude. Normally the reaction begins in a low viscosity liquid. With steadily increasing molecular size, the viscosity increases and finally a polymeric solid is produced. The changes in mechanical properties can be measured using the DMA. The isothermal curing reaction of the DGEBA-DDM epoxy-amine system serves as a good example.

4.2.6.1 固化反应中剪切模量的变化
Change of the shear modulus during the curing reaction

目的	测试等温固化反应中力学行为的变化	
Purpose	To measure the change in the mechanical behavior during an isothermal curing reaction.	
样品	DGEBA和DDM的化学当量混合物。	
Sample	Stoichiometric mixture of DGEBA and DDM.	
条件	测试仪器:DMA	**Measuring cell**:DMA
Conditions	样品夹具:液体剪切样品夹具	**Sample holder**:Shear sample holder for liquids
	样品制备:未反应的混合物(室温下液态)装入样品夹具。样品膜厚度为0.3mm。含有样品的样品夹具装入预热平衡的炉体中。	**Sample preparation**:The unreacted mixture (liquid at room temperature) was transferred to the sample holder. The sample film thickness was 0.3mm. The sample holder with sample was mounted in the preheated, equilibrated furnace.
	DMA测试:90℃(样品温度)等温测试120min。	**DMA measurement**:Isothermal measurement at 90℃ (sample temperature) for 120min.

频率:1Hz,
最大位移振幅:70μm,
最大力振幅:20N。

Frequency: 1Hz,
Maximum displacement amplitude: 70μm,
Maximum force amplitude: 20N.

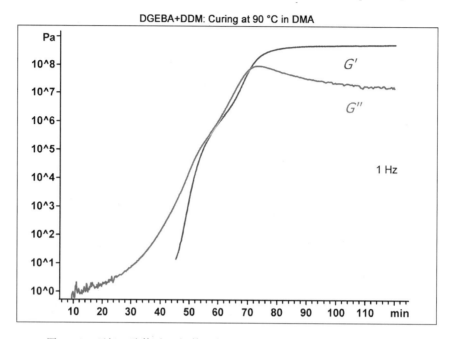

图 4.38 环氧-胺体系 90℃ 等温交联反应期间测试的贮存和损耗模量。

Fig. 4.38 Storage and loss moduli measured during the isothermal crosslinking reaction at 90℃ of an epoxy-amine system.

解释 下列方程给出黏度与剪切模量之间的关系:

Interpretation The relationship between the viscosity and the shear modulus is given by the equation:

$$G'' = \omega \eta'$$

式中 G'' 是损耗模量,$\omega = 2\pi f$ 是角频率,f 是频率,η' 是动态黏度的实部。

where G'' is the loss modulus. $\omega = 2\pi f$ is the angular frequency and f the frequency. η' is the real part of the dynamic viscosity.

图 4.38 以化学当量的 DGEBA-DDM 反应混合物为实例,表示这种体系在加聚反应中的力学行为变化。

Figure 4.38 shows the change in the mechanical behavior of such a system during the polyaddition reaction using a stoichiometric DGEBA-DDM reaction mixture as an example.

为了用转化率代替横坐标上的时间,可利用 4.2.5.3 节的 DSC 测试及那里描述的计算。假定 DSC 中和 DMA 中的转化以完全相同的方式进行,这首先要求准确的样品温度校准,其次,为了能够在温度平衡阶段中准确测量温度,在装入样品前,样品温度热电偶必须放入空的剪切样品夹具中。

To replace the time on the abscissa by the conversion, the DSC measurements from Section 4.2.5.3 can be used with the evaluation described there. This assumes that the conversion in the DSC and in the DMA proceed in exactly the same way. This first of all requires accurate calibration of the sample temperature. And second, the sample temperature thermocouple must be positioned in an empty shear sample holder before the sample is inserted in order to allow proper temperature measurement during the temperature equilibration phase.

结果示于图 4.39。图中的小图示意性地代表分子结构。黑线段代表树脂,红点代表交联剂或固化剂。

The result is shown in the Figure 4.39. The small diagrams in the figure schematically represent molecular structures. The black bars are symbols for the resin and the red points the crosslinker or hardener.

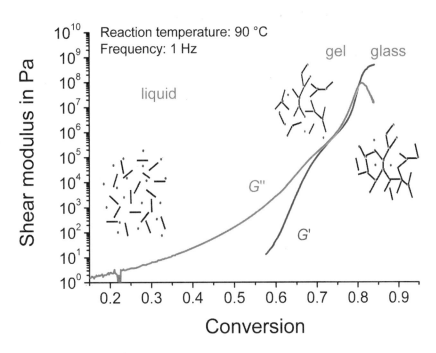

Fig. 4.39 Curves of the shear and loss modulus from Figure 4.38, normalized to conversion according to Figure 4.35 in Section 4.2.5.3.

Before the reaction begins, the sample consists of a low molecular weight mixture of resin and hardener. This is heated as quickly as possible to the reaction temperature. Initially, relatively small molecules are formed in the reaction and the material behaves like a Newtonian liquid. This means that only the real part of the viscosity exists. For the shear modulus, it follows that the storage modulus is practically non-existent. The loss modulus is less than 100 Pa, which corresponds to a viscosity of less than 16 Pa s. As the molecules become larger, the viscosity increases.

At a conversion of above 50%, the molecules are so large that the liquid loses its Newtonian behavior due to the stronger hydrodynamic interaction. The storage component of the shear modulus is now significant, but still much smaller than the loss modulus. The material therefore still retains the characteristic behavior of a liquid ($G'' > G'$).

At about 70% conversion, the molecules begin to assume macroscopic dimensions. Storage and loss moduli are of the same order of magnitude and finally G' becomes larger than G''. The material becomes dimensionally stable and has the properties of a solid. A gel is formed. The gel point can be defined as the time at which the ratio of the storage and loss modulus (i.e. the loss factor, tanδ) is independent of frequency. In a stoichiometric system this is the intersecttion of the two curves G' and G''. With

的交点。贮存模量约为 1MPa 时，材料具有弹性体的模量，但是因为聚合度较低，所以损耗模量较大。

转化率约 80% 时，G' 有一个与 G″ 曲线上出峰一致的台阶式增加。贮存模量达到约 1GPa 的值。这个行为描述了化学引发的玻璃化转变，分子运动因聚合物网状结构的形成而受到严格限制。最后材料在约 85% 转化率时玻璃化，因为扩散受阻，所以实际上停止反应。

a storage modulus of about 1Mpa, the material has the modulus of an elastomer, but because of the lower degree of polymerization, the loss modulus is larger.

At a conversion of about 80% there is a stepwise increase in G' that coincides with a peak in the G″ curve. The storage modulus reaches a value of about 1GPa. This behavior describes a chemically induced glass transition because molecular mobility is severely restricted through the formation of the polymer network. Finally the material vitrifies at a conversion of about 85%, and because diffusion is hindered the reaction practically stops.

结论 梅特勒-托利多 DMA 仪器能够跟踪交联反应过程中材料的力学行为。实际上一次单独的实验就能测试从低分子量的液体通过凝胶到玻璃态的整个范围。性能变化超过 9 个数量级。

为了获得玻璃态时的正确模量值，可修正反应过程中由于样品收缩而发生的几何形状变化。

Conclusions The METTLER TOLEDO DMA instrument allows the mechanical behavior of a material to be followed during a crosslinking reaction. Practically the entire range from a low molecular mass liquid, via a gel, to the glassy state can be measured in one single experiment. The properties change by more than 9 orders of magnitude.

To obtain the correct modulus value in the glassy state, geometry changes can be corrected that occur during the reaction due to sample shrinkage.

4.2.6.2 固化反应中剪切模量的频率依赖性
Frequency dependence of the shear modulus during a curing reaction

目的 **Purpose**	测试等温固化反应中不同频率下力学行为的变化。 Measurement of the change of mechanical behavior during an isothermal curing reaction at different frequencies.	
样品 **Sample**	DGEBA 和 DDM 的化学当量混合物。 Stoichiometric mixture of DGEBA and DDM.	
条件 **Conditions**	测试仪器：DMA 样品夹具：液体剪切样品夹具 样品制备： 未反应的混合物（室温下液态）装入样品夹具。样品膜的厚度为 0.3mm。装有样品的样品夹具装入预热平衡的炉中。 DMA 测试： 不同频率下 100℃（样品温度）等温测试 120min。 频率：0.1、1、10、100、470Hz。 最大位移振幅：低频下 70μm，100Hz 下 30μm，470Hz 下 15μm。	Measuring cell：DMA Sample holder：Shear sample holder for liquids Sample preparation：The unreacted mixture (liquid at room temperature) was transferred to the sample holder. The sample film thickness was 0.3mm. The sample holder with sample was mounted in the preheated, equilibrated furnace. DMA measurement：Isothermal measurement at 100℃ (sample temperature) for 120min at individual frequencies. Frequencies：0.1, 1, 10, 100, 470Hz Maximum displacement amplitude：70μm at low frequencies, 30μm at 100Hz and 15μm at 470Hz.

最大力振幅：20N.   Maximum force amplitude：20N.

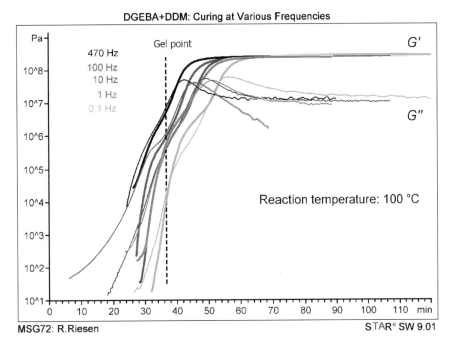

图 4.40　环氧—胺体系等温固化过程中测试的贮存和损耗模量

Fig. 4.40　Storage and loss modulus measured during the isothermal curing of an epoxy-amine system.

解释　每一次测试都用新制备的样品。可见在测试时间短时，频率较高下模量较大。这可以用黏度 η 与剪切模量 G'' 之间的关系来解释：

$$G''=2\pi f\eta'$$

式中 G'' 是损耗模量，$\omega=2\pi f$ 是角频率，f 是频率，η' 是动态黏度的实部。30min 后，假塑性高至能测试贮存模量。37min 后，贮存和损耗模量具有相同值。两条模量曲线的交点总是发生在与频率无关的同一反应时间后。损耗和贮存模量比（即 $\tan\delta$）与频率无关是判断凝胶点的一个标准。较高频率（大于 100Hz）下两条模量曲线的交点不出现。如参考文献所示，在凝胶点关系：

当指数 $n=0.5$ 时是有效的。当满足以下条件时,此指数定律是正确的：

在双轴都为对数刻度的图中得到一条直线。为了证实之，将凝胶点

Interpretation　A fresh sample was used for each measurement. It can be seen that at short measurement times, the modulus is greater at higher frequency. This can be explained by the relationship between viscosity, η', and the shear modulus, G'':

$$G''=2\pi f\eta'$$

where G'' is the loss modulus. $\omega=2\pi f$ is the angular frequency and f the frequency. η' is the real part of the dynamic viscosity. After about 30min, the pseudoplasticity is so high that the storage modulus can be measured. After 37min, the storage and loss moduli have the same value. This point of intersection of the two modulus curves always occurs after the same reaction time independently of the frequency. The frequency independence of the ratio of the loss and storage moduli (i. e. of $\tan\delta$) is a criterion for the gel point. At higher frequencies (above 100Hz) a point of intersection of the two modulus curves does not occur. As shown in Reference [1], at the gel point the relationship

$$G'(\omega)=G''(\omega)=a\omega^{n}$$

with an exponent $n=0.5$ is valid. This power law is true for the condition

$$G'(f,T_{\text{react}})=G''(f,T_{\text{react}})$$

A straight line is obtained in a diagram where both axis are shown on a logarithmic scale. To confirm this, the corresponding modulus values of the gel point were plotted as a function of frequency.

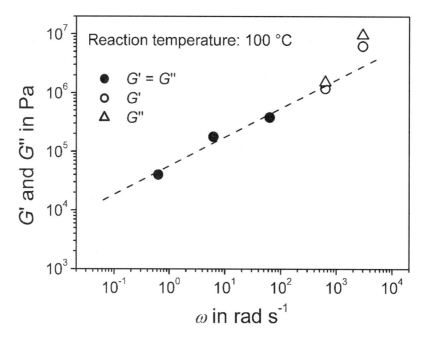

Fig. 4.41 Storage and loss moduli as a function of frequency. The filled circles correspond to the points of intersection of the modulus curves. With the open symbols, there is no point of intersection.

If the value at 470 Hz is not taken into account, a fit of the measured values gives $n = 0.49$ and $a = 463$ $(kPa\ s)^{-n}$.

After the gel point, the storage modulus exhibits a small "shoulder", whose duration decreases with increasing frequency. Afterward G' increases further and G'' exhibits a peak. This behaveior is characteristic for a relaxation process. In this case, it is the dynamic glass transition during curing. Because of the frequency dependence of the glass transitions, at higher frequencies the maximum of G'' is measured after shorter reaction times. This shift is also the reason why a gel point is no longer measured at high frequencies. The gelation process is then overlapped by the glass process.

Vitrification then begins after the dynamic glass process.

Conclusions The METTLER TOLEDO DMA instrument can be used to measure mechanical behavior during crosslinking over a wide frequency range. This yields detailed information about the gel point and relaxation behavior during a chemically induced glass processes.

参考文献　H. H. Winter and F. Chambon, J. Rheol., 30 (1986) 367.

Reference　H. H. Winter and F. Chambon, J. Rheol., 30 (1986) 367.

4.3　贮存效应 Storage effects

4.3.1　贮存后的后固化 Postcuring after storage

目的　仅用若干个DSC测试来测量树脂体系可用于加工的时间。
对冷固化体系进行催化以使其在低温下固化,例如甚至低于室温。双组分体系的组分混合后,通常仅有很短的时间用于树脂固化。

Purpose　To measure the time available for processing a resin system using just a few DSC measurements.
Cold-curing resin systems are catalyzed so that they cure at low temperatures, for example even below room temperature. After mixing the components of a two-component system, only a short time is usually available to apply the resin mixture.

样品　冷固化双组分环氧树脂体系,低于40℃温度固化的用于制造适用于结构增强制件的板的层压树脂。

Sample　Cold-curing two-component EP resin system, laminating resin used to make sheets for structural reinforcement elements for curing at temperatures below 40℃.

条件
测试仪器:DSC
坩埚:40μL铝坩埚,盖钻孔
样品制备:树脂-硬化剂体系在冰水浴中混合,搅拌约5min,之后在冰水浴中继续放置5min。然后在室温下快速装入铝坩埚。

在10℃三个单独的样品等温固化4、20和100h,以模拟低温贮存和加工期间的行为。立即测试第一个样品。

DSC测试:
以10K/min从0℃升温至220℃。
气氛:干燥空气,50mL/min

Conditions
Measuring cell:DSC
Pan:Aluminum 40μL with pierced lid
Sample preparation:The resin-hardener system was mixed in an ice-water bath, stirred for about 5min and then left for a further 5min in the ice-water bath. The aluminum crucibles were then quickly filled at room temperature.

Three individual samples were cured isothermally at 10℃ for 4, 20 and 100 hours to simulate the behavior during low temperature storage or processing. One sample was measured immediately.

DSC measurement:Heating from 0℃ to 220℃ at 10K/min
Atmosphere:Dry air, 50mL/min

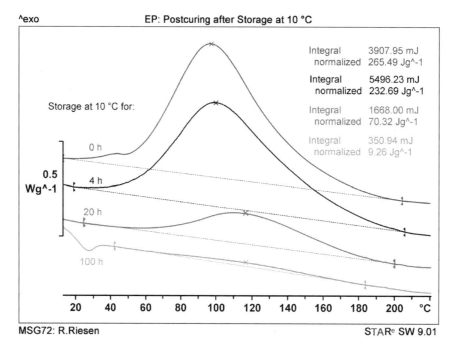

图 4.42　Fig. 4.42

最上面的 DSC 曲线(红色)表示未固化起始材料的固化反应。其他曲线表示材料在10℃下贮存不同时间(4、20 和100h)后的后固化反应曲线。由这些曲线可以计算材料的转化率。

该树脂体系在 10℃ 下贮存或加工的时间窗口为约 4h,因为在这段时间内转化率小于 10%。另一方面,冷固化树脂体系在 10℃ 下 100h 后几乎已完全固化。该曲线上约 25℃ 处的吸热效应是玻璃化转变处的焓松弛峰。在长时间的反应中,玻璃化转变温度已经慢慢提高到反应温度,可能已导致玻璃化。因此测定的该样品的后固化反应焓可能并不准确。当然用温度调制 DSC 可以获得更好的准确性。

The uppermost DSC curve (red) shows the curing reaction of the uncured starting material. The other curves show the postcuring reaction curves after the material has been stored for different periods (4, 20, and 100h) at 10℃. These curves allow the conversion of the material to be calculated.

The storage or processing time window for this resin system at 10℃ is about 4 h because in this period the conversion is less than 10%. On the other hand, the cold-curing resin system has almost completely cured after about 100h at 10℃. The endothermic effect in this curve at about 25℃ is the enthalpy relaxation peak at the glass transition. During the long reaction time, the glass transition temperature has slowly increased to the reaction temperature and has possibly led to vitrification. The postcuring reaction enthalpy determined for this sample might therefore be inaccurate. Better accuracy could of course be obtained using temperature modulated DSC.

计算
Evaluation

贮存时间,h Storage time in h	反应焓,J/g Reaction enthalpy in J/g	转化率,% Conversion in %
0	265.5	0
4	232.7	12
20	70.3	73
100	9.3	96

结论 DSC 能用来测定完全固化树脂所需要的后固化反应焓。如果新制备的树脂体系的反应焓是已知的或已测得的,那么就能计算已经发生的转化率。这提供了关于加工或固化条件的重要信息——例如,如果树脂贮存在特定的温度下,经多久还能加工。此时,10%以下的转化率是尚可接受的。

Conclusions DSC can be used to determine the postcuring reaction enthalpy needed to completely cure a resin. If the reaction enthalpy of the freshly prepared resin system is known or measured, the conversion that has already occurred can be calculated. This provides important information about processing or curing conditions-for example, how long the resin can still be processed if stored at a particular temperature. Here, a conversion up to 10% can be tolerated.

4.3.2 环氧树脂—碳纤维:贮存对预浸料的影响
EP-CF: Influence of storage on preprgs

目的 说明如何用 DSC 来跟踪测试作不同温度贮存试验的预浸料,测定贮存多久后仍可使用。在正常加工中,预浸料的加工性只能保证一定时间(所谓的"外置寿命")。一方面,必须详细说明贮存条件。另一方面,实际应用前应可检查预浸料是否仍然适合使用。实际上,剩余化学活性(后固化反应焓)的快速测试就能表征材料的化学历史。

Purpose To show how storage tests at different temperatures followed by DSC measurements can determine how long prepregs can be used.

The processibility of prepregs in normal processes is only guaranteed and possible for a certain period (the so-called "out-life"). On the one hand, the conditions for storage must be specified. On the other hand, before the actual application, one should be able to check whether the prepreg is still suitable for use. In practice, a rapid measurement of the residual chemical reactivity (postcuring reaction enthalpy) can characterize the chemical history of a material.

样品 EP-CF,由碳纤维和环氧树脂基体制备的适合于飞行器结构制件的预浸料(纤维含量约为重量 50%)。

生产两星期后收到的材料,然后在不同温度下贮存了不同的时间。

Sample EP-CF, prepreg made of carbon fibers and an EP resin matrix (fiber content about 50% by weight) for aircraft constructional elements.

Material received two weeks after production, then stored at different temperatures for different times.

条件
测试仪器:DSC
坩埚:40μL 铝坩埚,盖钻孔
样品制备:收到的材料:用刀从预浸料切出样品。
贮存条件:预浸料在 −20℃ 和 23℃ 下贮存了 10 个月和 15 个月的时间。
样品质量介于 3 到 11mg 之间。
DSC 测试:
以 10K/min 从 25℃升温至 350℃。
气氛:干燥空气,50mL/min

Conditions
Measuring cell:DSC
Pan:Aluminum 40μL with pierced lid
Sample preparation:Material as received: A sample was cut out of the prepreg using a knife.
Storage conditions:The prepregs were stored at −20℃ and 23℃ for a period of 10 months and 15 months.
Samples masses between 3 and 11 mg.
DSC measurement:
Heating from 25℃ to 350℃ at 10K/min
Atmosphere:Dry air, 50mL/min

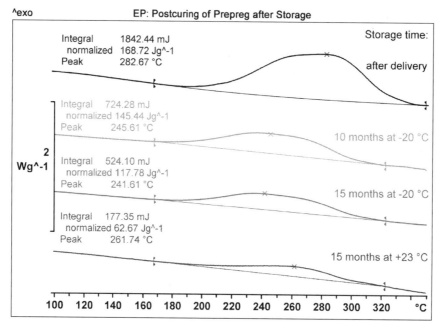

图 4.43　Fig. 4.43

解释　后固化反应的比焓随着贮存温度和贮存时间的增加而降低。甚至在 −20 ℃ 下贮存时间仍是重要的。然而应该注意的是,归一化的焓值不等于基体树脂真实的剩余后固化反应焓,因为它们参照的是总样品的质量而不只是聚合物含量。不过,可进行定性的比较或者确定重要的趋势。

Interpretation　The specific enthalpy of the postcuring reaction decreases with increasing storage temperature and increasing storage time. Even the storage time at −20 ℃ is important.
It should however be noted that the normalized enthalpies do not correspond to the actual residual postcuring reactions of the matrix resin because they refer to the total sample mass and not just the polymer content. Qualitative comparisons can however be made or important trends established.

计算
Evaluation

储存时间 Storage time	后固化反应的比焓, J/g Specific enthalpy of the postcuring reaction in J/g	峰温, ℃ Peak in ℃
收到材料时 as received	168.7	283
−20 ℃ 10 个月 10 months at −20 ℃	145.4	246
−20 ℃ 15 个月 15 months at −20 ℃	117.8	242
+23 ℃ 15 个月 15 months at +23 ℃	62.7	262

结论　贮存稳定性是预浸料加工性的重要参数。借助于简单的 DSC 测试,可定量分析贮存效应和限定材料可加工的期限。所定义的后固化焓的限值与不同的因素有关,例如材料的黏性或柔性。

Conclusions　The storability is an important parameter for the processibility of prepregs. With the aid of simple DSC measurements, storage effects can be quantitatively analyzed and limits defined for the processibility of the material. The limits defined for postcuring enthalpy depends on different factors such as tackiness or the flexibility of the material.

4.4 填料和增强纤维 Fillers and reinforcement fibers

4.4.1 玻璃化转变温度和"固化因子"按照 IPC-TM-650 的 DSC 测定
Glass transition temperature and "Cure Factor" by DSC according to IPC-TM-650

目的 按照 IPC-TM-650 No. 2.4.25 标准方法测试玻璃化转变温度 T_g 和固化因子。
这些测试适合于预浸料、覆金属箔或非覆金属箔层压板和印制电路板。规定了某些类型材料的相对固化度或固化因子(CF)的测定。本应用说明如何进行这些测定。

Purpose To measure the glass transition temperature, T_g, and the cure factor according to standard method IPC-TM-650 No. 2.4.25.
These tests are suitable for prepreg, metallic clad or unclad laminates, and printed circuit boards. It also provides a determination of relative degree of cure, or Cure Factor (CF) for some types of materials. The application shows how these determinations are performed.

样品 没有铜导线的印制电路板(含环氧树脂基体的玻璃织物)。
Sample Printed circuit board without copper conduction tracks (glass fabric with epoxy matrix).

条件 测试仪器:DSC
Conditions **Measuring cell**:DSC

样品设备:用小锯从提供的材料上切出样品,在105℃干燥2h,然后贮存在干燥器内。
样品放入盖钻孔的 40μL 铝坩埚。

Sample preparation: Samples were cut from the material supplied using a small saw, dried for 2h at 105℃ and then stored in a desiccator.
The sample was put in a 40μL Al crucible with pierced lid.

DSC 测试:第一次和第二次升温均以 20K/min 从 35℃ 至 185℃,两次之间冷却不受控。

DSC measurement: 1st and 2nd heating runs at 20K/min from 35℃ to 185℃ with uncontrolled cooling between runs.

气氛:静态空气
Atmosphere:Static air

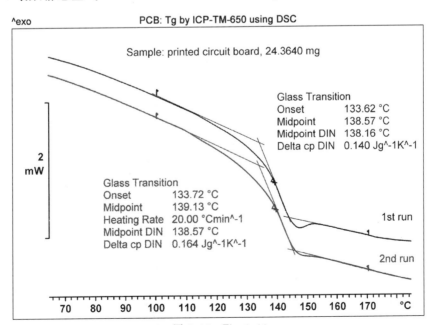

图 4.44 Fig. 4.44

解释 按照 IPC-TM-650 测试方法，样品升温到至少高于玻璃化转变温度 30K 两次，本例中为 185℃。在两次测试之间，快速冷却样品并立即开始第二次测试。

两条曲线的玻璃化转变出现在几乎相同的温度。这意味着经过加热材料到 185℃，任何可能的后固化反应几乎没有引起任何改变。这依次说明材料已经完全固化。

Interpretation According to the IPC-TM-650 test method, the sample is heated twice to a temperature that is at least 30K higher than the glass transition, in this case 185℃. Between runs, the sample is cooled rapidly and the second measurement immediately started.

The glass transition occurs at practically the same temperature in both curves. This means that any possible postcuring reaction through heating the material to 185℃ makes almost no change. This in turn indicates that the material was already completely cured.

计算 按照 IPC-TM-650 2.4.25 测试方法，DSC 曲线的玻璃化转变温度计算为半台阶高度的中点。

"固化因子"为两次玻璃化转变温度之差（第二次升温的减去第一次升温的），在本例中几乎为零。

结果汇总结于下表：

Evaluation The glass transition temperatures of the DSC curves were calculated as the midpoint at half step height according to IPC-TM-650 2.4.25.

The "Cure Factor" is the difference between the two glass transition temperatures (2nd heating run minus the 1st heating run) and is in this case practically zero.

The results are summarized in the following table:

	DSC
T_g（第一次升温） T_g (1st heating run)	138.2℃
T_g（第二次升温） T_g (2nd heating run)	138.6℃
CF ($T_{g2} - T_{g1}$)	0.4 K

结论 IPC-TM-650 规定了测定玻璃化转变温度的实用的 DSC 测试方法。用 DSC，第二次升温中玻璃化转变温度的提高可用来计算所谓的固化因子。后固化反应程度越大，CF 就越大。在本特例中，几乎没有检测到后固化反应，这说明材料已经完全固化（参见 4.2.1 节实例"重复后固化对玻璃化转变的影响"）。

Conclusion IPC-TM-650 provides a practical DSC test method for the determination of the glass transition temperature. With DSC, the increase of the glass transition temperature in the second heating run can be used to calculate a so-called Cure Factor. The larger the postcuring reaction, the larger the CF. In this particular case, practically no postcuring reaction was detected, which indicates that the material was already completely cured (see the example in Section 4.2.1 Effect of repeated postcuring on the glass transition).

4.4.2 玻璃化转变温度和 z-轴热膨胀按照 IPC-TM-650 的 TMA 测定
Glass transition temperature and z-axis thermal expansion by TMA according to IPC-TM-650

目的 按照 IPC-TM-650 No. 2.24.4 测试方法测试玻璃化转变温度和 z-轴热膨胀系数 CTE。这些试验适合于印制线路板中使用的介电材料。

本应用描述了如何进行这些测试。

Purpose	To measure the glass transition temperature and the coefficient of thermal expansion, CTE, in the z-axis according to the IPC-TM-650 No. 2.24.4 test method. These tests are suitable for dielectric materials used in printed boards. The application describes how these tests are performed.
样品 Sample	无铜导线的印制线路板(含环氧树脂基体的玻璃织物)。 Printed circuit board without copper tracks (glass fabric with epoxy matrix).
条件 Conditions	测试仪器:TMA,球点探头 Measuring cell: TMA Probe with ball-point probe tip 样品制备:用小锯从收到的材料上切出样品,在105℃下干燥2h,然后贮存在干燥器内。 Sample preparation: Samples were cut from the material received using a small saw, dried for 2h at 105℃ and then stored in a desiccator. TMA 测试: TMA measurement: The probe was positioned directly on the sample. 探头直接放在样品上。 TMA 负载:0.2N TMA load: 0.2N 第一次和第二次升温均以20K/min从35℃至185℃,两次升温之间冷却不受控。 First and second heating runs at 20K/min from 35℃ to 185℃, with uncontrolled cooling between the two runs. 气氛:静态空气 Atmosphere: Static air

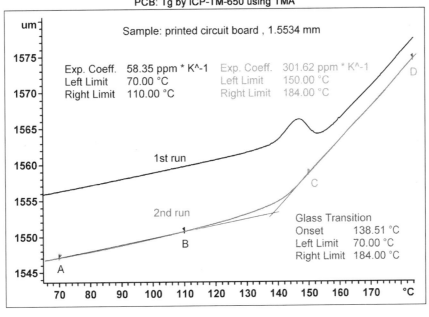

图 4.45 Fig. 4.45

解释 图示为第一次和第二次升温的 TMA。在玻璃化转变区域曲线的斜率发生变化。在第一次升温中,在玻璃化转变区域可见由于样品的热历史和力学历史所形成的一个峰。第二次升温用来测定玻璃化转变温度 T_g,得到 138.5℃ 的值。玻璃化转变温度以下测得的平均热膨胀系数为 58.3×10^{-6}/K,玻璃化

Interpretation The figure shows the 1st and 2nd TMA heating runs. The slopes of the curves change in the glass transition region. In the 1st run, a peak can be seen in the glass transition region that originates from the thermal and mechanical history of the sample. The 2nd heating run was used to determine the glass transition temperature, T_g. A value of 138.5℃ was obtained. The mean thermal expansion coefficient measured below the glass transition was 58.3ppm/K and above the glass transition 301.2ppm/K.

转变温度以上为 $301.2 \times 10^{-6}/K$。

| 计算 | 按照 IPC-TM-650 No. 2.4.24(TMA)测试方法测定的玻璃化转变温度为通过 AB 和 CD 的直线的交点。玻璃化转变前后的热膨胀系数 CTE 从 A 至 B 之间和 C 至 D 之间样品的高度(厚度)的变化计算。 结果汇总结于下表。|

Evaluation The glass transition temperature was determined according to the IPC-TM-650 No. 2.4.24 (TMA) test method as the point of intersection of the straight lines drawn through points AB and CD. The coefficient of thermal expansion, CTE, before and after the glass transition was calculated from the change in height (thickness) of the sample between A and B, and C and D.

The results are summarized in the following table.

	TMA
T_g(第一次升温) T_g (1st heating run)	~138℃
T_g(第二次升温) T_g (2nd heating run)	138.7℃
玻璃化转变以下的 CTE CTE below the glass transition	$58.3 \times 10^{-6}/K$
玻璃化转变以上的 CTE CTE above the glass transition	$301.6 \times 10^{-6}/K$

结论 IPC-TM-650 是测定玻璃化转变温度和热膨胀的实用的 TMA 测试方法。

在 TMA 测试中,因为松弛效应,用第一次升温曲线测定 T_g 常常是困难的,因而只能用第二次升温进行标准计算。除了温度信息,TMA 测试也提供印制线路板 z-方向的膨胀系数(线性热膨胀系数)信息。

Conclusion IPC-TM-650 is a practical TMA test method for the determination of the glass transition temperature and thermal expansion.

In TMA measurements, it is often difficult to determine T_g in the first heating curve because of relaxation effects, so that a standard evaluation can only be performed using the second heating run. Besides the temperature information, the TMA measurement also provides information on the expansion coefficient (linear coefficient of thermal expansion) in the z-direction of the printed circuit boards.

4.4.3 印制线路板,纤维取向对膨胀行为的影响
Printed circuit boards, influence of fiber orientation on expansion behavior

目的	说明若干个 TMA 实验如何能测试印制线路板在焊接中或正常使用时的尺寸变化。 电子工业中的高性能印制线路板是由玻璃织物增强的特殊环氧树脂基体制成的。使用玻璃织物的一个目的是保持印制线路板的热膨胀系数尽可能低。然而织物导致板的力学性能出现各向异性,即板在 x-(宽)、y-(长)和 z-(高或厚)方向呈现不同的膨胀系数。
Purpose	To show how a few TMA experiments can measure dimensional changes when printed circuit boards are soldered or in normal use. High performance printed circuit boards in the electronics industry are made from a special epoxy resin matrix reinforced with glass fabric. One purpose of the glass fabric

is to keep the thermal expansion of the printed circuit board as low as possible. The fabric however leads to mechanically anisotropic behavior of the board, that is, the board exhibits different expansion coefficients in the x-(width), y-(depth), and z-(height or thickness) directions.

样品 Sample	由环氧树脂和玻璃织物组成的无铜箔的印制线路板 FR4。 Printed circuit board without copper foil consisting of epoxy resin and glass fabric, FR4.
条件 Conditions	测试仪器:TMA　　　　　　　　　　　Measuring cell:TMA 探头:球点探头　　　　　　　　　　Probe:Ball-point tip 样品制备:用钻石刀从印制线路板　　Sample preparation:The samples were cut out 上切下样品,在105℃下干燥2h,然　　from the printed circuit board using a diamond 后贮存在干燥箱中。　　　　　　　　saw, dried at 105℃ for 2h and stored in a 　　　　　　　　　　　　　　　　　desiccator. 样品尺寸:　　　　　　　　　　　　Sample dimensions: 宽(x): 4.989mm　　　　　　　　　width (x): 4.989mm 长(y): 2.677mm　　　　　　　　　depth (y): 2.677mm 高(z): 1.584mm　　　　　　　　　height (z): 1.584mm TMA 程序:　　　　　　　　　　　　TMA program: 以 3K/min 从 30℃升温至 150℃　　　Heating from 30 to 150℃ at 3K/min 显示第二次升温测试,即热历史已　　The second heating measurement is shown, i.e. 被消除。　　　　　　　　　　　　　the thermal history has been eliminated. 气氛:氮气,50mL/min　　　　　　　Atmosphere:Nitrogen, 50mL/min

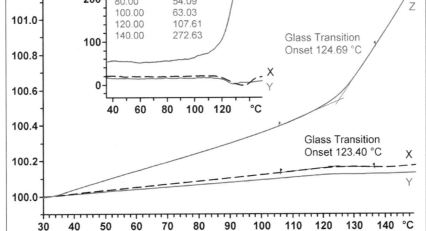

图 4.46　Fig. 4.46

解释　图示为样品三个轴方向上测试的 TMA 曲线。在 x-和 y-方向与玻璃纤维平行,基体树脂的膨胀严格受限,至玻璃化转变仅膨胀约 0.1%。插图显示,样品在 x-和 y-

Interpretation　The figure shows the TMA curves measured in the three axial directions of the sample. Parallel to the glass fibers in the x-and y-directions, the expansion of the matrix resin is very restricted and is only about 0.1% up to the glass transition. The inserted diagram shows that the expansion

方向至玻璃化转变的膨胀系数与铜的相等,铜通常用作导电材料(17×10^{-6}/K)。在玻璃化转变处,这两个方向上可观察到轻微的收缩。

由于纤维对 x- 和 y- 方向膨胀的限制,导致玻璃纤维不能在限制膨胀的 z- 方向发生热膨胀,在玻璃化转变以下,膨胀系数大于 x- 和 y- 方向约四倍(见插图)。

通常玻璃化转变温度从 z- 方向测试的曲线计算,因为难以准确地画 x- 和 y- 方向曲线上切线。玻璃化转变后,TMA 测试曲线的斜率或膨胀系数在 z- 方向显著增加了五倍,而在 x- 和 y- 方向它甚至降低为负值。

coefficient of the sample in the x- and y-directions up to the glass transition corresponds to that of copper, which is generally used as the conducting material (17ppm/K). At the glass transition, a slight shrinkage is observed in these two directions.

The limitation of expansion in the x- and y-directions through the fibers leads to the thermal expansion occurring in the z-direction where the glass fibers do not restrict expansion. Below the glass transition, the expansion coefficient is about four times greater than in the x- and y-directions (see inserted diagram).

Usually the glass transition temperature is evaluated from the curve measured in the z-direction because of the difficulty of drawing the tangents accurately to the curves in the x- and y-directions. After the glass transition, the slope of the TMA measurement curve or expansion coefficient increases markedly in the z-direction by a factor five, whereas in the x- and y-directions it decreases or is even negative.

结论 各向异性材料(例如印制线路板)的热膨胀系数和绝对尺寸变化可通过样品 x、y 和 z 方向的简单 TMA 测试测定。结果可用于质量控制或计算板在加工或操作使用中升温时的机械应力。

Conclusions The thermal expansion coefficient and absolute dimensional changes of anisotropic materials (e. g. printed circuit boards) can be determined by simple TMA measurements in the x-, y- and z-directions of the sample. The results can be used for quality control or to calculate the mechanical stress of the board during processing or through heating in operational use.

4.4.4 碳纤维增强树脂玻璃化转变的测定
Determination of the glass transition of CF-reinforced resins

用纤维高填充或增强的复合材料通常基体树脂含量低。这就使得难以用 DSC 测定玻璃化转变温度,因为不易测试热容的微少变化。此外,碳纤维的比热容与温度有相当大的关系,因此 DSC 曲线不是水平的,这使玻璃化转变温度的测定就更困难。

Composite materials that are highly filled or reinforced with fibers usually have a only low matrix resin content. This makes it difficult to determine the glass transition temperature by DSC because the small change in the heat capacity is difficult to measure. In addition, the specific heat capacity of carbon fibers exhibits relatively large temperature dependence. The result of this is that the DSC curve is not horizontal, which makes the determination of glass transition temperature even more difficult.

目的 测定增强复合材料的玻璃化转变。在这种情况下,通过在参比坩埚中加入适量的填料或增强纤维可使计算变得简单。这可补偿样品中这些材料的热容。

Purpose To determine the glass transition of reinforced composite materials. In such cases, the evaluation can be simplified by filling the reference crucible with an appropriate amount of filler or reinforcement fibers. This compensates the heat capacity of these materials in the sample.

样品 EP-CF,来自适合于结构增强制件的单向碳纤维增强环氧树脂(纤维含量约60%重量)的

	两块层压板。	
	两块层压板用同一树脂体系,然而碳纤维类型(尺寸影响)和固化不同。	
Sample	EP-CF, two laminates from unidirectional carbon-fiber reinforced EP resins (fiber content about 60% by weight) for constructional reinforcement elements. The same resin system was used for both laminates. The laminates differ however in the type of carbon fiber (influence of size) and in curing.	

| 条件 Conditions | 测试仪器:DSC 坩埚:40μL 铝坩埚,盖钻孔 样品设备:用刀从固化的层压板上切下样品。 在有碳纤维的实验中,切割量等于60%样品质量的碳纤维,称重后装入坩埚(约为复合材料中60%纤维含量)。 DSC 测试: 空参比坩埚: 第一次:−20℃恒温 1min,以 10K/min 从−20℃升温至220℃,接着以10K/min 降温到−20℃,然后第二次:以10K/min 从−20℃升温至270℃。 参比坩埚含碳纤维: 第一次:−20℃恒温 2min,以 10K/min 从−20℃升温至170℃,接着以10K/min 降温至−20℃,然后第二次:以10K/min 从−20℃升温至270℃。 气氛:干燥空气,50mL/min | Measuring cell:DSC Pan:Aluminum 40μL with pierced lid Sample preparation:Samples were cut from the cured laminates with a knife. In the experiments with carbon fibers, a quantity of carbon fibers equivalent to 60% of the sample mass was cut and weighed into the crucible (about 60% fiber content in the composite). DSC measurement: With empty reference crucible: 1st run:1min at −20℃, heating from −20℃ to 220℃ at 10K/min, followed by cooling to −20℃ at 10K/min then 2nd run: Heating from −20℃ to 270℃ at 10K/min With carbon fibers in the reference crucible: 1st run:2min at −20℃, heating from −20℃ to 170℃ at 10K/min, followed by cooling to −20℃ at 10 K/min then 2nd run: Heating from −20℃ to 270℃ at 10K/min Atmosphere:Dry air, 50mL/min |

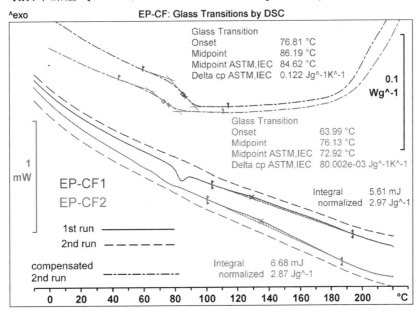

图 4.47   Fig. 4.47

| 解释 | 图的下面部分所示为样品对 | Interpretation | The lower part of the figure shows the four |

空参比坩埚测试的无补偿的第一次和第二次升温的四条 DSC 曲线。图的上部所示为从补偿测试获得的第二次升温（点-划线）。用空参比坩埚的测试，环氧基体系的玻璃化转变温度在第一次或第二次测试中是不易观察见的；玻璃化转变的位置和外形与固化度（第一次测试）和碳纤维类型（尺寸的影响）有关。

当碳纤维用于参比坩埚时，则玻璃化转变变得清晰得多。用空参比坩埚测试的曲线和用碳纤维填充的参比坩埚测试的曲线之间的不同是由于碳纤维比热的温度依赖性。参比坩埚中碳纤维的使用补偿了这些效应，因此环氧基体的玻璃化转变更容易看到。

当样品升温到玻璃化转变以上时，对于第一次升温，观察到一个弱的放热固化反应。因此树脂的固化度提高了，这导致了第二次升温中 T_g 的值较高。在补偿测试的第二次升温中，T_g 的提高较少，因为在第一次测试中样品仅加热到 170℃（曲线未显示）。参比坩埚中没有碳纤维的测试中，样品升温到 220℃。

计算 两个样品的后固化焓几乎相同，但是玻璃化转变温度明显不同，差约 10K。这与尺寸的影响有关。虽然基体树脂存在的量近似相等，但是两个样品的比热容变化还是明显不同的。这可以由可能不同的树脂/纤维比来解释。

uncompensated 1st and 2nd DSC heating runs, where the samples were measured against an empty reference crucible. The upper part of the figure displays the 2nd heating runs obtained from compensated measurements (dot-dash lines). In the measurements using empty reference crucibles, the glass transition temperatures of the EP matrix system are not readily apparent in the 1st or the 2nd runs; the location and the appearance of the glass transition depends on the degree of cure (1st run) and the CF type (influence of sizing).

When carbon fibers are used in the reference crucible, the glass transition becomes much clearer. The difference between the curves measured with an empty reference crucible and those measured with a reference crucible filled with carbon fibers is due to the temperature dependence of the specific heat capacity of the carbon fibers. The use of carbon fibers in the reference crucible compensates these effects so that the glass transitions of the EP matrix are easier to see.

When the samples are heated to above the glass transition, for the first time, a weak exothermic curing reaction is observed. The degree of cure of the resin thereby increases, which results in a higher T_g in the 2nd heating run. The increase in T_g is less in the 2nd heating runs of the compensated measurements because the sample was only heated to 170℃ in the 1st run (curves are not shown). In the measurements without carbon fibers in the reference crucible, the samples were heated to 220℃.

Evaluation The postcuring enthalpy is practically the same for both samples but the glass transition temperatures differ significantly by about 10K. This has to do with the influence of the size. Although approximately the same amount of matrix resin should be present, the change in the specific heat capacity of the two samples is also clearly different. This can be explained by the possibility of different resin/fiber ratios.

参比坩埚 Reference crucible	空的 Empty		含碳纤维的 With carbon fibers	
	EP-CF1	EP-CF2	EP-CF1	EP-CF2
样品质量,mg Sample mass in mg	1.89	2.33	6.11	7.06

参比坩埚 Reference crucible	空的 Empty		含碳纤维的 With carbon fibers	
	EP-CF1	EP-CF2	EP-CF1	EP-CF2
T_g, ℃ 　第一次测试 　1st run 　第二次测试 　2nd run	79 99	69 89	78.0 84.6	70.0 72.9
ΔC_p, J/gK 　第一次测试 　1st run 　第二次测试 　2nd run	— —	— —	0.12 0.12	0.06 0.08
后固化, J/g Postcuring in J/g	3.0	2.9	—	—

结论 当用 DSC 测试时，含碳纤维的复合材料的玻璃化转变不总是清晰的。首先 40% 的树脂含量是相对小的，其次，碳纤维的热容随着温度而显著提高。如果在参比坩埚中用相近数量的碳纤维进行测试，玻璃化转变能观察得清楚多，因而能更准确地测定。

Conclusions The glass transitions of composite materials containing carbon fibers are not always so clear when measured by DSC. First, the resin content of 40% is relatively small and second, the heat capacity of the carbon fibers increases significantly with temperature. The glass transition can be much more clearly observed and hence more accurately determined if the measurement is performed with a similar amount of carbon fiber in the reference crucible.

4.4.5 复合材料纤维含量的热重分析测定
Determination of the fiber content of composites by thermogravimetric analysis

目的 用热重分析测定复合材料中纤维材料的含量——质量控制应用。

为了改善力学和热学性能，经常用非常不同类型的材料填充或增强树脂。有机填料和增强材料（例如木粉）可提高塑料的韧性。添加纤维可使刚度和结构强度显著提高。除了天然有机纤维如黄麻和剑麻，合成无机纤维（例如玻璃和碳纤维）和有机纤维如芳纶纤维也广泛用于增强目的。芳纶纤维由聚对苯二甲酰对苯二胺组成，因高拉伸强度和约 550℃ 相对高的分解温度而备受关注。

Purpose To determine the content of fiber material in a composite using thermogravimetric analysis -a quality control application.

Resins are frequently filled or reinforced with very different types of material in order to improve their mechanical and thermal properties. Organic fillers and reinforcing materials (e. g. wood powder) increase the toughness of a plastic. The addition of fibers can result in a major increase in stiffness and structural strength. Besides natural organic fibers such as jute and sisal, synthetic inorganic fibers (e. g. glass and carbon fibers) and organic fibers such as aramid are widely used for reinforcement purposes. Aramid fibers consist of poly-p-phenylene terephthalamide and are remarkable for their high tensile strength and their relatively high decomposition temperature of about 550℃.

样品 Sample	复合材料，单独的环氧树脂和芳纶纤维。 Composite material, separate epoxy resin and the aramid fibers.	
条件 Conditions	测试仪器：TGA 坩埚：70μL 铝坩埚，无盖 样品设备：从复合材料上锯下 54.2mg 重的一片。 9.57mg 芳纶纤维被剪短以便能塞入坩埚。 35.7mg 环氧树脂直接装入坩埚。 TGA 测试： 以 20K/min 从 25℃升温至 600℃ 气氛：氮气，200mL/min	Measuring cell：TGA Pan：Aluminum 70μL, without lid Sample preparation：A piece weighing 54.2mg was sawn out of the composite material. 9.57mg aramid fibers were cut to short lengths so that they could be packed into a crucible. 35.7mg epoxy resin was filled directly into the crucible. TGA measurement： Heating from 25 to 600℃ at 20K/min Atmosphere：Nitrogen, 200mL/min

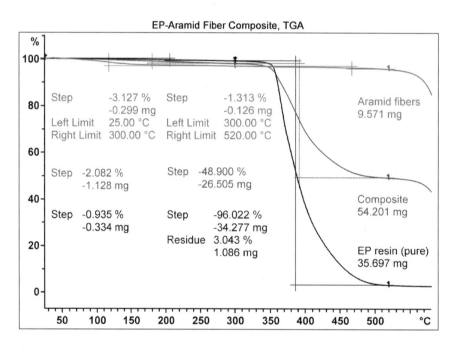

图 4.48　Fig 4.48

解释　与玻璃纤维不同，芳纶纤维在 520℃以上分解，如图中 TGA 曲线所示。因此纤维含量不能从树脂完全分解后的残余物直接测定。芳纶纤维开始时失去 3.1%的水分；在 300℃至 520℃之间失重 1.3%。
纯树脂首先失去水分，然后在 300℃至 520℃之间以单一台阶过程热解。
复合材料首先由于失水失重 2%，然后，正如预期的，以两个台阶分解，高至 520℃的树脂热解和高于

Interpretation　In contrast to glass fibers, aramid fibers decompose above 520℃ as the TGA curve in the figure shows. The fiber content cannot therefore be directly determined from the residue after complete decomposition of the resin. The aramid fibers initially lose 3.1% moisture; between 300℃ and 520℃ the mass loss is 1.3%.
The pure resin first loses moisture and then pyrolyzes in a single step process between 300 and 520℃.
The composite material first suffers a mass loss of about 2% due to loss of moisture and then, as expected, decomposes in two steps, the pyrolysis of the resin up to 520℃ and the

此温度的纤维分解。

计算 表格汇总了至520℃的各个阶段的失重。300℃以后由于热解发生的失重是基于300℃时的干质量计算的。用这些值来计算树脂含量。

Evaluation The table summarizes the individual mass losses up to 520℃. The mass losses through pyrolysis from 300℃ onward are calculated based on the dry mass at 300℃. These values are used to calculate the resin content.

样品 Sample	热重台阶,% Thermogravimetric steps in %		
	室温至300℃ RT to 300℃	300 至 520℃ 300 to 520℃	300 至 520℃,对300℃时的干含量 300 to 520℃, with respect to the dry content at 300℃
芳纶纤维 Aramid fiber	3.13	1.31	1.35
复合材料 Composite	2.08	48.90	49.93
纯树脂 Pure resin	0.94	96.02	96.92

复合材料的树脂含量可从300℃至520℃之间的热解台阶计算。因为树脂本身不是完全分解的且纤维也有少量损失,所以在计算中必须考虑这些。

因此复合材料干物质的树脂含量为:

The resin content of the composite material can be calculated from the pyrolysis step between 300℃ and 520℃. Since the resin itself does not completely decompose and the fibers also show a small loss, this has to be taken into account in the calculation.

The resin content of the composite dry substance is therefore given by:

$$树脂含量 = (49.93\% - 1.35\%)/(0.9692 - 0.0135) = 50.83\%$$

$$\text{Resin content} = \frac{49.93\% - 1.35\%}{0.9692 - 0.0135} = 50.83\%$$

归一到原始样品质量计算得到的复合材料实际水分含量:

 水分:2.08%
 树脂:49.77%
 芳纶纤维:47.14%(因无其他填料存在,可认为与100%略有差别)。

Calculated with respect to the original sample mass with the actual moisture content of the composite yields:

 Moisture: 2.08%
 Resin: 49.77% and
 Aramid fiber: 47.14% (as difference to 100% because no other filler is present)

结论 本例表明,即使有纤维分解,热重分析也是一种快速和准确测定复合材料中树脂和纤维含量的很好方法。然而,树脂和纤维的热解不可发生在同样的温度范围内。无机填料也可通过热解残留物的燃烧测定。

Conclusions This example shows that thermogravimetric analysis is an excellent method for the rapid and accurate determination of the resin and fiber content in composite materials, even if the fibers decompose. The pyrolysis of the resin and the fibers must not, however, occur in the same temperature range. Inorganic fillers could also be determined by combustion of the pyrolysis residues.

分解温度提供了有关组分鉴定的附加的定性信息。

The decomposition temperatures provide additional qualitative information concerning the identity of the components.

4.4.6 预浸料中的碳纤维含量 Carbon fiber content in prepregs

目的 Purpose	如果树脂经历完全的燃烧且没有其他的无机填料存在,那么就容易测定矿物填料含量。本应用的目的是说明可以测定碳纤维的含量,即使树脂在热解条件下生成一定量的碳。 The determination of the content of a mineral filler is easy if the resin undergoes complete combustion and no other inorganic fillers are present. The purpose of the application is to show that the content of carbon fibers can be determined even if the resin produces a certain amount of carbon under pyrolytic conditions.
样品 Sample	由环氧树脂和碳纤维制成的预浸料织物 Prepreg fabric made of epoxy resin and carbon fibers
条件 Conditions	测试仪器:TGA　　　　　　　　Measuring cell:TGA 坩埚:30μL 氧化铝坩埚　　　　Pan:Alumina 30μL 样品设备:从预浸料上切出约 8mg 重的一小片。　　　　　　　Sample preparation:A small piece weighing about 8mg was cut out of the prepreg material. TGA 测试:　　　　　　　　　TGA measurement: 以 5K/min 从 30℃ 升温至 700℃,以 20K/min 降温至 200℃,均在氮气中(热解条件),接着切换到空气吹扫,以 5K/min 升温至 1000℃。 Heating from 30℃ to 700℃ at 5K/min and cooling to 200℃ at 20K/min, all in nitrogen (pyrolytic conditions) followed by switching to air purge and heating to 1000℃ at 5K/min. 气氛:氮气和空气,70mL/min　　Atmosphere:Nitrogen and air,70mL/min

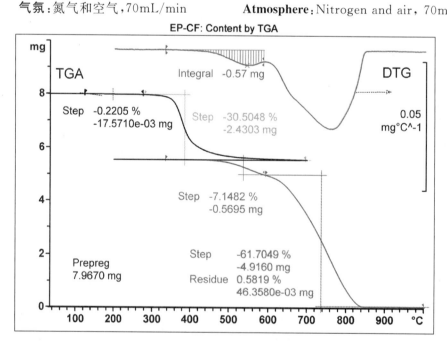

图 4.49　Fig. 4.49

解释　图示为当样品在热解条件下加热到 700℃ 时得到的曲线:产生 30.5% 的失重。因为必须认为有部

Interpretation　The diagram shows the curve obtained when the sample is heated under pyrolytic conditions up to 700℃: a mass loss of 30.5% results. Since it must be assumed that part of the resin carbonizes (see "Content determination"), the atmosphere cannot be switched directly to air or oxygen. In this case, the combustion would only show the sum of the different types of carbon. The sample is therefore first cooled to 200℃ still under nitrogen, switched to air, and then heated to 1000℃. The different types of carbon then burn off sequentially. Active carbon with a large surface area (e. g. carbon black) combusts at lower temperatures than carbon with a greater packing density (e. g. carbon fibers, graphite).

This is the case here: before the carbon fibers burn from about 600℃ onward, a combustion loss of 7.1% is observed. The separation of the steps is most easily seen in the DTG curve. The carbon black produced in the pyrolysis step can thus be separated from the carbon of the carbon fibers.

Evaluation The mass loss through combustion of carbon black from the pyrolysis step is 7.14% (corresponding to 0.569 mg). It can also be calculated by integration of the DTG peak. This method is sometimes more accurate with steps that are difficult to define because all the different baseline types that can be used for peak integration in DSC are available. The resin content is 37.65% (30.50% + 7.15%). The residue of 0.58% shows that there are small amounts of noncombustible materials in the composite. Hence, the carbon fiber content is 61.77% (100% − 37.65% − 0.58%).

Conclusions The determination of the carbon fiber content in a prepreg (here 62%) requires a special temperature program, similar to that used for the content determination of elastomers (for example ISO 9924). Pyrolysis of the matrix resin produces active carbon which in contrast to carbon fibers burns at lower temperatures. If the matrix resin is available in the pure form, the method described in Section 4.4.5 can be used.

Reference ISO 9924-2, Rubber and rubber products-Determination of the composition of vulcanizates and uncured compounds by thermogravimetry-Part 2: Acrylonitrile-butadiene and halobutyl rubbers.

4.5 材料性能的检测 Checking material properties

4.5.1 印制线路板生产中的质量保证
Quality assurance in the production of printed circuit boards

目的	说明玻璃化转变温度测定的可重复性。这对质量控制是非常重要的。 印制线路板的固化度在质量保证中是常规测定的,主要为确保板的力学性能。没有完全固化的线路板在随后的焊接过程中会进一步交联并伴随收缩。这可导致导线升离。 固化度是用两次升温测试由玻璃化转变温度进行测定的。因为材料是高度填充的,所以DSC仪器必须非常灵敏,以便也能自动计算玻璃化转变。
Purpose	To demonstrate the repeatability of the determination of the glass transition temperature. This is very important for quality control. The degree of cure of printed circuit boards is routinely determined in quality assurance and is mainly responsible for the mechanical properties of the board. A board that is not properly cured undergoes further crosslinking with accompanying shrinkage in the soldering process that follows. This can lead to conduction tracks lifting off. The degree of cure is determined from the glass transition temperature using two heating runs. Since the materials are highly filled, the DSC instrument must be very sensitive so that the glass transitions can also be evaluated automatically.
样品	印制线路板 FR4。
Sample	Printed circuit board, FR4.

条件 Conditions

测试仪器:DSC	Measuring cell: DSC
坩埚:40μL 铝坩埚,盖钻孔	Pan: Aluminum 40μL with pierced lid
样品制备:从板上锯下重约 30mg 的片。用砂纸将边缘磨光,以便样品与坩锅底部接触良好。	Sample preparation: Pieces weighing about 30mg were sawn out of the board. The edges were smoothed with abrasive paper so that the samples made good contact with the bottom of the crucible.
DSC 测试: 第一次和第二次升温以 20K/min 从 60℃至 190℃。将样品从炉中取出,迅速冷却到室温,在初始条件下再次放入(例如用自动进样器)。	DSC measurement: 1st and 2nd heating runs from 60℃ to 190℃ at 20 K/min. The samples were removed from the furnace, quickly cooled to room temperature and inserted again at the starting conditions (e.g. automatically with the sample robot).
气氛:静态空气	Atmosphere: Static air

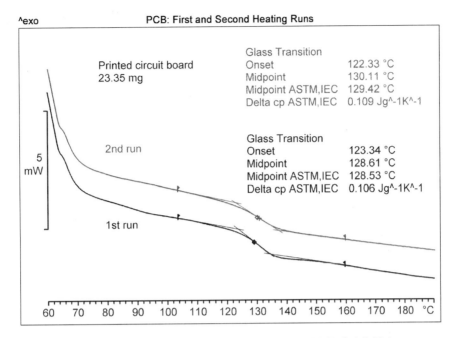

图 4.50 印制线路板第一次和第二次 DSC 升温测试的玻璃化转变。

Fig. 4.50 Glass transitions as measured by the 1st and 2nd DSC heating runs of a printed circuit board.

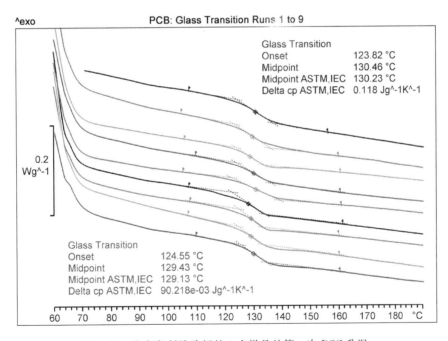

图 4.51 取自印制线路板的 9 个样品的第一次 DSC 升温。

Fig. 4.51 DSC 1st heating runs of 9 specimen taken from a printed circuit board.

解释 图 4.50 所示为第一次和第二次 DSC 升温测试。玻璃化转变之间的不同是非常小的,因此几乎不发生后固化,所以材料满足质量标准。第一次升温几乎看不到松弛焓,在 90℃和 100℃之间可以看见一个由于失去

Interpretation Figure 4.50 shows the 1st and 2nd DSC heating runs. The difference between the glass transitions is very small. From this it follows that very little postcuring occurred so that the material satisfies the quality criteria. In the 1st heating run practically no relaxation enthalpy can be seen. A small endothermic effect is visible between 90℃ and 100℃ due to the

吸附水导致的小的吸热效应；在第二次升温中，不再观察到该效应。

loss of adsorbed moisture; in the 2nd heating run, this effect is no longer observed.

图 4.51 所示为取自同一块板的样品（沿着边缘 10cm 长）的一系列测试，以评估测试的重复性。在这些 DSC 曲线上也能看到小的汽化效应。在一个测试中，吸热假象出现在 165℃，这干扰了自动分析，如果将界限设在 160℃。不过，对每一条曲线单独选择了计算界限，这样就可使切线就与曲线最佳匹配。

Figure 4.51 shows a series of measurements of samples taken from the same board (over a length of 10 cm along the edge) to estimate the repeatability of the determination. The small vaporization effects can also be seen in these DSC curves. In one measurement, an endothermic artifact occurs at 165℃. This interferes with the automatic evaluation if the limit is set to 160℃. The evaluation limits were however chosen individually for each curve so that the tangents optimally fitted the curves.

表 4.4 按照 ASTM 测试方法计算的图 1 所示曲线的中点温度。
Table 4.4 Evaluation of the midpoint temperature of the curves shown in Figure 1 according to the ASTM test method.

	T_g 第一次升温 1st heating run	T_g 第二次升温 2nd heating run
按照 ASTM 的中点温度 Midpoint temperature according to ASTM	128.5℃	129.5℃

用界限 110℃ 和 160℃ 的自动计算与手工设限进行的计算作了比较。

The automatic evaluations using limits of 110℃ and 160℃ are compared with evaluations performed using manually set limits.

表 4.5 本表总结了图 4.51 中所示的测试系列的玻璃化转变温度（未显示第二轮加热）。
Table 4.5 The table summarizes the glass transition temperatures of the measurement series shown in Figure 4.51 (2nd runs are not shown).

玻璃化转变，℃ Glass transition in ℃	第一轮，自动 1st run auto	第一轮，手动 1st run manual	第二轮，自动 2nd run auto	第二轮，手动 2nd run manual
1	130.4	130.5	128.0	127.1
2	129.4	129.0	128.6	127.9
3	127.9	127.7	127.8	127.2
4	129.4	129.2	129.3	129.2
5	131.5	129.8	128.9	128.3
6	128.0	128.6	130.1	126.9
7	129.2	128.6	128.0	127.2
8	128.5	128.2	128.3	127.8
9	130.0	129.1	129.3	128.7
平均值 Mean value	129.4	129.0	129.0	127.8
标准偏差 Standard deviation	1.16	0.84	1.03	0.79

计算 当手工选择切线位置（优化

Evaluation The standard deviation is smaller when the position

定位)时,标准偏差较小。玻璃化转变处的松弛效应和其他假象会干扰第一次升温实验的计算,产生不准确的结果。在自动计算中,必须注意所画的切线。

of the tangents is chosen manually (optimum positioning). Relaxation effects at the glass transition and other artifacts can interfere with the evaluation of the 1st heating run and produce inaccurate results. In the automatic evaluation, attention must be paid to the drawn tangents.

结论 比较第一次和第二次升温的 T_g 值可快速表明实际上是否满足质量标准:差值不可大于3K。所测试的样品满足这个标准。

T_g 测定的重复性或再现性可用九个测试的标准偏差进行表征。从实验室间测试的结果可见[Schmid等],1K的偏差值对同一个人和同一台仪器(重复性)是典型的。不同实验室测定 T_g 值的偏差为3K。为了获得好的再现性,除了注意环境的影响外,必须仔细检查切线的位置,如有必要则优化之。

Conclusions Comparison of the T_g values of the 1st and 2nd heating runs quickly shows whether the quality criterion is satisfied in practice: the difference must not be greater than 3K. The samples measured satisfy this criterion.

The repeatability or reproducibility of the T_g determination can be characterized by the standard deviation using nine measurements. A value of 1 K for one person and one instrument (repeatability) is typical as was seen from the results of an interlaboratory test [Schmid et. al.]. The T_g determination measured by different laboratories (reproducibility) gave a value of 3K. To achieve good reproducibility, the location of the tangents must be carefully checked and if necessary optimized, besides paying attention to environmental influences.

参考文献 M. Schmid, A. Ritter, S. Affolter,"用差示扫描量热仪(DSC)对聚合物的实验室间测试",Macromol. Mater. Eng.,286 (2001) p. 605—610.

Reference M. Schmid, A. Ritter, S. Affolter, "Interlaboratory Tests on Polymers by Differential Scanning Calorimetry (DSC): Determination of Glass Transition Temperature", Macromol. Mater. Eng., 286 (2001) p. 605—610.

4.5.2 碳纤维增强热固性树脂的玻璃化转变测定
Determination of the glass transition of carbon fiber reinforced thermosets

对于高刚度和低质量的结构部件,例如在航空和航天工业,越来越多地使用碳纤维增强的复合材料。这些复合材料是轻质材料,表现出优于如铝这样的传统材料的力学性能。在航空和航天应用中安全使用复合材料的一个重要参数是树脂的玻璃化转变温度。出于安全原因,它应该至少超过最高操作温度以上50K。在航空和航天工业中,玻璃化转变温度传统上是通过用DMA单悬臂梁模式测试与温度有关的模量来测定的。作为选择,越来越多地采用温度调制DSC。后者技术的优势是可将玻璃化转变和可能重叠的后固化过程相互分开。

For construction elements of high stiffness and low mass, for example in the aerospace industry, more and more composites with carbon fiber reinforcement are being used. These composites are lightweight materials and exhibit mechanical properties superior to those of traditional materials such as aluminum. An important parameter for the safe use of composites in aerospace applications is the glass transition temperature of the resin. For safety reasons, this should be at least 50K above the maximum operating temperature. In the aerospace industries, the glass transition is traditionally determined from the temperature-dependent modulus measured by DMA in single cantilever mode. As an alternative, temperature-modulated DSC is being increasingly used. The advantage of the latter technique is that it can separate the glass transition and potentially overlapping postcuring processes from one another.

目的	用温度调制 DSC 和 DMA 测定玻璃化转变并比较获得的温度,甚至在可能有后固化的情况下。
Purpose	To determine the glass transition by temperature-modulated DSC and DMA and compare the temperatures obtained, even in cases where postcuring is possible.
样品	不同类型的碳纤维环氧预浸料: 层压板 1:8 层经纱的 Fibredux 914C,180℃固化。 层压板 2:8 层经纱的 Fibredux 924C,180℃固化。 层压板 3:7 层经纱的 Fibredux 6268C,130℃固化。
Sample	Different types of carbon-fiber epoxy prepregs: Laminate 1: Fibredux 914C with 8-ply warp yarns, cured at 180℃ Laminate 2: Fibredux 924C with 8-ply warp yarns, cured at 180℃ Laminate 3: Fibredux 6268C with 7-ply warp yarns, cured at 130℃

| 条件 | 测试仪器:DSC
 DMA
 DSC 坩埚:20μL 铝坩埚
 样品制备:
 样品在 70℃下干燥 48h。
 DSC 测试:切出重约 16mg 的小片。
 DMA 测试:长度约 22.5mm;典型厚度 2.3mm;典型宽度 10mm。

DSC 测试:
 以平均速率 5K/min 从 30℃升温至 260℃的 ADSC,正弦振动振幅 1K,周期 1min。
 30mL/min 氮气
 DMA 测试:
 单悬臂,应用的夹紧转矩为 40~60 Ncm
 以 5K/min 从 30℃升温至 270℃
 最大位移振幅 16μm
 频率 1Hz
 气氛:50mL/min 氮气 | **Measuring cell**:DSC
 DMA
 Pan:Aluminum 20μL
 Sample preparation:
 Samples were dried at 70℃ during 48h.
 DSC measurements: small pieces weighing about 16 mg were cut out. DMA measurements: free length 22.5mm; thickness typically 2.3mm, width typically 10mm

 DSC measurement:
 ADSC from 30 to 260℃ at a mean rate of 5 K/min, amplitude of sinusoidal oscillation 1K, period 1min.
 30mL/min nitrogen
 DMA measurement:
 Single cantilever, applied clamping torque 40~60Ncm
 30℃ to 270℃ at 5K/min
 Maximum displacement amplitude 16μm
 Frequency 1Hz
 Atmosphere:50mL/min nitrogen |
| **Conditions** | | |

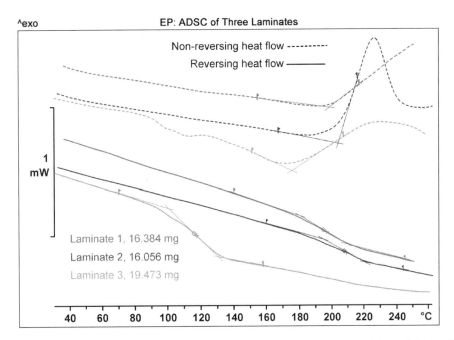

图 4.52 三块层压板 ADSC 测试的玻璃化转变和后固化。(由可逆热流曲线可清楚地确定玻璃化转变。在不可逆热流曲线上可见样品的后固化。在常规 DSC 测试中,后固化和玻璃化转变区域会重叠,即不可能鉴别和正确计算玻璃化转变)

Fig. 4.52 ADSC measurements of the glass transition and postcuring of three laminates. The glass transitions can be clearly identified in the reversing heat flow curves. The postcuring of the samples is visible in the non-reversing heat flow curves. In conventional DSC measurements, the postcuring and glass transition regions would overlap, that is, it would not be possible to identify and properly evaluate the glass transition.

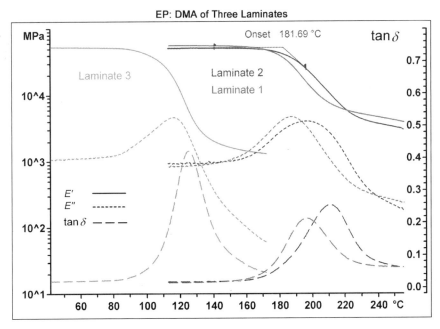

图 4.53 三块不同层压板单悬臂模式测试的 DMA 曲线(样品尺寸如下(宽度、高度)mm;层压板 1:10.00、2.50;层压板 2:10.00、2.58;层压板 3:10.06、2.34)

Fig. 4.53 DMA curves of three different laminates measured in single cantilever mode. The dimensions of the samples were as follows (width, height) in mm: Laminate 1:10.00, 2.50; Laminate 2:10.00, 2.58; Laminate 3:10.06, 2.34.

解释 ADSC：ADSC 技术可将热容变化的效应与重叠后固化现象分开。在可逆热流曲线上能清晰地确定玻璃化转变。不可逆热流曲线呈现在玻璃化转变后立即偏离基线的放热。这是后固化。在常规 DSC 测试中,后固化反应会与玻璃化转变区域重叠,从而不可能准确的测定 T_g。

DMA：DMA 测试的玻璃化转变可表征为贮存模量台阶式变化的起始点,或者损耗模量或损耗因子(tanδ)的峰值。对于层压板,通常关注材料的最初开始软化。因此贮存模量台阶变化的起始温度被广泛接受作为这类材料的玻璃化转变温度。DMA 测试的后固化反应比 DSC 测试的不明显得多。实际上,只有层压板 2 呈现一些后固化效应的迹象,因为与其他两个样品的峰形相比,它的 E'' 峰看起来略宽些。

计算 为了研究玻璃化转变温度的再现性,用 ADSC 测试了每一块层压板的两个样品,用 DMA 测试了每块板的三个样品。结果汇总于下表。ADSC 的玻璃化转变温度用可逆热流曲线计算,按照 ASTM E1356 测定其中点。经比较证实,用不同计算方法和不同实验技术测定的样品的玻璃化转变温度之间存在显著差异。最低值通常是线性刻度的 E' 的起始点,最高值是 tanδ 的峰值。

Interpretation ADSC：The ADSC technique allows effects due to changes in heat capacity to be separated from overlapping postcuring phenomena. The glass transition can be clearly identified in the reversing heat flow curves. The non-reversing heat flow curves exhibit exothermic deviations from the baseline immediately after the glass transition. This indicates postcuring. In a conventional DSC measurement, the postcuring reaction would overlap the glass transition region making a reliable determination of T_g impossible.

DMA：The glass transition in a DMA measurement can be characterized by the onset in the step-like change of the storage modulus or as a peak in the loss modulus or the loss factor (tanδ). With laminates, one is usually interested in the very beginning of softening of the material. The onset temperature of the step change in the storage modulus is therefore widely accepted as the glass transition temperature in these types of material. The postcuring reaction is much less obvious in the DMA measurements than in the DSC measurements. In fact, only Laminate 2 shows some indications of postcuring effects because the E'' peak appears slightly broadened compared to the shape of the peaks of the other two samples.

Evaluation To investigate the reproducibility of the glass transition temperature, two samples of each laminate were measured by ADSC and three samples of each by DMA. The results are summarized in the following table. The glass transition temperature from ADSC was evaluated using the reversing heat flow curve and determining the midpoint according to ASTM E1356. The comparison confirms the significant differences between glass transition temperatures of samples determined by different evaluation procedures and experimental techniques. The lowest values are usually the onsets of E' on a linear scale, the highest the peak values of tanδ.

样品 Sample	T_g,℃ ADSC 测试 by ADSC	Δc_p J/gK	T_g,℃ E'(线性)起始点 Onset E' (linear)	T_g,℃ E'(对数)起始点 Onset E' (logarithmic)	E'' 最大值 Max E''℃	tanδ 最大 Max tanδ℃
层压板 1 Laminate 1	194.5 193.3	0.08 0.09	170.1 173.0 171.5	175.6 176.6 175.9	186.4 186.2 187.7	195.6 195.0 197.4

续表

样品 Sample	T_g,℃ ADSC 测试 by ADSC	Δc_p, J/gK	T_g,℃ E'(线性)起始点 Onset E' (linear)	T_g,℃ E'(对数)起始点 Onset E' (logarithmic)	E''最大值 Max E''℃	tanδ 最大 Max tanδ℃
层压板 2 Laminate 2	207.5 205.6	0.07 0.09	171.1 171.0 169.7	181.7 178.2 181.2	196.0 196.5 196.3	210.5 209.9 210.7
层压板 3 Laminate 3	115.8 114.8	0.17 0.16	97.7 97.7 97.7	108.6 108.6 108.6	115.9 115.9 115.9	124.8 125.0 124.9

结论 用 ADSC 或 DMA 方法可容易地测试玻璃化转变温度。ADSC 对检测后固化反应远优于 DMA。玻璃化转变温度与测试技术密切相关,对于 DMA,与频率和计算方法有关。因此只有基于来自可比技术用相同计算方法的结果才可能进行比较。

对于复合材料,通常将线性刻度作图的贮存模量台阶变化的起始温度作为有关的玻璃化转变温度。

Conclusions Glass transition temperatures can readily be measured by ADSC or DMA methods. ADSC is far superior to DMA with respect to the detection of postcuring reactions. The glass transition temperatures strongly depend on the measuring technique and with DMA on the frequency and evaluation procedure. Comparisons should therefore only be made based on results from comparable techniques using the same evaluation procedure.

For composites, the onset temperature of the step change of the storage modulus plotted on a linear scale is usually accepted as the relevant glass transition temperature.

4.5.3 按照 ASTM 标准 E1641 和 E1877 求解分解动力学和长期稳定性
Decomposition kinetics and long-term stability according to ASTM standards E1641 and E1877

目的 按照 ASTM E1641 测试方法用动态 TGA 测量法测定分解反应的活化能。在 ASTM E1877 测试方法中,这被用来估算材料的等温长期稳定性(耐热性),例如在操作温度即焊接槽温度下。

如果将印制线路板加热到高温,则基体树脂开始分解并逸出气体。这个过程引起印制线路板层结构的分层并导致板的最终毁坏。为了防止燃烧发生,用含溴化单体单元(四溴双酚 A)的专门环氧树脂作为基体树脂制板。这些添加剂防止燃烧但不阻止板分层。

对于板的生产过程和以后的操作使用,了解板在发生一定程度的降解前对特定温度能耐多久是重要的。

Purpose To determine the activation energy of the decomposition reaction using dynamic TGA measurements according to the ASTM E1641 test method. This is used in the ASTM E1877 test method for estimating the isothermal long-term stability (thermal endurance) of the material, e. g. at the operating temperature or the soldering bath temperature.

If printed circuit boards are heated to high temperatures, the matrix resin begins to decompose and gases are evolved. This process causes delamination of the layered structure of the printed circuit board and leads to ultimate destruction of the board. To prevent combustion occurring, the board is made flame resistant by using a special epoxy resin containing brominated monomer units (tetrabrom-bisphenol A) as the matrix resin. These additives prevent combustion but not delamination of the board.

For the manufacturing process of a board and its operational use later on, it is important to know how long a board can withstand a particular temperature before a certain amount of degradation occurs.

样品 由环氧树脂和玻璃织物组成的无铜箔的印制线路板(FR4)。
Sample Printed circuit board without copper foil consisting of epoxy resin and glass fabric (FR4).

条件 测试仪器：TGA **Measuring cell**：TGA
Conditions 坩埚：30μL 氧化铝坩埚 **Pan**：Alumina 30μL

样品制备：用钻石锯从印制线路板上锯下样品。样品尺寸约为 2mm×2mm×1.5mm；样品质量约为 22mg。使用的样品尺寸和质量尽量接近以获得好的再现性。

Sample preparation：The samples were cut out of the printed circuit board using a diamond saw. Sample dimensions about 2mm×2 mm×1.5mm; sample mass about 22mg. A narrow size and mass range was used to achieve good reproducibility.

TGA 测试：
以 1、2、5 和 10K/min 从 200℃升温至 380℃

TGA measurement：
Heating from 200℃ to 380℃ at 1, 2, 5 and 10K/min

气氛：氮气，20mL/min **Atmosphere**：Nitrogen, 20mL/min

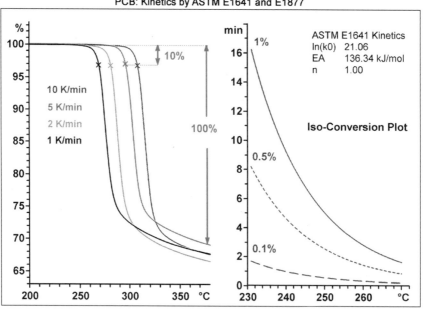

图 4.54   Fig 4.54

解释 为了按照 ASTM 1641 来测定分解反应的活化能，以 1、2、5 和 10K/min 的升温速率将四个样品充分加热到分解温度以上，失重约 30%（见图左边）。在四条曲线上确定对应于特定转化率的温度（这里

Interpretation To determine the activation energy of the decomposition reaction according to ASTM 1641, four samples were heated to temperatures well above their decomposition temperatures at heating rates of 1, 2, 5 and 10 K/min. The mass loss is about 30% (see diagram on the left). The temperature corresponding to a certain conversion (here 10% of the mass loss

失重为至 380℃ 时失重的 10%）。算得的活化能在分解反应开始阶段与反应级数无关（在 ASTM E1641 中假定为 1）。然后将所用的升温速率的对数与 1000/T 作图。除了其他动力学参数，还可从得到的直线的斜率迭代计算出 Arrhenius 活化能。

按照 ASTM E1877"从热重分解数据计算耐热性的标准方法"，活化能是确定与等温使用温度 T_f 有关的寿命 t_f 的必需参数（见右边的图，等转化率曲线）。在本例中，寿命的计算是根据假设在印制线路板变为不可用之前分解可达到 0.1%、0.5% 或 1.0%。选择的温度为焊接槽中可能发生的温度。如果设定 1% 为极限，那么板在 250℃ 下加热不可超过 5min。

结论 按照 ASTM E1641 测试方法，只用四个代表性动态 TGA 测试，就能测定一级分解反应的活化能。该参数用来估算材料在不同温度下的寿命，适用于质量控制目的或材料规格（ASTM E1877）。

本例的计算表明，几秒钟的焊接短时间，甚至在 260℃（接近焊接槽温度）下对印制线路板不引起分解导致的任何严重损坏（失重<0.1%）。用实际测试来检查根据动力学预测的寿命是非常重要的。

up to 380℃) was determined for all four TGA curves. The activation energy to be calculated is only independent of the reaction order (in ASTM E1641 assumed to be one) at the beginning of the decomposition reaction. The heating rates used are then plotted logarithmically versus 1000/T. Besides other kinetic parameters, the Arrhenius activation energy can then be iteratively calculated from the slope of the straight line obtained.

This activation energy is the parameter necessary to determine the lifetime t_f as a function of the isothermal temperature of use T_f according to ASTM E1877 "Standard Practice for Calculating Thermal Endurance of Materials from Thermogravimetric Decomposition Data" (see diagram on the right, Iso-Conversion plot). In this case, the lifetime was calculated on the assumption that the decomposition can reach 0.1%, 0.5% or 1.0% before the printed circuit board becomes unusable. The temperatures were chosen that would occur in a soldering bath. If 1% is set as the limit, the board must not be heated for longer than 5min at 250℃.

Conclusions With just four representative dynamic TGA measurements, the activation energy of a first order decomposition reaction can be determined according to the ASTME1641 test method. This parameter is used to estimate the lifetime of a material at different temperatures for quality control purposes or for the specification of a material (ASTM E1877).

The calculations in the example show that short soldering times of a few seconds, even at 260℃ (the approximate soldering bath temperature) do not cause any serious damage (<0.1% mass loss) to the printed circuit board through decomposition.

It is very important to check lifetimes predicted on the basis of kinetics by performing practical tests.

4.5.4　印制线路板的老化 Aging of printed circuit boards

目的　用 DSC 测试的的玻璃化转变温度，用作检查印制线路板老化测试后可能变化的质量标准。印制线路板在使用中受温度大幅变化的影响。条件一定不要对力学稳定性或其他性能有不利影响。这可通过进行许多极端温度循环来检查和用热分析来研究各种变化。

Purpose　The glass transition temperature is measured by DSC and used as quality criterion to check printed circuit boards for possible changes after aging tests.

Printed circuit boards are subjected to large changes in temperature in use. The conditions must not adversely affect the mechanical stability or other properties. This can be checked by performing any number of extreme temperature cycles and

	investigating the changes by thermal analysis.	
样品 **Sample**	EP-GF：新的和人工老化的玻璃纤维增强环氧树脂含铜导线印制线路板的片。 EP-GF: pieces of new and artificially aged glass-fiber reinforced epoxy printed circuit boards with copper tracks.	
条件 **Conditions**	测试仪器：DSC 坩埚：40 μL 铝坩埚，盖钻孔 样品制备：含铜导线的环氧-玻纤印制线路板有差别地老化（一个老化周期：−65℃ 下 15min，125℃ 下 15min，空气中）。用刀从印制线路板的同一部位切下重约 23mg 由环氧-玻纤和铜组成的样品。 DSC 测试： 第一次测试：25℃下恒温 5min，以 20K/min 从 25℃升温至 200℃，在 200℃ 下恒温 5min，接着以 20K/min降温至 25℃，25℃下恒温 5min，然后第二次测试：以 25K/min 从 25℃升温至 200℃ 气氛：干燥空气，50mL/min	**Measuring cell**：DSC **Pan**：Aluminum 40 μL with pierced lid **Sample preparation**：EP-GF printed circuit boards with copper tracks were aged differently (one aging cycle: 15 min at −65℃, 15min at 125℃, in air). Specimen consisting of EP-GF and copper weighing about 23 mg were cut from the same area of the printed circuit board using a knife. **DSC measurement**： 1st run: 5min at 25℃, heating from 25℃ to 200℃ at 20K/min, 5min at 200℃, followed by cooling to 25℃ at 20K/min, 5min at 25℃ and 2nd run: Heating from 25℃ to 200℃ at 25K/min **Atmosphere**：Dry air，50mL/min

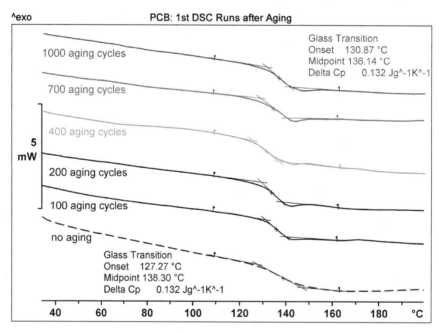

图 4.55 Fig. 4.55

解释 图示为进行了高达 1000 次老化循环的样品的第一次升温。为了检测任何可能的老化效应，比较了第一次升温的曲线。所有的样品显示近似相同的玻璃化转变温度。在这种情况下，用 DSC 检测不到玻

Interpretation The figure shows the 1st heating runs of samples that had undergone up to 1000 aging cycles. The first heating curves were compared in order to detect any possible aging effects. All the samples exhibit approximately the same glass transition temperature. In this case, no systematic shift of the glass transition temperatures and therefore no aging effects

璃化转变温度的系统性移动，因此无老化效应。然而，经过大量老化循环的样品的确显示了焓松弛。这对在仅低于玻璃化转变温度下经历较长时间的树脂模塑料是典型的。这里没有显示第二次升温。因为样品在第一次中都被升温到200℃，所以所有的样品实际上具有相同的固化度，第二次升温中的玻璃化转变温度几乎是相同的。与第一次测试相比，玻璃化转变温度只显示轻微的增加，这说明原来状态下是高度固化的。

could be detected by DSC. Samples that had undergone a larger number of aging cycles did however exhibit enthalpy relaxation. This is typical for resin molding compounds that spend longer times at temperatures just below the glass transition temperature.

The 2nd heating runs are not shown here. Since the samples were all heated to 200℃ in the first run, all the samples have practically the same degree of cure. The glass transition temperatures in the 2nd run are practically the same. Compared to the 1st run, the glass transition temperatures show only a slight increase, which indicates that the degree of cure in the original state was high.

计算
Evaluation

表 4.6　第一次升温测试。
Table 4.6　1st heating run.

老化[循环次数] Aging [number of cycles]	0	100	200	400	700	1.000
玻璃化转变温度,℃ Glass transition temperature in ℃	138	135	134	131	134	136
Δc_p, J/gK Δc_p in J/gK	0.13	0.12	0.13	0.13	0.13	0.13

表 4.7　第二次升温测试。
Table 4.7　2nd heating run.

老化[循环次数] Aging [number of cycles]	0	100	200	400	700	1.000
玻璃化转变温,℃ Glass transition temperature in ℃	143	141	140	139	140	140
Δc_p, J/gK Δc_p in J/gK	0.15	0.14	0.15	0.14	0.14	0.14

结论　用DSC测定玻璃化转变温度可获得关于老化现象的信息。在本例中，印制线路板满足要求，因为高达1000次循环没有检测到老化效应。

Conclusions　The determination of glass transition temperatures by DSC allows information to be obtained about aging phenomena. In this case, the printed circuit boards satisfy the requirements because no aging effects could be detected up to 1000 cycles.

4.5.5 分解产物的 TGA-MS 分析
Analysis of decomposition products by TGA-MS

质谱(MS)能对在 TGA 中进行分解的样品逸出的挥发性分解产物进行定性和定量分析。具有足够高蒸汽压的化合物从 TGA 进入质谱仪,被离子化,最终离子化的碎片离子按其质荷比 m/z 被分析。与红外光谱法不同,MS 也能检测单原子或对称双原子气体如氮气或氯气。而且它能区别原子的同位素,例如溴的同位素。

MS 通过一根加热石英毛细管与 TGA 连接,毛细管的开口靠近 TGA 中测试的样品。在 TGA 测试过程中,在选择的质量范围内以快速的时间间隔记录质谱,或者以选定的 m/z 值连续记录信号强度。

A mass spectrometer (MS) allows volatile decomposition products evolved from a sample undergoing decomposition in a TGA to be qualitatively and quantitatively analyzed. Compounds with a sufficiently high vapor pressure pass from the TGA into the MS where they are ionized and the resulting ionized fragment ions analyzed based on their mass-to-charge ratio, m/z. In contrast to infrared spectroscopy, MS can also detect monatomic or symmetrical diatomic gases such as nitrogen or chlorine. It can furthermore distinguish between istopes of atoms, e. g. of bromine.

The MS is connected to the TGA via a heated quartz capillary whose opening is mounted close to the sample under investigation in the TGA. During the TGA measurement, mass spectra are recorded at rapid intervals over a selected mass range or the signal intensity continuously recorded at selected m/z values.

目的	说明当将样品升温并进行分解时,连接在线 MS 分析的 TGA 测试如何能快速鉴定从 PCB 板逸出的气体的性质。	
Purpose	To show how a TGA measurement with online MS analysis can rapidly identify the nature of gases evolved from a PCB when a sample is heated and undergoes decomposition.	
样品	印制线路板 FR4。	
Sample	Printed circuit board, FR4.	
条件	测试仪器:	**Measuring cell**:
	耦联 Pfeiffer Thermostar™ 质谱的 TGA。	TGA coupled to a Pfeiffer Thermostar™ mass spectrometer
Conditions	坩埚:70μL 氧化铝坩埚,无盖	**Pan**: Alumina 70μL, without lid
	样品制备:从板上锯下重约 30mg 的试片。	**Sample preparation**: Pieces weighing about 30mg were sawn out of the board.
	TGA 测试:	**TGA measurement**:
	以 10 K/min 从 35℃ 升温至 260℃	Heating from 35℃ to 260℃ at 10K/min
	气氛:氩气,60mL/min	**Atmosphere**: Argon, 60mL/min

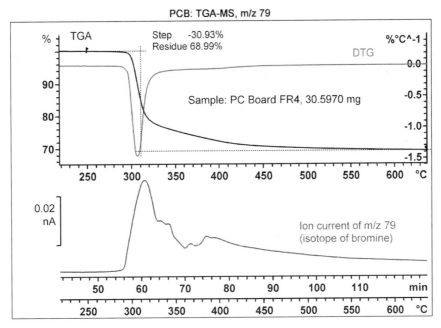

图 4.56　Fig 4.56

解释　图的上部表示 TGA 和 DTG 曲线。

逸出气体的质谱显示质荷比 m/z 为 79、81、94 和 96 的碎片。这些可容易地归属于溴的两个同位素 79（自然界丰度 50.5%）和 81（约 49.5%）和甲基溴。如果所关注的碎片的质荷比已经从试验性实验或从文献了解，则它们可被连续地测试，并与 TGA 曲线一起作图，如这里下部的图所示。

m/z 79 离子流曲线表明，含溴分解产物（例如溴化氢、甲基溴等）在第一个分解台阶逸出。

计算　氩气中的热解表明，至 650℃ 31% 的失重和主要由于玻纤增强物产生的 69% 剩余物。

结论　在产品开发、失效分析和对竞争性产品的研究中，需要全面的分析。对热稳定性和分解产物性质的分析数据，包括在相对低的温度（低于 300℃）下逸出的气体的分析就是一个实例。同时，印制线路板的纤维含量也可测定。

Interpretation　The upper diagram shows the TGA and DTG curves.

The mass spectra of the gases evolved exhibit fragments with mass-to-charge ratios, m/z, of 79, 81, 94 and 96. These can easily be assigned to the two isotopes of bromine 79 (natural abundance 50.5%) and 81 (about 49.5%) and methyl bromide. If the mass-to-charge ratios of fragments of interest are already known from trial experiments or from the literature, they can be continuously measured and plotted together with the TGA curves, as shown here in the lower diagram.

The m/z 79 ion current curve shows that decomposition products containing bromine (e.g. hydrogen bromide, methyl bromide, etc.) are evolved in the first decomposition step.

Evalclusion　The pyrolysis in argon shows a mass loss of 31% up to 650℃ and a residue of 69% that is mainly due to the glass-fiber reinforcement.

Conuation　Comprehensive analyses are needed in product development, failure analysis and for the investigation of competitive products. Analytical data on thermal stability and the nature of decomposition products including gases that are evolved at relatively low temperatures (below 300℃) is one such example. At the same time the fiber content of the printed circuit board can be determined.

当在较高的温度暴露较长的时间,例如在焊接槽中、操作期长或万一着火,印制线路板的基体树脂会分解,形成有毒的和有害环境的气体。

The matrix resin of a printed circuit board decomposes when it is exposed to higher temperatures over a longer period, e. g. in a soldering bath, during long operational periods, or in case of fire. Toxic and environmentally hazardous gases are formed.

4.5.6 印制线路板分层的 TMA-EGA 测量
Delamination of printed circuit boards by TMA-EGA

目的	TMA 是一种非常灵敏的方法,可用来测定最初的不可逆变化(分层)的温度,同时鉴定逸出气体的性质。	
	如果印制线路板暴露在过热下,例如在焊接槽中,板层有分离的危险(分层)。这是几乎看不见的,但能破坏电路连接。如果热应力太高,则会继续分解,此时引起进一步破坏,有气体逸出。	
Purpose	To determine the temperature of the first irreversible change (delamination) using TMA as a very sensitive method and at the same time identify the nature of the gases evolved.	
	If printed circuit boards are exposed to excessive heat, e. g. in a soldering bath, there is the risk that the layers of the board separate (delaminate). This is hardly visible but it can destroy the electrical connections. If the thermal stress is too high, decomposition can continue, in which case gases are evolved that cause further damage.	
样品	由含环氧基体树脂的玻璃纤维和阻燃剂制造的印制线路板。	
Sample	Printed circuit board made from glass fiber with an epoxy matrix resin and flame retardant.	

条件 / **Conditions**

测试仪器:耦联 Pfeiffer Thermostar™ 质谱仪的 TMA。

Measuring cell: TMA coupled to a Pfeiffer Thermostar™ mass spectrometer

样品制备:重约 57mg 直径 4mm 的圆片放在直径 6mm 厚 0.5mm 的石英圆片上。

Sample preparation: Disks of 4 mm diameter weighing approx. 57mg were placed on a fused silica disk, 6mm in diameter and 0.5mm thick.

TMA 测试:
3mm 球点探头直接放在样品上。施加给样品的负载为 0.05N。样品首先加热到 100℃ 以消除由样品制备或预处理产生的"记忆效应"。然后以 20K/min 将样品从 30℃ 升温至 650℃。

TMA measurement: The 3mm ball point probe was positioned directly on the sample. The load applied to the sample was 0.05 N. The samples were first heated to 100℃ to remove any "memory effects" resulting from sample preparation or pretreatment. The sample was then heated from 30℃ to 650℃ at 20K/min.

气氛:氮气,10mL/min

Atmosphere: Nitrogen, 10mL/min

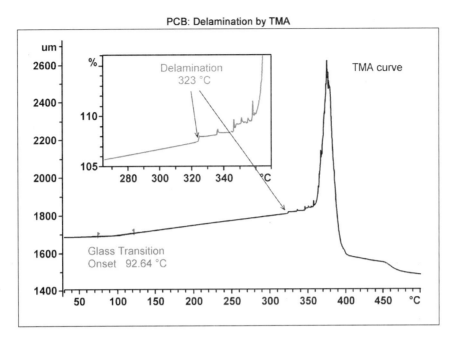

图 4.57　Fig. 4.57

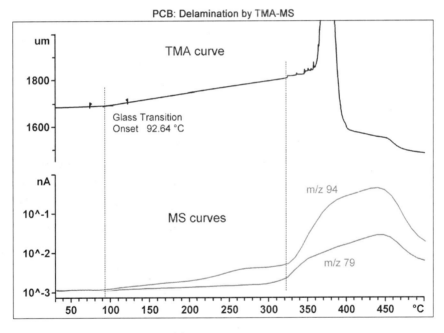

图 4.58　Fig. 4.58

解释　图 4.57：TMA 曲线记录高至 500℃时 PCB 板的尺寸变化。在 92℃处 TMA 曲线的斜率变化等于样品的玻璃化转变。323℃以上突然的尺寸变化是由于样品的分层，如放大曲线（红色）更详细的显示。

图 4.58：m/z 79 和 m/z 94 离子的 MS 信号强度是溴和甲基溴的特征。在玻璃化转变以上检测到含溴

Interpretation　Figure 4.57：The TMA curve records the dimensional changes of the PCB up to 500℃. The change in the slope of the TMA curve at 92℃ corresponds to the glass transition of the sample. The sudden dimensional changes above 323℃ are due to delamination of the sample, as the zoomed curve (red) shows in more detail.

Figure 4.58：The MS signal intensities of the m/z 79 and m/z 94 ions are characteristic of bromine and methyl bromide. Products containing bromine were detected from the glass transition

的产物。这些产物的浓度稳定地增加,然后在 200℃ 以上更为显著。最后,约 300℃ 以上,分层和全部分解开始,甚至伴随更多的溴化产物的形成。

temperature onward. The concentration of these products increases steadily and then more markedly above 200℃. Finally, above about 300℃, delamination and full degradation begins with the formation of even more brominated products.

结论 样品尺寸的变化伴随着微量气体物质的释放。在印制线路板软化至分层过程中,清晰地鉴别出含溴化合物。得到的结果与测得的尺寸变化相关。
虽然很少使用,但是 TMA 和 MS 的联用对于同步定量和定性表征分层过程确是一种有用的技术。

Conclusion Changes in the dimensions of the sample were accompanied by the release of traces of gaseous substances. Bromine-containing compounds were clearly identified during the softening and delamination of a printed circuit board. The results obtained were correlated with the measured dimensional changes.
Although seldom used, the combination of TMA with MS is a useful technique for the simultaneous quantitative and qualitative characterization of delamination processes.

4.5.7 印制线路板分层时间按照 IPC-TM-650 的 TMA 测定
Time to delamination of printed circuit board by TMA according to IPC-TM-650

目的	IPC-TM-650 No. 2.4.24.1 测试方法描述了测定层压板和印制线路板分层发生时间的 TMA 方法。本实验说明如何进行旨在质量控制的测定。
Purpose	The IPC-TM-650 No. 2.4.24.1 test method describes a TMA method for the determination of the time up until which delamination of laminates and printed circuit board occurs. The experiment shows how such a determination is performed, e. g. for quality control purposes.
样品	无铜导线的印制线路板(含环氧基体的玻璃纤维织物)。
Sample	Printed circuit board without copper tracks (glass-fiber fabric with epoxy matrix).
条件	测试仪器:TMA,球点探头
Conditions	样品制备:用小锯从提供的材料上锯下样品。样品在 105℃ 干燥 2h,然后贮存在干燥器中。 TMA 测试:以 20K/min 从 35℃ 升温至 260℃ 或 288℃,接着在结束温度恒温 15min。 负载:0.2N 气氛:静态空气
	Measuring cell:TMA with ball-point probe **Sample preparation**:Samples were cut from the material supplied using a small saw. The samples were dried at 105℃ for 2h and then stored in a desiccator. **TMA measurement**:Heating at 20K/min from 35℃ to 260℃ or 288℃ followed by a 15min isothermal period at the end temperature. Load:0.2N **Atmosphere**:Static air

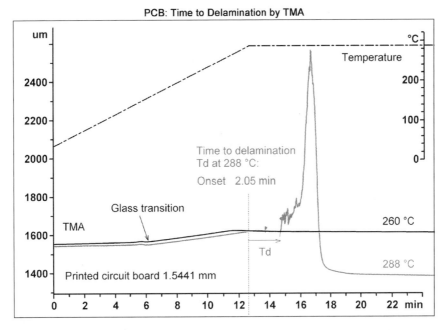

图 4.59　Fig. 4.59

解释　图示为对两个不同结束温度测定分层时间的 TMA 测试曲线。对高至 288℃测试的温度曲线以点划线表示。垂直的红点直线标示测试中从升温到等温的切换，以 288℃为结束温度。

在 260℃ 15min 的等温期间（黑色 TMA 曲线），没有观察到分层过程。因此对第二个样品测试提高了等温温度。

在 288℃，不可逆效应发生在约 2.0min 后（红色 TMA 曲线）。这不可能由于水分，因为样品在制备时已干燥过。甚至当 0.2N 的相对大的力施加到样品末端上时，仍能检测到分层。通常用 0.005N 的力。

两次升温都可用来测定玻璃化转变温度，约为 136℃。松弛效应干扰了测定。

计算　在 IPC-TM-650 No. 2.4.24.1 测试方法中，分层时间测定为等温过程的开始到破坏的时间。破坏表现为 TMA 厚度信号产生不可逆变化的任何效应。这里为在 288℃下厚度突然增加的起始点，时

Interpretation　The figure shows the TMA measuring curves for the determination of the time to delamination for the two different end temperatures. The temperature profile for the measurement up to 288℃ is shown by the dash-dot curve. The vertical red dotted line indicates the switch-over from heating to isothermal in the measurement with an end temperature of 288℃.

No delamination processes were observed during the 15min isothermal period at 260℃ (black TMA curve). The isothermal temperature was therefore increased for a second sample measurement.

At 288℃, an irreversible event occurs after about 2.0min (red TMA curve). This cannot be due to moisture because the sample was dried during sample preparation. The delamination was detected even when a relatively large force of 0.2N was applied to the sample tip. Normally a force of 0.005N is used.

Both heating runs can be used to determine the glass transition temperature, which was found to be about 136℃. A relaxation effect interferes with the determination.

Evaluation　In the IPC-TM-650 No. 2.4.24.1 test method, the time to delamination is determined as the time from the beginning of the isothermal period to failure. Failure is any event which causes an irreversible change in the TMA thickness signal. The time determined here as the onset of the sudden increase of thickness at 288℃ is 2.0min.

间测定为 2.0 min。

结论　IPC-TM-650 测试方法是测定分层时间的实用标准方法,容易应用,因此可用于质量控制。对于稳定的基体树脂,必须使用所推荐的较高等温温度。

Conclusion　The IPC-TM-650 test method is a practical standard procedure for the determination of the delamination time. It is easy to apply and can therefore be used for quality control. With stable matrix resins, the recommended higher isothermal temperature must be used.

4.5.8　质量保证,黏结层的失效分析
Quality assurance, failure analysis of adhesive bonds

目的　本实验的目的是研究金属件之间的黏结层是否已充分固化。这可用 DSC 测试容易地完成。测试一个好的和一个差的黏结层以作比较。

DSC 具有测试只用少量材料的优势。因此从黏接层上取下少量黏合剂样品后就能分析有缺陷的黏接层。

一种双组分环氧树脂被用来黏接两块金属件。当使用这些制件时,很清楚一些黏接层在稍高于室温下失效。

Purpose　The purpose of the experiment was to investigate whether the adhesive bonds between metal parts had been sufficiently cured. This can be easily done using DSC measurements. For comparison, a good and a bad bond were measured.

DSC has the advantage that only small amounts of material are used for measurements. Faulty adhesive bonds can therefore be analyzed afterward by taking a small sample of the adhesive from the bond.

A two-component epoxy resin was used to bond two metal parts. When the parts were used, it became clear that some of the bonds failed at slightly elevated room temperatures.

样品　来自两个不同黏接层的环氧树脂黏合层(双组分黏合剂);在生产中组分被混合和固化。

黏接层 1:好,满足要求

黏接层 2:差

Sample　EP resin adhesive layer (two-component adhesive) from two different adhesive bonds; components were mixed and cured in production.

Adhesive bond 1: good, satisfies requirements

Adhesive bond 2: bad

条件　测试仪器:DSC
Conditions

坩埚:40 μL 铝坩埚,敞口

样品制备:

用小刀从两个开口黏接层切下重约 3mg 的样品(层 1 和层 2)。

DSC 测试:第一次:以 20K/min 从 −30℃升温至 160℃,在 160℃下恒温 1min,接着以 20K/min 降温到 −30℃,在 −30℃下恒温 1min,然后第二次:以 10K/min 从 −30℃升温

Measuring cell: DSC

Pan: Aluminum 40 μL, open

Sample preparation: Samples weighing approx. 3mg (Layers 1 and 2) were taken from the two adhesive bonds using a small knife.

DSC measurement: 1st run: Heating from −30℃ to 160℃ at 20 K/min, 1 min at 160℃ followed by cooling to −30℃ at 20K/min, 1 min at −30℃, then 2nd run: Heating from −30℃ to 250℃ at 10K/min.

至 250℃。

气氛：干燥空气，50mL/min   **Atmosphere**：Dry air，50mL/min

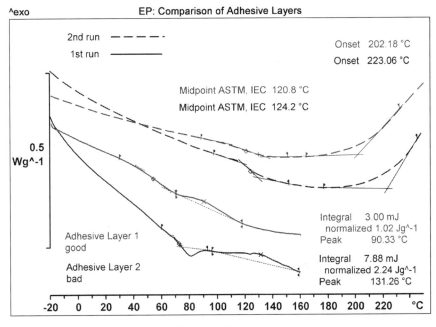

图 4.60　Fig. 4.60

解释　两个黏合层都没有完全固化，在第一次升温中呈现放热效应。而且，层 2 显示在玻璃化转变区域呈现焓松弛。当固化发生在略低于 T_g 下长时间而不是在更高温度下短时间时，就可以观察到这个焓松弛。

在第二次测试中，两个黏合层均呈现明显更高但不同的玻璃化转变温度以及不同的分解起始值。两个结果间接地表明，两个黏合层中使用了不同的树脂-硬化剂比。比较第一次测试的与第二次测试的玻璃化转变温度表明，层 1 不如层 2 固化得好。

Interpretation　Both adhesive layers were not fully cured and exhibit exothermic effects in the 1st heating run. Furthermore, Layer 2 shows enthalpy relaxation in the glass transition region. This enthalpy relaxation is observed when curing takes place over a long period slightly below T_g instead for a short time at a higher temperature.

In the 2nd runs, both adhesives layers exhibit significantly higher but different glass transition temperatures as well as different onset values for decomposition. Both results suggest that different resin-hardener ratios were used in the two adhesive layers. Comparison of the glass transition temperatures from the 1st run with those of the 2nd run indicate that Layer 1 was less-well cured than Layer 2.

计算
Evaluation

表 4.8 第一次升温测试。
Table 4.8 1st heating run.

	玻璃化转变 Glass transition		后固化 Postcuring
	中点温度,℃ Midpoint temperature in ℃	焓 Enthalpy J/g	峰,℃ Peak in ℃
黏合层 1(好) Adhesive Layer 1 (good)	52	1.0	90
黏合层 2(差) Adhesive Layer 2 (bad)	73	2.2	131

表 4.9 第二次升温测试。
Table 4.9 2nd heating run.

	玻璃化转变 Glass transition	分解 Decomposition
	中点温度,℃ Midpoint temperature in ℃	起始点,℃ Onset in ℃
黏合层 1(好) Adhesive Layer 1 (good)	121	202
黏合层 2(差) Adhesive Layer 2 (bad)	124	223

结论 同一样品的第一次和第二次动态 DSC 测试对于黏合层的比较研究是有用的,可得到关于体系的玻璃化转变温度、后固化和树脂/硬化剂比的信息。

Conclusions First and second dynamic DSC measurements of the same sample are useful for comparative investigations of adhesive layers and allow information to be obtained about the glass transition temperature, postcuring and the resin/hardener ratio of the system.

4.5.9 油与增强环氧树脂管的相互作用
Interaction of oil with a reinforced EP resin pipe

目的 说明 DSC 测试如何能表明油是否起了增塑剂的作用和树脂是否在其他方面发生了变化。如果玻璃纤维增强环氧树脂管用于输送液体,那么必须研究它们之间可能的相互作用。液体能够扩散进入基体,并可能起增塑剂的作用。增塑剂降低玻璃化转变温度,并且树脂会溶涨。取决于使用的温度,这会降低管道的稳定性,因而可能产生破坏。

Purpose To demonstrate how DSC measurements can show whether an oil acts as a plasticizer and whether the resin has otherwise changed.
If glass-fiber reinforced epoxy resin pipes are used to transport liquids, their possible interaction must be investigated. The liquids can diffuse into the matrix and possibly

act as a plasticizer. Plasticizers lower the glass transition temperature and the resin can swell. Depending on the temperature of use, this can reduce the stability of piping and so give rise to possible damage.

样品	EP-GF,由玻纤增强环氧树脂制造的复合管(纤维含量约72%)。在使用中,管道充满了油。	
Sample	EP-GF, laminated piping made of glass-fiber reinforced EP resin (fiber content about 72%). In operation, the pipe was filled with oil.	
条件 Conditions	测试仪器:DSC 坩埚:40μL 铝坩埚,盖钻孔 样品制备: 用锯切割管子,从内表面和管壁中部切下样品。考虑到玻璃纤维含量高,样品用量相对大。 DSC 测试:第一次:以 20K/min 从 25℃升温至 190℃,在 190℃下恒温 2min,接着以 20K/min 降温至 25℃,然后在 25℃下恒温 2min。 第二次:以 20K/min 从 25℃升温至 350℃。 气氛:干燥空气,50mL/min	Measuring cell:DSC Pan:Aluminum 40μL, with pierced lid Sample preparation:The pipe was cut through with a saw and a sample cut out from the inside surface and from the middle part of the wall of the pipe. The sample mass used was relatively large to take into account the high glass fiber content. DSC measurement:1st run: Heating from 25℃ to 190℃ at 20K/min, 2min at 190℃, followed by cooling to 25℃ at 20K/min, then 2min at 25℃. 2nd run: Heating from 25℃ to 350℃ at 20K/min. Atmosphere:Dry air, 50mL/min

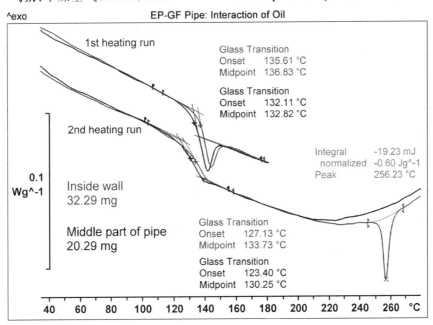

图 4.61　Fig 4.61

解释 在每个样品的第一次升温中,玻璃化转变区域显示由于样品热历史的原因产生的清晰的焓松弛峰。在第二次升温中,松弛峰不再存在,这是由于用了相同的降温(第	Interpretation The glass transition region in the 1st run of each sample shows a clear enthalpy relaxation peak due to the thermal history of the sample. In the 2nd run, the relaxation peaks are no longer present. This is a result of using the same cooling (1st run) and heating rates (2nd run).

一次测试)和升温速率(第二次测试)的结果。

玻璃化转变温度在第一和第二次测试中是相同的,所以固化看来是完全的。此外,管壁内表面和管中部的曲线之间只检测到非常小的差别。

然而,取自管壁内表面的样品大约250℃至260℃处,出现一个明显的吸热峰。此峰在取自管壁中部的样品的曲线上不出现。显然,无疑这是一个与扩散进了管壁内表面的油的低分子量组分的挥发有关的表面现象。峰面积是相对小的,用油的汽化焓约600J/g,求得含量约为0.1%。

管壁中部样品树脂开始放热分解的温度比内表面样品的略低。

The glass transition temperatures are identical in the 1st and 2nd runs so that curing seems to be complete. In addition, only very minor differences can be detected between the curves of the inside surface of the wall and the middle part of the pipe.

An endothermic peak at about 250 to 260℃ of the sample taken from the inside surface of the wall of the pipe is however very noticeable. This peak does not occur in the curve of the sample taken from the middle part of the wall of the pipe. Apparently it is a surface phenomenon that no doubt has to do with the vaporization of low molecular mass constituents of the oil that diffused into the inside surface of the wall of the pipe. The peak area is relatively small and corresponds to a content of about 0.1% using a enthalpy of vaporization of oil of about 600J/g.

The exothermic decomposition of the resin begins at a somewhat lower temperature for the sample from the middle part of the wall of the pipe than for the sample from the inside surface.

计算 Evaluation

管壁 Pipe wall	内表面 Inside surface	中部 Middle
玻璃化转变的中点温度 Midpoint temperature of the glass transition		
第一次测试,℃ 1st run in℃	137	133
第二次测试,℃ 2nd run in℃	134	130
第二次测试的吸热峰,℃ Endothermic peak in 2nd run in℃	256	—

结论 与接触聚合物的液体的扩散有关的相互作用现象,可以用取自紧靠表面和更深位置的样品进行分析。

本特例测试表明,油不起增塑剂作用,所以在这方面材料满足要求。

Conclusions Interaction phenomena that have to do with the diffusion of liquids in contact with the polymer can be analyzed by taking samples from positions close to the surface and deeper down.

In this particular case, the measurements showed that the oil did not act as a plasticizer so that in this respect the material satisfies the requirements.

5 不饱和聚酯树脂 Unsaturated polyester resins

5.1 进货控制：固化特性和玻璃化转变
Incoming goods control: curing characteristics and glass transition

目的 说明 DSC 如何能快速测定树脂体系是否满足要求。
加工前检查固化特性对复杂树脂体系是特别重要的。这能快速鉴别与通常的反应进程的偏差，并可避免一些意外的发生。

Purpose To show how DSC measurements can quickly determine whether a resin system satisfies requirements.
It is particularly important with complex resin systems to check the curing characteristics before processing. This enables deviations from the usual reaction course to be rapidly identified and allows unpleasant surprises to be avoided.

样品 不饱和聚酯，适用于玻纤增强塑料管的四组分树脂。
Sample UP, four-component resin for glass-fiber reinforced plastic pipes.

条件 测试仪器：DSC
Conditions 坩埚：40μL 铝坩埚，盖钻孔
样品制备：将树脂体系混合，坩埚中放入 25mg 的树脂混合物。
DSC 测试：第一次：25℃恒温 2min，以 15K/min 从 25℃升温至 240℃，240℃恒温 2min，接着以 15K/min 降温至 25℃，25℃恒温 2min，然后第二次：以 15K/min 从 25 升温至 240℃
气氛：干燥空气，50mL/min

Measuring cell: DSC
Pan: Aluminum 40μL, with pierced lid
Sample preparation: The resin system was mixed and the crucibles filled with 25mg of the resin mixture.
DSC measurement: 1st run: 2min at 25℃, heating from 25℃ to 240℃ at 15K/min, 2min at 240℃ followed by cooling to 25℃ at 15K/min, 2min at 25℃ then 2nd run: Heating from 25 to auf 240℃ at 15K/min
Atmosphere: Dry air, 50mL/min

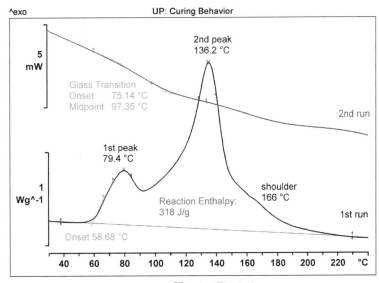

图 5.1　Fig. 5.1

解释 第一次升温曲线显示固化反应,第二次升温曲线显示完全固化树脂的玻璃化转变区域。固化曲线呈现两个具有特征峰温的峰。这是由于含有树脂、硬化剂、抑制剂和促进剂的四组分体系的复杂配方产生的。

Interpretation The first heating curve shows the curing reaction and the second curve the glass transition region of the completely cured resin. The curing curve shows two peaks with characteristic peak temperatures. This is due to the complex formulation of the four-component system containing resin, hardener, inhibitor and accelerator.

计算 总反应焓为 318.0J/g,反应的起始点为 59℃。第一个峰的最大值出现在 79℃,第二个的在 136℃。在 166℃ 可以看到一个肩。固化体系的第二次测试的玻璃化转变温度在 97℃。

Evaluation The total reaction enthalpy is 318.0J/g. The onset of the reaction is at 59℃. The first peak maximum occurs at 79℃, the second at 136℃. A shoulder is visible at 166℃.

The 2nd run of the cured system shows a glass transition temperature at 97℃.

结论 第一次升温表征不饱和聚酯树脂体系的固化反应。完全固化树脂的玻璃化转变温度可在第二次升温测定。为了避免树脂可能的分解,第一次升温的最终温度不可太高。峰温、反应焓和玻璃化转变温度可在相同的条件下作为参考值进行比较。而且,曲线的形状应该相同。

Conclusions The 1st heating run characterizes the curing reaction of the UP resin system. The glass transition of the fully cured resin can then be determined in the 2nd run. The final temperature must not be too high in the first heating run in order to avoid possible decomposition of the resin. The peak temperatures, reaction enthalpy and glass transition temperature can be compared under identical conditions as reference values. Furthermore, the shape of the curves should be the same.

5.2 不饱和聚酯:促进剂含量的影响
UP: Influence of the accelerator content

目的 用一系列 DSC 测试研究促进剂含量对固化反应的影响。
促进剂的浓度对固化反应的进程有较大的影响。在多组分体系的开发中,必须找到获得最佳固化度的促进剂含量。

Purpose To investigate the influence of the accelerator content on the curing reaction using a series of DSC measurements.
The concentration of the accelerator has a large influence on the course of the curing reaction. In the development of a multicomponent system, the accelerator content must be found that gives the optimum degree of cure.

样品 不饱和聚酯,适用于精细涂层(胶衣、面漆)的三组分树脂。
Sample UP, three-component resin for fine coatings (Gelcoats, Topcoats).

条件 测试仪器:DSC
Conditions 坩埚:40μL 铝坩埚,盖钻孔
样品制备:
树脂体系与不同量的促进剂混合,将 19 到 22mg 的树脂混合物放入坩埚,添加促进剂后 9min 左右开

Measuring cell:DSC
Pan:Aluminum 40μL, with pierced lid
Sample preparation:The resin system was mixed with different amounts of accelerator, the crucibles filled with 19 to 22 mg of the resin mixture, and the measurement started about 9

始测试。
DSC 测试：
以 10K/min 从 25℃升温至 280℃
气氛：干燥空气，50mL/min

min after adding the accelerator.
DSC measurement：
Heating from 25℃ to 280℃ at 10 K/min
Atmosphere：Dry air，50mL/min

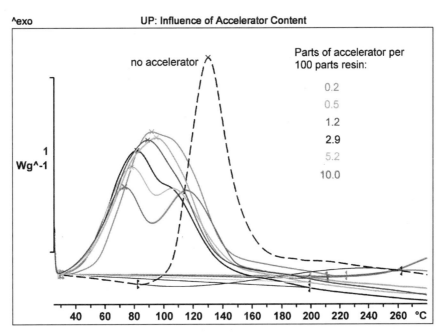

图 5.2　Fig. 5.2

解释　促进剂含量越高，反应开始得越早。一些曲线呈现肩和双峰，这意味着反应是复杂的，分几步进行。正如预计的，随着促进剂含量增加，第一个反应峰向低温移动。同时，反应焓不断降低。
曲线的变化表明促进剂含量是如何改变反应的进程和反应机理的。

Interpretation　The higher the accelerator content, the sooner the reaction starts. Some of the curves exhibit shoulders and double peaks. This means that the reaction is complex and proceeds in several steps. As expected, the first reaction peak shifts to lower temperature with increasing accelerator content. At the same time, the reaction enthalpies continually decrease. The changes in the curves show how the accelerator content changes the course of the reaction and the reaction mechanism.

计算
Evaluation

促进剂 [phr] Accelerator [parts per 100 parts resin]	0	0.2	0.5	1.2	2.9	5.2	10.0
第一个峰，℃ 1st peak ℃	131	92	95	90	82	79	74
第二个峰，℃ 2nd peak in ℃	—	—	—	—	—	106	116
反应焓，J/g Reaction enthalpy in J/g	289	318	283	279	266	254	225

结论　用 DSC 表征多组分不饱和聚酯树脂的固化可得出关于配方

Conclusions　The characterization of curing of multicomponent UP resins by DSC allows conclusions to be drawn about their

的结论。使用最佳的促进剂浓度，可使树脂满足生产过程的需要。DSC 测试可用作生产过程的附加质量保证标准。

formulation. The resin can be adapted to the production processes using optimum accelerator concentration. The DSC measurements can be used as an additional quality assurance criterion for the production process.

5.3 不饱和聚酯：硬化剂含量的影响
UP: Influence of the hardener content

目的 用一系列 DSC 测试说明硬化剂含量对反应进程的影响。

过氧化物这样的硬化剂的浓度影响反应的进程，因而能控制生产条件、优化产品性能。因此对于多组分体系，每一个反应成分的浓度必须单独设定。DSC 能特别好地测试每一种组分对反应动力学的影响，因此能快速确定最佳条件。如同上例，用一系列曲线来说明这些。

Purpose To demonstrate the influence of the hardener content on the course of the reaction using a series of DSC measurements.

The concentration of hardeners such as peroxides influences the course of the reaction and so enables production conditions to be controlled and product properties optimized. With multicomponent systems, the concentration of each reaction partner must therefore be individually set. DSC can measure the effect of each component on the reaction kinetics especially well and can therefore quickly determine the best conditions. As in the previous example, this is shown using a series of curves.

样品 不饱和聚酯，适用于精细涂层（胶衣、面漆）的三组分树脂。
Sample UP, three-component resin for fine coatings (Gelcoats, Topcoats).

条件 测试仪器：DSC
Conditions 坩埚：40μL 铝坩埚，盖钻孔
样品制备：树脂体系与不同量的硬化剂混合，将 19 到 23mg 的树脂混合物放入坩埚，添加硬化剂后 9min 左右开始测试。
DSC 测试：以 10K/min 从 25℃ 升温至 280℃
气氛：干燥空气，50mL/min

Measuring cell: DSC
Pan: Aluminum 40μL with pierced lid
Sample preparation: The resin system was mixed with different amounts of hardener, the crucibles filled with 19 to 23mg of the resin mixture, and the measurement started about 9 min after adding the hardener.
DSC measurement: Heating from 25℃ to 280℃ at 10K/min
Atmosphere: Dry air, 50mL/min

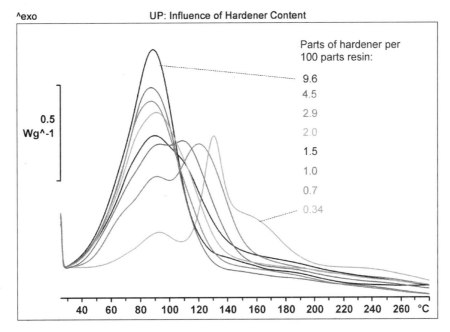

图 5.3　Fig. 5.3

解释　硬化剂含量越高，第一个峰越大，即在90℃下的反应随着硬化剂含量的增加而加快进行。对于低硬化剂含量，曲线呈现双峰。第二个峰随着硬化剂含量的增加向低温移动。

曲线的形状不同表明，反应机理和网状结构是不同的。

Interpretation　The higher the hardener content, the larger the first peak, that is, the reaction at 90℃ proceeds faster with increasing hardener concentration. With low hardener contents, the curves exhibit two peaks. The second reaction peak shifts to lower temperatures with increasing hardener content.

The very different curve shapes indicate that the reaction mechanisms and network structures are different.

计算
Evaluation

硬化剂 [phr] Hardener [parts per 100 parts resin]	0.34	0.7	1.0	1.5	2.0	2.9	4.5	9.6
第一个峰，℃ 1st peak in ℃	93	91	94	90	91	88	88	89
第二个峰，℃ 2nd peak in ℃	131	121	110	—	—	—	—	—

结论　DSC测试表征多组分不饱和聚酯树脂的固化过程。这能得出关于最佳硬化剂量的结论。通过使用最优化的硬化剂浓度也能改进网状结构，从而使性能得到改善。在多组分体系中，当然也必须研究其他成分的影响。

Conclusions　The DSC measurements characterize the curing processes in multicomponent UP resins. This allows conclusions to be drawn about the optimal amount of hardener. Network structures and hence properties can also be improved by using the optimum hardener concentration. In multicomponent systems, the influence of other components must of course also be investigated.

5.4 抑制剂对等温固化的影响
Influence of the inhibitor on isothermal curing

目的	用一系列 DSC 测试说明抑制剂如何能延迟固化反应。 抑制剂,例如苯醌或更有效的 TBC(4-叔丁基邻苯二酚),确保在给定工艺条件下有足够长的适用期。无论是与温度的关系还是与这里讨论的浓度的关系,DSC 均是证实反应延迟的较适合的方法。
Purpose	To show how an inhibitor can retard a curing reaction using a series of DSC measurements. Inhibitors, e. g. quinone or the more effective TBC (4-tert-butylcatechol), ensure that pot-life is long enough for the given processing conditions. DSC is an excellent method to demonstrate the retardation of the reaction, whether as a function of temperature or, as discussed here, concentration.
样品	不饱和聚酯树脂,适用于玻璃纤维塑料管的四组分树脂
Sample	UP, four component resin for glass-fiber plastic pipes.
条件 Conditions	测试仪器:DSC **Measuring cell**: DSC 坩埚:40μL 铝坩埚,盖钻孔 **Pan**: Aluminum 40μL with pierced lid 样品制备: **Sample preparation**: The resin system was mixed with different amounts of inhibitor, the crucibles filled with 17 to 21mg resin mixture, and the measurement started about 9 min after adding the inhibitor. 树脂体系与不同量的抑制剂混合,将 17 到 21mg 的树脂混合物,放入坩埚添加抑制剂后 9min 左右开始测试。 DSC 测试:45℃下等温长达 600min **DSC measurement**: Isothermal at 45℃ for up to 600min 气氛:干燥空气,50mL/min **Atmosphere**: Dry air, 50mL/min

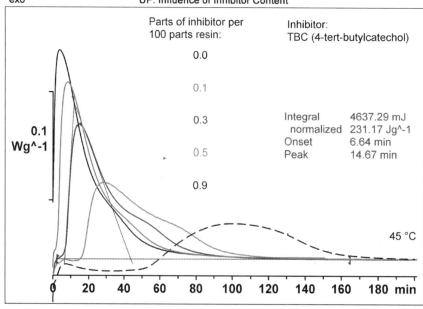

图 5.4   Fig. 5.4

| 解释 | 抑制剂含量越高,反应开始所需时间越长。峰最大值向更长的时间移动。随着抑制剂含量的增加,反应峰变宽、变低,并直至抑制剂含量每100份树脂0.5份之前都显示峰肩。在更高抑制剂浓度,肩消失。对于等温反应,不对称反应峰是典型的。每百份0.9份抑制剂的DSC曲线(虚线)显示一个吸热效应,这可能是由于等待期间反应开始前苯乙烯的汽化产生的。 | **Interpretation** The higher the inhibitor content, the longer it takes for the reaction to start. The peak maxima shift to longer times. The reaction peak becomes broader and lower with increasing inhibitor content and shows a shoulder up until an inhibitor content of 0.5 parts per 100 parts resin. The shoulder disappears at higher inhibitor concentrations. The asymmetrical reaction peak is typical for isothermal reactions. The DSC curve for 0.9 parts per hundred (phr) (dashed line) exhibits an endothermic effect. This could be due to the vaporization of styrene during the waiting time before the reaction starts. |

计算 如图中对0.3phr浓度的曲线所示,曲线的外推起始点计算为反应开始。

Evaluation The reaction begin was evaluated as the extrapolated onset as shown in the diagram for a concentration of 0.3 phr.

抑制剂 [份数/百份树脂] Inhibitor [parts per 100 parts resin]	0	0.1	0.3	0.5	0.9
起始点,min Onset in min	0	2	7	16	56
峰,min Peak in min	4	8	15	28	98

结论 等温DSC测试直接表明与不饱和聚酯树脂多组分体系配方有关的固化行为。这能用作树脂混合物的附加质量保证标准。

当比较快速冷却固化体系时,必须注意确保样品制备时间始终是相同的并尽可能缩短。

Conclusions Isothermal DSC measurements directly show the curing behavior which depends on the formulation of UP multicomponent systems. This can be used as an additional quality assurance criterion for the resin mixture.

When comparing fast cold curing systems, care must be taken to ensure that the sample preparation time is always the same and as short as possible.

5.5 不饱和聚酯:贮存后的固化行为
UP: Curing behavior after storage

目的 用一系列DSC测试说明树脂混合物贮存期间的固化。

活性液态树脂体系不能长期贮存。制备了一系列在一定温度贮存不同时间的样品来测定加工极限:剩余的固化反应越小,贮存期间体系已固化的越多。超过适用期时的转化率,即从该时起树脂不能再用了,与过程有关。通常,贮存期间只允许有低的固化度。

Purpose To show the curing of a resin mixture during storage using a series of DSC measurements.

Reactive liquid resin systems cannot be stored for long periods. A series of samples that had been stored at a certain temperature for different times was prepared to determine the processing limits: the smaller the residual postcuring reaction, the more the system has cured during storage. The conversion at which the pot life is exceeded, that is, from when on the resin can no longer be used depends on the process. In general, only a low degree of curing can be tolerated during storage.

样品 Sample	不饱和聚酯树脂,适用于玻璃纤维塑料管的四组分树脂。 UP, four-component resin for glass-fiber plastic pipes.
条件 Conditions	测试仪器:DSC / **Measuring cell**:DSC 坩埚:40μL 铝坩埚,盖钻孔 / **Pan**:Aluminum 40μL with pierced lid 样品制备: / **Sample preparation**:The resin system was mixed, the crucibles filled with 25 to 30mg of the resin system and stored at room temperature for different times (0, 1, 3, 5, 18 and 100h). 混合树脂,将 25 至 30mg 树脂体系装入坩埚,在室温贮存不同的时间(0、1、3、5、18 和 100h)。 DSC 测试:25℃ 恒温 2min,以 15K/min 从 25℃升温至 240℃ / **DSC measurement**:Isothermal at 25℃ for 2min, heating from 25℃ to 240℃ at 15 K/min 气氛:干燥空气,50mL/min / **Atmosphere**:Dry air,50mL/min

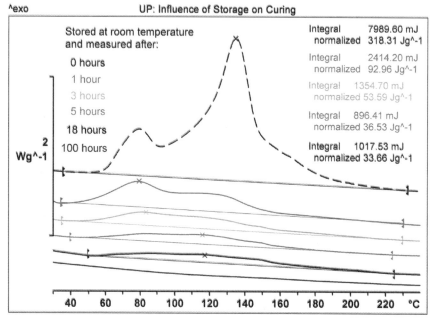

图 5.5　Fig. 5.5

解释 立即测试的样品的参比曲线(贮存时间 0h,黑色虚线)表示新制备的四组分树脂混合物的完全固化反应。室温下混合物贮存后,曲线形状显著变化。反应焓随着贮存时间的增加而下降。发生在贮存期间的转化率可以通过将树脂混合物的剩余(即后固化)反应焓与参比混合物的反应焓(立即测试的)进行比较而得到。结果表明,树脂混合物在室温下贮存100h后完全固化。

Interpretation The reference curve of the sample immediately measured (storage time 0 hours, black dashed line), shows the complete curing reaction of the freshly prepared four-component resin mixture. The curve shape changes significantly after storing the mixture at room temperature. The reaction enthalpies decrease with increasing storage time. The conversion that occurred during storage can be determined by comparing the residual (i. e. postcuring) reaction enthalpy of the resin mixtures with the reaction enthalpy of the reference mixture (immediately measured). The results show that the resin mixture is completely cured after a storage time of 100 h at room temperature.

计算
Evaluation

储存时间,h Storage time in h	反应焓,J/g Reaction enthalpy in J/g	转化率,% Conversion in %
0	318	0
1	93	71
3	54	83
5	37	88
18	34	89
100	不能计算 cannot be evaluated	> 95

结论 DSC 测试可用来测定贮存树脂配方的（剩余）反应焓和转化率。这为继续加工提供了关于不饱和聚酯树脂体系最长贮存时间的重要信息。在这里描述的实例中，最长贮存时间显然远小于 1h。

Conclusions DSC measurements can be used to determine the (residual) reaction enthalpy and the conversion of stored resin formulations. This provides important information for further processing with regard to the maximum storage times of UP resin systems. In the case described here, the maximum storage time was clearly much less than one hour.

5.6 乙烯基酯树脂：由促进剂引起的固化温度的移动
VE: Shift of curing temperature due to the accelerator

目的 说明不同量的促进剂对乙烯基酯树脂固化温度的影响。
促进剂浓度对固化温度影响大。这能优化生产条件,例如使树脂适应现有的生产设备。通过比较固化曲线也可在加工前检查新进树脂的组成。

Purpose To show the influence on the curing temperature caused by different amounts of accelerator in vinyl ester resins.
The accelerator concentration has a great influence on the curing temperature. This allows production conditions to be optimized, for example to adapt a resin to existing production equipment. A comparison of curing curves also allows the composition of incoming resin to be checked before processing.

样品 乙烯基酯树脂:适合于化工厂建筑用玻璃纤维增强塑料模塑制件的三组分树脂。
Sample VE, three-component resin for glass-fiber reinforced plastic molded parts for chemical processing plant construction.

条件 测试仪器：DSC
Conditions 坩埚：40μL 铝坩埚,盖钻孔
样品制备：树脂与不同量的促进剂混合,将 18 至 22mg 的树脂混合物装入坩埚。
DSC 测试：以 10K/min 从 25℃升温至 280℃
气氛：干燥空气,50mL/min

Measuring cell：DSC
Pan：Aluminum 40μL with pierced lid
Sample preparation：The resin was mixed with different amounts of accelerator and the crucibles filled with 18 to 22mg of the resin mixture.
DSC measurement：Heating from 25℃ to 280℃ at 10K/min
Atmosphere：Dry air, 50mL/min

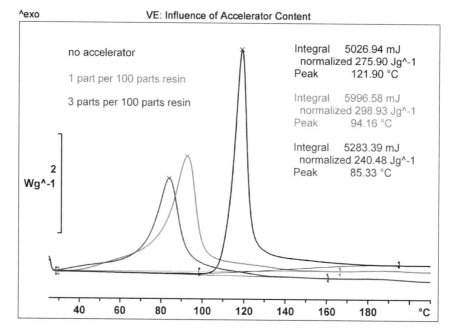

图 5.6　Fig. 5.6

解释　固化峰随着促进剂浓度的增加而向较低温度移动。DSC 峰变宽，由反应产生的热在 50℃ 已清晰可见。反应焓变化显著。这是因为反应在室温就已开始。也有可能是由于混合质量引起。

Interpretation　The curing peak shifts to lower temperature with increasing accelerator content. The DSC peaks become broader and the generation of heat through the reaction can already be clearly seen at 50℃. The reaction enthalpies vary noticeably. This is because the reaction has already begun at room temperature. It may possibly also be due to mixing quality.

计算
Evaluation

促进剂 [phr] Accelerator [parts per 100 parts resin]	0	1	3
峰温, ℃ Peak temperature in ℃	122	94	85

结论　用 DSC 升温测试表征多组分乙烯基酯树脂的固化过程可得出关于配方的结论。这能用作树脂混合物的附加质量保证标准。通过合适选择促进剂浓度能优化生产工艺。

Conclusions　The characterization of curing processes in multicomponent VE resins using DSC heating measurements allows conclusions to be drawn about the formulation. This can be used as an additional quality assurance criterion for the resin mixture. The production processes can be optimized through the right choice of the accelerator concentration.

5.7　乙烯基酯—玻璃纤维：使用后管材的固化度
VE-GF: Degree of cure of a pipe after use

目的　由乙烯基树脂制造的玻璃纤维增强管可能在使用中发生后固化，导致性能改变。本应用中能通过比较内外管壁的 DSC 实验进行研究。

Purpose　Glass-fiber reinforced pipes made of vinylester resin can undergo postcuring during use, resulting in a change of properties. The application shows how this can be investigated in a

DSC experiment in which the inside and outside walls of the pipe are compared.

样品 Sample	乙烯基酯－玻璃纤维：使用一段时间后的由玻璃纤维增强乙烯基酯树脂制造的复合管。 VE-GF, laminated piping made of glass-fiber reinforced VE resin after a period of use.	
条件 Conditions	测试仪器：DSC	Measuring cell：DSC
	坩埚：40μL 铝坩埚，盖钻孔	Pan：Aluminum 40μL with pierced lid
	样品制备：用钻石锯锯下一块乙烯基酯－玻璃纤维管材，然后从含丰富树脂的内外管表面用刀割出 6 至 10mg 重的样品。	Sample preparation：A piece was cut out of the VE-GF pipe using a diamond saw. Samples weighing 6 to 10 mg were then cut out of the resin-rich inside and outside surfaces of the pipe using a knife.
	DSC 测试：	DSC measurement：
	以 20K/min 从 25℃ 升温至 220℃	Heating from 25℃ to 220℃ at 20K/min
	气氛：干燥空气，50mL/min	Atmosphere：Dry air，50mL/min

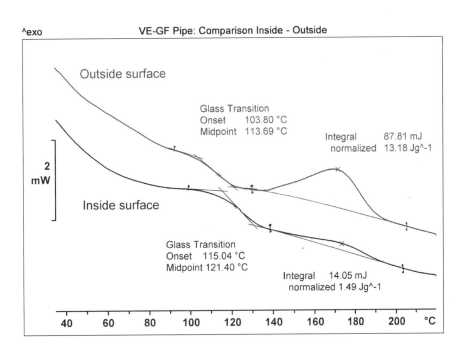

图 5.7 Fig. 5.7

解释 管里边的材料比外边的材料的固化要好得多。从玻璃化转变看这是显然的：里边样品的玻璃化转变温度明显高于外边样品。里边样品的后固化焓也明显小于外边的。对样品质量归一化的反应焓（"表观"焓）与真实的后固化反应并不一致，因为它参照的是整个样品质量而不只是聚合物含量。由于不知道样品的准确玻璃纤维含量，所以不能测定聚合物的后固化比焓。要得到绝对值，尚需对被测试的样品进行 TGA 分析。

Interpretation The inside material of the pipe is much better cured than the outside material. This is evident from the glass transitions: the glass transition temperature of the sample from the inside is significantly higher than that of the sample from the outside. The postcuring enthalpy of the sample from the inside is also appreciably smaller than that from the outside. The reaction enthalpies normalized to sample mass ("apparent" enthalpies) do not correspond to the actual postcuring reaction because it refers to the total sample weight and not just the polymer content. Since the exact glass fiber content of the samples is not known, the polymer specific postcuring enthalpy cannot be determined. To obtain absolute values, the samples measured would have to be analyzed afterward by TGA.

计算 Evaluation	玻璃化转变 Glass transition		后固化 Postcuring	
	起始点,℃ Onset in ℃	中点,℃ Midpoint temperature in ℃	表观焓,J/g Apparent enthalpy in J/g	峰,℃ Peak in ℃
管的里边 Inside of pipe	115	121	1.5	173
管的外边 Outside of pipe	104	114	13.2	171

结论 玻璃化转变温度和后固化焓的比较可定性表明管壁是否均一地从外边向里边固化。本实例中,显然不是这种情况,因为管里边和外边的玻璃化转变温度和后固化焓呈现明显的不同。通过比较同一批未使用的管子会表明,这些不同是否是在管子的使用过程中产生的或是否已经在制造后就业已存在。

Conclusions A comparison of the glass transition temperatures and the postcuring enthalpies shows qualitatively whether the wall of the pipe is uniformly cured from the outside through to the inside. In this example, this was clearly not the case because the glass transition temperatures and postcuring enthalpies of the inside and the outside of the pipe showed significant differences. A comparison of the measurements from an unused pipe of the same lot would show whether these differences arose through using the pipe or whether they were already present after manufacturing.

5.8 粉末涂料的紫外光固化
Curing of powder coatings using UV light

今天,粉末涂料技术广泛应用于不同材料(木材、塑料、金属等)。这种涂料除了性能优越,其使用还具有重要的生态优势。例如,与液体油漆不同,由于不使用溶剂所以只有数量可忽略不计的挥发性有机化合物(VOC)释放到大气中。粉末涂料通常是喷涂到制件上形成涂层并固化。于是固化或交联过程可在炉中等温进行(典型的在 180℃)也可在低温(例如 110℃)通过紫外光进行。

用紫外光固化的优势是对温度灵敏的材料(如木材或塑料制品)能用高价值涂料涂层。

实际上,联用的红外/紫外加工线用于粉末涂料的紫外光固化。在红外区,粉末在红外(IR)光的作用下"熔化",在要涂层的基板上形成均一的薄膜。液态薄膜形成后在几秒或几分内就在紫外区固化。

Today, powder coating technology is applied to a wide range of different materials (wood, plastics, metals, etc.). Besides the excellent properties of such coatings, their use also offers important ecological advantages. For example, unlike liquid paints, no solvents are used so that only negligible amounts of volatile organic compounds (VOCs) are released into the atmosphere. The powder coating is usually sprayed onto the parts to be coated and then cured. The curing or crosslinking process is then performed either thermally in an oven (typically at about 180℃) or by means of UV light at lower temperatures (e.g. at 110℃).

Curing with UV light has the great advantage that materials sensitive to temperature (such as wood or plastic products) can be coated with high value coatings.

In practice, a combined IR/UV processing line is used for the UV curing of powder coatings. In the IR zone, the powder "melts" under the influence of the infrared (IR) light and forms a homogeneous film on the substrate to be coated. The liquid film so formed is then cured in the UV zone within seconds or minutes.

目的	本应用的目的是说明 DSC 如何能用来研究粉末涂料的紫外光固化行为。 与紫外光固化有关的三个基本问题： 1. 温度和紫外光对固化过程有什么影响？ 2. 样品必须在紫外光下曝光多久才能获得足够的固化交联度？ 3. 固化反应的最佳参数是什么？ 前两个问题在下面部分详述，后一点在最后的结论中讨论。
Purpose	The purpose of this application is to show how DSC can be used to investigate the UV curing behavior of powder coatings. Three basic questions arise in connection with UV curing: 1. What influence do temperature and UV light intensity have on the curing process? 2. How long does the sample have to be exposed to UV light to achieve an adequate degree of cure or crosslinking? 3. What are the optimum parameters for the curing reaction? The first two questions are investigated in detail in the following sections. The latter point is discussed in the final conclusions.
样品	研究的样品是作为商品用作木材纤维板和碎木板表层透明底漆的粉末涂料。
Sample	The sample investigated was a commercially available powder coating material used as a clear primer for wood fiberboard and chipboard surfaces.
条件 **Conditions**	测试仪器： 带光量热附件的 DSC 坩埚：无盖 40μL 铝坩埚 样品制备：典型的，8.5mg 原料均匀地涂满坩埚的底部，形成厚约 0.8mm 的一层。 DSC 测试： 每例样品在进行固化的温度下等温保持 25min。在该时间段中，样品对三个不同的紫外光强度之一曝光 15min。 然后通过用 10K/min 的单独升温实验测试样品的玻璃化转变温度来测定固化度。 气氛：静态空气	**Measuring cell：** DSC with photocalorimetry accessory **Pan：**Aluminum 40μL without lid **Sample preparation：**Typically 8.5mg of the material was spread evenly over the bottom of the crucible forming a layer about 0.8mm thick. **DSC measurement：** In each case, the samples were held isothermally for 25 min at the temperature at which curing was to be performed. During this time interval, the samples were exposed to one of three different UV light intensities for 15 min. The degree of cure was afterward determined by measuring the glass transition temperature of the sample in a separate heating experiment at 10K/min. **Atmosphere：**Static air

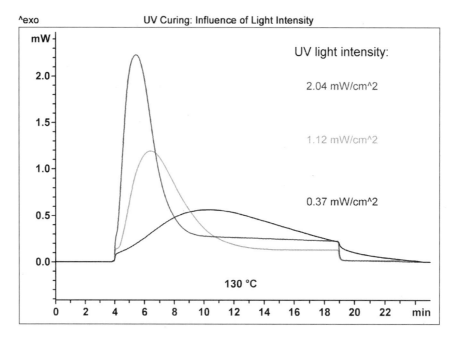

图 5.8　130℃下光强度对固化的影响。

Fig. 5.8　Influence of light intensity on curing at 130℃.

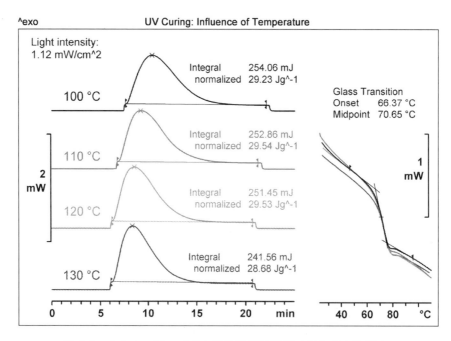

图 5.9　在 1.12mWcm^{-2} 的入射光强度下温度对固化反应的影响。

Fig. 5.9　Influence of temperature on the curing reaction at an incident light intensity of 1.12mWcm^{-2}.

解释　为了研究温度和紫外光强度对固化反应的影响,用三个不同的紫外光强度在四个不同的温度测试等温 DSC 曲线。图 5.8 表示在不同入射光强度下测试的 DSC 曲线的进程。在 130℃下测试中,光在 4min 时开始

Interpretation　To investigate the influence of temperature and UV light intensity on the curing reaction, isothermal DSC curves were measured at four different temperatures using three different UV light intensities. Figure 5.8 shows the course of the DSC curves measured at different incident light intensities. In the measurements at 130℃, the light was switched on at 4

照射,曝光持续了 15min。在 0.37 mWcm^{-2} 的弱光强度下,曲线表明,固化反应在 15min 后尚未完成。甚至在紫外光切断后,反应还在继续并逐渐停下来。

图 5.9 左边的曲线表示当粉末涂料体系在 1.12mWcm^{-2} 的恒定光强下在不同温度固化时测得的热流。样品在所示温度下等温保持 25min,对紫外光曝光都为 15min。反应焓表明,在这个光强水平上温度对固化度没有任何明显的影响。然而,固化峰的宽度和高度在不同温度下确实不同。在较高温度下,最大热流增加而峰宽下降,即固化发生较快。

图右边的小图所示为在固化后立即测试的 DSC 升温曲线。它表明固化后测得的玻璃化转变温度与温度无关。

min and the exposure lasted 15 min. At the weak light intensity of 0.37 mWcm^{-2}, the curve shows that the curing reaction is not complete after 15 min. Even after the UV light has been switched off, the reaction continues and gradually dies down.

The curves on the left side of Figure 5.9 show the heat flows measured when the powder coating system was cured at different temperatures at a constant light intensity of 1.12 mWcm^{-2}. The samples were held isothermally for 25 min at the temperatures indicated and exposed to UV light for a total of 15 min.

The reaction enthalpies show that temperature does not have any appreciable influence on the degree of cure at this level of light intensity. The widths and heights of the curing peaks do, however, differ at the different temperatures. At higher temperatures, the maximum heat flow increases and the peak width decreases, i.e. curing occurs more rapidly.

The smaller diagram on the right of the figure shows the DSC heating curves measured immediately after curing. This means that glass transition temperatures measured after curing are independent of temperature.

表 5.8 所有测试的反应焓和峰温。
Table 5.8 Reaction enthalpies and peak heights for all measurements.

温度 Temperature℃	水平 1 Level 1 0.37 mWcm^{-2}		水平 2 Level 2 1.12 mWcm^{-2}		水平 3 Level 3 2.04 mWcm^{-2}	
	ΔH_{cure} J/g	峰 Peak mW	ΔH_{cure} J/g	峰 Peak mW	ΔH_{cure} J/g	峰 Peak mW
100	21.6	0.32	29.2	0.91	33.9	1.68
110	24.5	0.37	29.5	0.95	33.3	1.70
120	29.9	0.45	29.5	1.01	34.5	1.85
130	32.4	0.48	28.7	1.05	34.0	2.00

所用的样品质量在 8.4 至 8.8mg 之间。

在最低紫外光强度下反应焓随温度升高而增大完全是由于对达到完全固化来说时间太短了。因为活性基团的扩散随着温度而增加,所以在较高温度下发生较多的转化。这可说明由于层厚导致的反应受限,就是说,紫外光只作用在表面。

对于给定的入射紫外光强度,固化

The sample masses used were between 8.4 and 8.8 mg.

The increase in the reaction enthalpy with temperature at the lowest UV light intensity is above all due to the fact that the curing time is too short for complete curing. Since the diffusion of the reactive groups increases with temperature, more conversion occurs at higher temperatures. This could indicate a limitation to the reaction due to the thickness of the layer, that is, the UV light acts only at the surface.

For a given incident UV light intensity, the degree of cure and

度和玻璃化转变温度,主要与紫外光曝光时间有关。图 5.10 表示对不同曝光时间的等温固化以及在固化后立即对样品进行加热测试的 DSC 曲线。

hence the glass transition temperature depend mainly on the UV exposure time. Figure 5.10 displays the DSC curves for isothermal curing for different exposure times as well as heating measurements performed on the samples immediately after the curing.

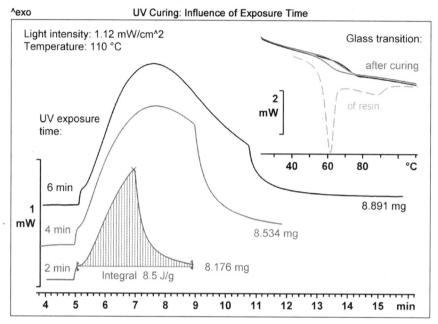

图 5.10　左图表示在 110℃的样品温度和 1.12 mW/cm² 的光强下紫外光曝光时间对固化度的影响。右图表示在 10K/min 升温速率下测得的部分固化样品和未固化树脂的玻璃化转变的 DSC 曲线。

Fig. 5.10　The left diagram shows the influence of UV exposure time on the degree of cure at a sampletemperature of 110℃ and light intensity 1.12mW/cm². The right diagram shows the DSC curves of the glass transitions of the partially cured samples and the uncured resin measured at a heating rate of 10K/min.

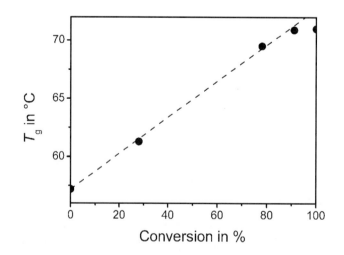

图 5.11　玻璃化转变温度与转化率的关系。

Fig. 5.11　Glass transition temperature, T_g, as a function of the conversion.

图 5.10 的测试结果能测定紫外光固化的转化率与相应的玻璃化转变温度之间的关系。转化率等于部分固化与完全固化样品的反应焓之比。根据表 5.8 和图 5.9,对 100% 紫外光固化,反应焓约为 30 J/g,相应的玻璃化转变温度为 71℃。

The results of the measurements in Figure 5.10 allow the relationship between the conversion on UV curing and the resulting glass transition temperature to be determined. The conversion corresponds to the ratio of the reaction enthalpies of a partially cured and fully cured sample. According to Table 5.8 and Figure 5.9, the reaction enthalpy is about 30 J/g and the corresponding glass transition temperature 71℃ with 100% UV curing.

计算 对于涂料充分固化的曝光时间可借助图 5.11 和图 5.12 测定。对于实际应用,涂料应该具有约 70℃ 的玻璃化转变温度。根据图 5.11,这在固化度即转化率 80% 时达到。图 5.12 表明,曝光时间约为 5min 后达到该转化率。

Evaluation The exposure time for adequate curing of the coating can be determined with the aid of Figures 5.11 and 5.12. For practical applications, the coating should have a glass transition temperature of about 70℃. According to Figure 5.11 this is achieved with a degree of cure or conversion of 80%. Figure 5.12 shows that this conversion is reached after an exposure time of about 5 minutes.

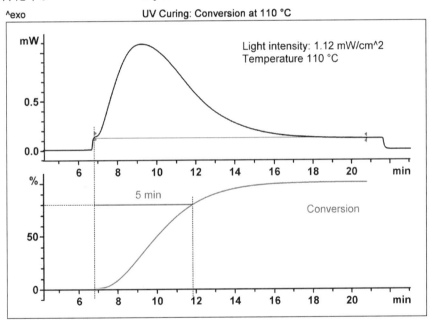

图 5.12 达到给定转化率即固化度例如 80%(在 110℃ 和 1.12 mW/cm² 入射光强度下)所必需的紫外光曝光时间可从 DSC 曲线测定。

Fig. 5.12 The UV exposure time necessary to achieve a given conversion or degree of cure, for example 80% (at 110℃ and an incident light intensity of 1.12 mW/cm²) can be determined from the DSC curing curve.

结论 对于这里研究的涂料体系,实验清楚地表明,如果入射光强度足够大,温度对实际固化过程的影响可忽略不计。然而,温度对未固化涂料的流动性能、因而对交联反应中生成的涂层的均一性确实具有重要的影响。其他许多研究已经

Conclusions For the coating system studied here, the experiments clearly showed that temperature has a negligible influence on the actual curing process if the incident light intensity is sufficiently large. Temperature does however have an important influence on the flow properties of the uncured coating material and hence on the homogeneity of the coating produced in the crosslinking reaction. A number of other studies have

确定,110℃是粉末涂料具有最佳流动行为的温度,接着可进行紫外光交联。

UV-DSC实验表明,在该温度下用 $1.12mW/cm^2$ 的光强度需要大约5min来获得足够的交联度。如果紫外光强度加倍,曝光时间下降至约3min。

不过紫外光曝光时间的下降有一定的限度:过高的光强度导致粉末涂料固化不完全。另一方面,过长的紫外光曝光时间引起基材表面温度的升高,这不可避免地导致涂层较差的力学性能。

established that 110℃ is the temperature for optimum flow behavior of the powder coating for the UV crosslinking process that follows.

UV-DSC experiments showed that about 5 minutes were needed to attain an adequate degree of crosslinking at this temperature with a light intensity of 1.12 mW/cm^2. If the UV light intensity is doubled, the exposure time is reduced to about 3 minutes.

There is however a limit to the possible reduction in UV exposure time: light intensities that are too high lead to incomplete curing of the powder coating. On the other hand, UV exposure times that are too long lead to an increase in temperature on the surface of the substrate, which inevitably results in the coating having poorer mechanical properties.

5.9 加工片状模塑料的模塑时间
Molding times for processing SMC

目的 说明DSC如何能用来研究模塑时间对片状模塑料固化的影响。
在基于不饱和聚酯树脂的片状模塑料预浸料树脂体系的开发中,须研究以下各点:
— 活性的控制和是否达到希望的固化度,即使配方改变;
— 成型和固化的最佳压缩时间;
— 预浸料的贮存稳定性。

Purpose To show how DSC can be used to study the effect of molding time on the curing of Sheet Molding Compounds.
In the development of a resin system for SMC prepregs based on UP resins, the following points have to be investigated:
— the control of reactivity and whether the desired degree of cure is achieved even if the formulation varies;
— the optimum compression time for molding and curing;
— the storage stability of the prepregs.

样品 在120℃一定压力下,片状模塑预浸料在加热的模中成型和固化同步完成。为此,模塑了四个不同时间的同一组成的许多测试板。

Sample The simultaneous molding and curing of this SMC prepreg is done at 120℃ under pressure in a heated compression mold. To do this, a number of test plates of the same composition were molded for four different times.

条件 测试仪器:DSC
Conditions 坩埚:40μL铝坩埚,盖钻孔
样品制备:
从模压板冲出重约30mg的圆柱形样品。

Measuring cell:DSC
Pan:Aluminum 40μL with pierced lid
Sample preparation:
Cylindrical samples weighing about 30mg were punched out from the molded plates.

DSC 测试:
以 20K/min 从 25℃升温至 300℃
气氛:氮气,50mL/min

DSC measurement:
Heating from 25℃ to 300℃ at 20K/min
Atmosphere:Nitrogen,50mL/min

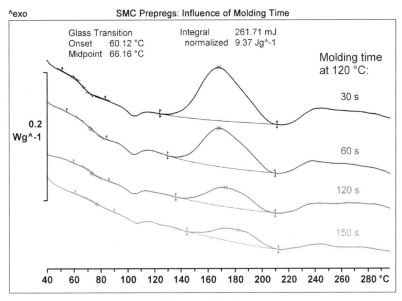

图 5.13 Fig. 5.13

解释 图 5.13 所示为测得的四个 DSC 曲线。随模塑时间增加而相继变小的放热峰清晰地表明了后固化反应。模塑期间的固化随着模塑时间增加越来越完全。与新制备树脂的反应焓(167J/g,不含织物)相比,测得的后固化反应焓小。120℃下后固化反应焓 ΔH_r(对数纵座标刻度)与模塑时间 t 之间的关系可示于图 5.14 中。

Interpretation Figure 5.13 displays the four DSC curves obtained. The postcuring reactions are clearly shown by exothermic peaks that become successively smaller with increasing molding time. Curing during molding becomes more and more complete with increasing molding time. The measured postcuring reaction enthalpies are small compared with the reaction enthalpy of the fresh resin (167 J/g without fabric). The relationship between the postcuring enthalpy, ΔH_r, (logarithmic ordinate scale) and the molding time, t, at 120℃ can is shown in Figure 5.14.

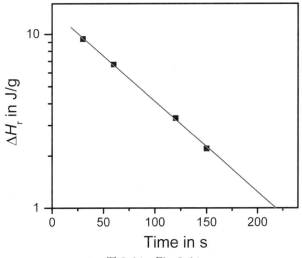

图 5.14 Fig. 5.14

对于实际上完全的固化,可假定后固化焓小于1J/g,因而是几乎不可测量的。曲线的外推可估算出树脂有效地完全固化的模塑时间约为220秒。

由于前述的主要反应(模塑过程)中较长模塑时间导致固化的增加也可在玻璃化转变温度 T_g 的提高中观察到(见表)。

计算 此表表示模塑时间对玻璃化转变温度和后固化焓 ΔH_{post} 的影响。

For practically complete curing, it can be assumed that the postcuring enthalpy is less than 1J/g and is therefore hardly measurable. Extrapolation of the curve allows a molding time of about 220s to be estimated at which the resin is effectively fully cured.

The result of increased cure due to longer molding times in the preceding main reaction (the molding process) can also be seen in the increase in the glass transition temperature, T_g, (see table).

Evaluation The table shows the influence of molding time on the glass transition temperature and the postcuring enthalpy, ΔH_{post}.

120℃下模塑时间,秒 Molding time at 120℃ in s	30	60	120	150
T_g,℃ T_g, in ℃	65	71	72	75
ΔH_{post}, J/g ΔH_{post} in J/g	9.4	6.7	3.3	2.2

反应焓参照样品重量即树脂加织物。

The reaction enthalpies refer to the sample weight, i.e. resin and fabric.

结论 预浸料成型和固化所必需的时间能用不同的模塑时间先进行几次试验性模塑来测定。然后用DSC测量法来测试在不同模塑时间后的样品后固化反应焓。

Conclusions The time necessary for the molding and curing of prepregs can be determined by first performing several trial molding runs using different molding times. DSC measurements are then used to measure the postcuring reaction enthalpy of samples after different molding times.

6 甲醛树脂 Formaldehyde resins

6.1 酚醛树脂：测试条件的影响
PF: Influence of measurement conditions

目的 说明不同的实验条件如何影响 DSC 曲线。
对于酚醛树脂，水的吸热蒸发经常与放热固化反应重叠。于是无法清晰的分辨固化峰，动力学计算变为不可能。如果用高压坩埚或高压 DSC 测试固化反应，可抑制水的蒸发或移向更高温度。结果，固化和蒸发不再重叠，便可正确地记录放热反应。

Purpose To show how different experimental conditions affect the DSC curves.
With phenolic resins, the endothermic vaporization of water often overlaps the exothermic curing reaction. The curing peak is then no longer clearly defined and a kinetic evaluation becomes impossible. If the curing reaction is measured in a high-pressure crucible or in a high-pressure DSC, the vaporization of the water is suppressed or shifted to higher temperature. As a result, curing and vaporization no longer overlap and the exothermic reaction can be properly recorded.

样品 未固化的酚醛树脂，约 10mg。
Sample Uncured phenolic resin, approx. 10mg.

条件 Conditions		
测试仪器:DSC 和 HP DSC	**Measuring cell**:DSC and HP DSC
坩埚:	**Pan**:
40μL 铝坩埚	Aluminum 40μL
40μL 高压坩埚，镀金	High pressure aluminum 40μL, gold plated
DSC 测试:	**DSC measurement**:
以 10K/min 从 30℃升温至 300℃	Heating from 30℃ to 300℃ at 10K/min
气氛:	**Atmosphere**:
常压 DSC:氮气,50mL/min	DSC at normal pressure: nitrogen, 50mL/min
高压 DSC:氮气,3MPa	High pressure DSC: nitrogen, 3MPa

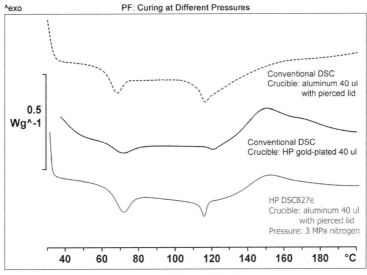

图 6.1 Fig. 6.1

解释 这里发生的效应是伴有焓松弛的玻璃化转变温度(65℃)、熔融(115℃)和固化反应(150℃)。在常规 DSC 中即常压下，放热固化峰很不清晰，因为水的吸热蒸发在反应起始时就已开始(虚线)。

如果在高压坩埚中进行测试，就能清晰得多地鉴别。固化峰由于等容(容积不变)条件，水的蒸发被抑止。不过，由于坩埚质量大和热传递较差，焓松弛的热效应和熔融较宽，分辨不太好(黑实线)。

如果在高压 DSC 3MPa 氮气压力的等压(恒定压力)条件下测试样品，水的蒸发过程移向较高温度。由于使用 40μL 铝坩埚，现在固化峰轮廓也很好并能得到良好的分辨率(红线)。

结论 独立于其他效应的反应峰对于研究固化过程是至关重要的。对于酚醛树脂，缩聚反应与生成水的蒸发同时发生。因此必须使两个热效应相互分离。在常规 DSC 中用高压坩埚或采用高压 DSC 仪器就能容易地做到，于是可防止水的蒸发或移向较高温度。

固化反应几乎不受条件改变的影响。固化和蒸发相互分离，能被正确地计算放热反应。

Interpretation The events that occur here are the glass transition with enthalpy relaxation (at 65℃), melting (at 115℃) and the curing reaction (at 150℃). In conventional DSC, i.e. at normal pressure, the exothermic curing peak is not at all clear because the endothermic vaporization of water begins as soon as the reaction starts (dotted curve).

If the measurement is performed in a high-pressure crucible, the curing peak can be much more clearly identified. Because of the isochoric (constant volume) conditions, the vaporization of water is prevented. The thermal effects of enthalpy relaxation and melting are, however, broader and less well resolved due to the large crucible mass and the poorer heat transfer (continuous black curve).

If the sample is measured under isobaric (constant pressure) conditions at a pressure of 3 MPa nitrogen in the high pressure DSC, the vaporization process of water is shifted to higher temperature. The curing peak is now also well defined and good resolution can be obtained through the use of the 40μL aluminum crucibles (red curve).

Conclusions A reaction peak free from other effects is crucial for studying curing processes. With phenolic resins, the polycondensation reaction and the vaporization of the formed water occur at simultaneously. The two thermal events therefore have to be separated from one another. This can easily be done using a high-pressure crucible in a conventional DSC or by means of a high pressure DSC equipment. The vaporization of water is thus prevented or shifted to higher temperature.

The curing reaction is hardly affected by the changed conditions. Curing and vaporization are separated from one another and the exothermic reaction can be properly evaluated.

6.2 酚醛树脂：用 TMA 区别完全和部分固化的酚醛树脂
PF:Differentiation between completely and partially cured phenolic resins by TMA

目的 未达最佳标准的工艺条件可导致性能不足的不完全固化树脂。本实验说明 TMA 如何能用来检测这种差异。

Purpose Suboptimal processing conditions can lead to incompletely cured resins with insufficient properties. The experiment shows how TMA can be used to detect such differences.

样品 适用于发动机部件的 40% 填料含量的完全和部分固化的酚醛树脂。在 190℃ 下进行短时间部分固化；通过后固化达到完全固化。

样品尺寸：约 3×3×3mm，从发动机部件切出。

Sample	Completely and partially cured phenolic resins with 40% filler content for motor parts. Partial curing was performed at 190℃ for a short time; complete curing was achieved through postcuring. Sample dimensions: about 3×3×3 mm, cut out from the motor part.
条件 **Conditions**	测试仪器:TMA　　　　　　　　　　**Measuring cell**:TMA 探头:球点探头　　　　　　　　　　**Probe**:Ball-point probe TMA 测试:　　　　　　　　　　　　**TMA measurement**: 以 10K/min 从 30℃升温至 300℃　　Heating from 30℃ to 300℃ at 10K/min 负载:0.1N　　　　　　　　　　　　Load:0.1N 气氛:静态空气　　　　　　　　　　Atmosphere:Static air

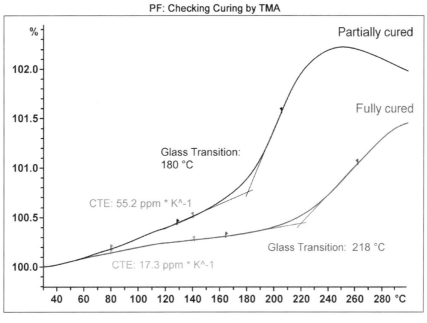

图 6.2　Fig 6.2

解释　完全固化的酚醛树脂呈现玻璃化转变温度 218℃。部分固化树脂的玻璃化转变温度在 180℃，明显较低。玻璃化转变前两个材料的平均膨胀系数(即 80℃至 140℃之间的 CTE_{mean})有三倍之差。玻璃化转变后，部分固化树脂的膨胀也明显更大。而且，该材料在比完全固化材料低得多的温度软化，如 TMA 曲线上的峰(在约 250℃)和变平处(在约 290℃)所示。在该点，探头开始穿透进入软样品材料。

TGA 测试表明，分解反应直至 400℃才开始。高至 300℃两个样品的失重小于 2%。

容易看到不完全固化的材料不满足

Interpretation　The completely cured phenolic resin exhibits a glass transition at 218℃. The glass transition temperature of the partially cured resin is significantly lower at 180℃. The mean expansion coefficients of the two materials before the glass transition (i.e. CTEmean between 80℃ and 140℃) differ by a factor three. After the glass transition, the expansion of the partially cured resin is also significantly larger. Furthermore, this material softens at a much lower temperature than the fully cured material, as the peak (at about 250℃) and the flattening off (at about 290℃) in the TMA curves show. At this point, the probe begins to penetrate into the soft sample material.

TGA measurements showed that the decomposition reaction does not begin until about 400℃. The mass loss for both samples up to 300℃ is less than 2%.

It is easy to see that in a technical application the incompletely

工业应用中的要求。

结论 固化度影响树脂的玻璃化转变温度和膨胀系数。不完全固化的发动机部件在太低的温度软化，或膨胀太多。这会导致该部件和其他部件发生故障。
TMA 测量法可以通过测试玻璃化转变温度和/或玻璃化转变前的膨胀系数来检查固化度。

Conclusions The degree of cure influences the glass transition temperature and the expansion coefficient of the resin. Incompletely cured motor parts soften at a temperature that is too low or expand too much. This can lead to failure of this and other components.
TMA measurements can be used to check the degree of the cure by measuring the glass transition temperature and/or the expansion coefficient before the glass transition.

6.3 酚醛树脂：树脂的软化行为 PF: Softening behavior of resins

未固化树脂的一个重要性能是它们的软化点。这是材料从坚硬的、脆性的状态变为黏性状态的温度。转变并不像熔点那样发生在热力学精确定义的温度，而是跨越一个与测试条件有关的温度范围。因此，为了得到能相互比较的结果，必须精确定义测试参数。ASTM D6090 测试方法"树脂软化点（Mettler 杯球法）"规定了这些条件。

An important property of uncured resins is their softening point. This is the temperature at which the material changes from the hard, brittle state to the viscous state. The transition does not occur at a thermodynamically precisely defined temperature, e. g. like the melting point, but stretches over a temperature range that depends on the measurement conditions. For this reason, the measurement parameters must be precisely defined in order to achieve results that can be compared with each other. The ASTM D6090 test method "Softening Point Resins (Mettler Cup and Ball Method)" specifies these conditions.

目的 / **Purpose**	用不同技术测定树脂的软化温度。 To determine the softening temperature of resins using different techniques.
样品 / **Sample**	用常规环球法（例如 ASTM E28）测定的软化点为 171℃ 的改性酚醛树脂。 Modified phenolic resin with a dropping point of 171℃, determined using the conventional ring and ball method (e. g. ASTM E28).

条件 / **Conditions**

测试仪器：
FP 滴点仪
TMA
DSC

样品支架：
FP：按照 ASTM D6090 软化点杯（孔 6.35mm）和置于满杯顶部的 2.77g 钢球。
TMA：树脂放在一铝箔薄片上，以便样品能容易地从样品支架上移出。
DSC：40μl 铝坩埚，盖钻孔。

Measuring cell：
FP dropping point apparatus
TMA
DSC

Sample holder：
FP: Cups for softening point (orifice 6.35 mm) and 2.77 g steel ball on top of the filled cup according to ASTM D6090.
TMA: The resin was placed on a thin piece of aluminum foil so that the sample could easily be removed from the sample support.
DSC: Aluminum 40μl with pierced lid.

预处理：

FP：预处理后的样品呈现相当高的黏度，这样无法用ASTM D6090建议的将"熔融"粉末样品装入杯子。改为将粉末压入杯子。

样品制备：

FP：树脂以纤细粉末一层层装入并用杆压实。

TMA：一片具有平行表面的树脂直接放在球点探头（直径3mm）下。

DSC：粉末和较小颗粒直接称量后放入，不压实。

FP 测试：

以 1K/min 从 165℃升温至 180℃

TMA 测试：

以 5K/min 从 35℃升温至 180℃

DSC 测试：

以 10K/min 从 -10℃升温至 250℃

气氛：静态空气

FP83HT测试原理：杯置于炉中，这样材料在软化时能垂直向下由孔流出。软化点定义为升温时样品在标准软化点样品杯中软化并从6.35mm开口流出20mm时的温度。

Pretreatment：

FP：The premelted samples showed a rather high viscosity, so the cups could not be filled by "melting" the powdered sample as proposed by ASTM D6090. Instead, the powder was pressed into the cup.

Sample preparation：

FP：The resin was filled in layers as a fine powder and compacted with a plunger.

TMA：A piece of resin with parallel surfaces was positioned directly under the ball-point probe (diameter 3 mm).

DSC：Powder and smaller pieces were weighed in directly without compacting.

FP measurement：

Heating from 165℃ to 180℃ at 1K/min

TMA measurement：

Heating from 35℃ to 180℃ at 5K/min

DSC measurement：

Heating from -10℃ to 250℃ at 10K/min

Atmosphere：Static air

FP83HT measurement principle：The cup is placed in a furnace so that the material can flow vertically downwards out of the orifice when it softens. The softening point is defined as the temperature at which the sample on slow heating softens in a standard softening point sample cup and flows 20 mm out of the 6.35mm opening.

图6.3 测试前样品顶部带有钢球的杯

Fig. 6.3 Cup before measurement with steel ball on top of the sample.

图6.4 软化点测试后的杯（带仍然软化的样品滴落物的杯放在台上。测得的软化点为175.7℃）

Fig. 6.4 Cup after the softening point determination. The cup with the sample drop that was still soft was placed on the table. The softening point measured was 175.7℃.

解释 TMA、DSC和TOPEM™测试曲线示于图6.5和图6.6中。第

Interpretation The TMA, DSC and TOPEM™ measurement curves are displayed in Figures 6.5 and 6.6. The first heating runs

一次升温用实线表示,第二次测试用虚线。

are shown by continuous lines, the second heating runs by dashed lines.

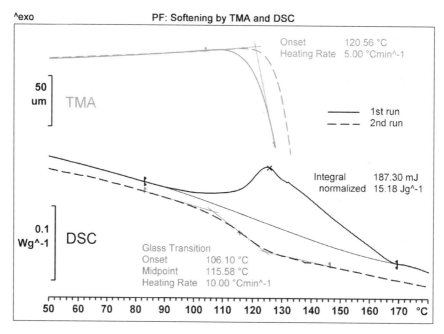

图 6.5　Fig. 6.5

FP 测试表明,粉末软化并流出杯口。样品上的球产生压力,保证可重复的软化点测定。因为黏度高,不形成单独的点滴。

将该流动行为与 TMA 中样品的行为作比较。这里,必须再次专门定义软化点,通常为外推起始点或探头穿透一定程度的温度(图 6.5)。第一次升温于 140℃ 停止,由于探头的一部分略已穿入样品。因为探头力分布在较大的面积上,所以记录的第二次升温的起始点稍迟些。

除了玻璃化转变温度,DSC 测试能检测其他的效应如熔融、反应或分解。第一次升温,放热效应与玻璃化转变重叠,无法直接测定。曲线上通常的信号台阶只能在第二次升温测试和计算。不可能确定放热效应是否是由于固化产生的,因而影响了树脂的玻璃化转变温度。不涉及失重。

The FP measurements show that the powder softens and flows out of the opening of the cup. The ball on the sample generates a pressure and ensures a reproducible softening point determination. Individual drops are not formed because the viscosity is high.

This flow behavior is compared with the behavior of the sample in the TMA. Here again, the softening point has to be specially defined. Normally this is the extrapolated onset or a certain penetration of the probe (Fig. 6.5). The first heating run was stopped at 140℃, since the probe already penetrated somewhat into the sample. The onset of the second heating run is recorded somewhat later because the probe force is distributed over a larger area.

Besides the glass transition, DSC measurements allow other events to be detected such as melting, reactions or decomposition. In the first heating run, an exothermic event overlaps the glass transition, which makes it impossible to determine directly. The usual signal step in the curve can only be measured and evaluated in the second heating run. It is not possible to determine whether the exothermic effect is due to curing and hence has an influence on the glass transition temperature of the resin. No loss of mass was involved.

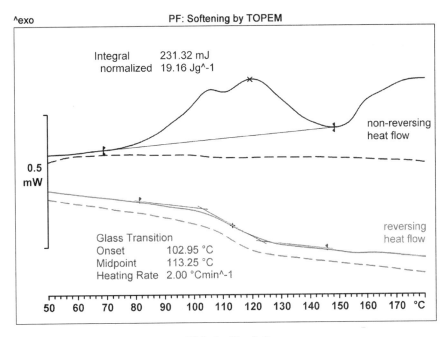

图 6.6　Fig. 6.6

与之不同，TOPEM™ 可在第一次升温中测定玻璃化转变温度（图 6.6）。与第二次升温的对比表明，玻璃化转变不受放热效应的影响。不可逆热流在放热效应后再次增加，这是因为粉末颗粒接合导致的从坩埚向样品的热传递变化的结果。

TMA、DSC 和 TOPEM™ 测试的比较清楚地说明，软化主要发生在已经达到玻璃化转变温度之后。

计算　软化点是完全可重复的，软化温度比用 ASTM E28 测定的值高约 4K。

In contrast, TOPEM™ allows the glass transition temperature to be determined in the first heating run (Fig. 6.6). The comparison with the second heating run shows that the glass transition is not influenced by the exothermic effect. The non-reversing heat flow increases again after the exothermic effect. This is a result of the change in heat transfer from the crucible to the sample because the powder particles coalesce.

A comparison of the TMA, DSC and TOPEM™ measurements clearly shows that softening occurs mainly after the glass transition temperature has been reached.

Evaluation　The softening points are very reproducible and show a softening temperature about 4 K higher than the value determined by ASTM E28.

表 6.1　用 FP83HT 自动测定的软化点 ℃。
Table 6.1　Softening points in ℃, determined automatically by FP83HT.

ASTM E28	171
FP83HT 测定值 Measured by FP83HT	175.7 175.9 174.8 175.1
平均值 Mean	175.4
标准偏差 Standard deviation	0.44

表 6.2 TMA、DSC 和 TOPEM™ 测试的结果。
Table 6.2 The results of the TMA, DSC and TOPEM™ measurements.

技术 Technique			第一次升温测试 1st heating run	第二次升温测试 2nd heating run
TMA	外推起始点 Extrapol. Onset	℃	120.6	127.6
DSC	T_g ΔH_{exo}	℃ J/g	— 15.2	115.5 —
TOPEM™	T_g ΔH_{exo}	℃ J/g	113.3 19.2	112.1 —

用 TOPEM™ 测试的放热峰，表明大于常规 DSC 测得的焓，因为在 DSC 测试中没有考虑比热容的变化。在比较玻璃化转变温度 T_g 时，应该注意到在 DSC 测试中用 10K/min 的升温速率，而 TOPEM™ 只用 2K/min。

结论 按照 ASTM D6090 测定未固化树脂的软化点是完全可重复的。
玻璃化转变温度比软化点低得多，由于在后一种情况下为低黏度流动。因为流动行为对软化点影响大，所以两个温度之间没有直接的相关性，然而软化的真实原因是玻璃化转变。
除了这个结果，DSC 还能检测玻璃化转变温度之上材料可能发生的固化或降解。

The exothermic peak measured using TOPEM™ shows a larger enthalpy than that obtained from the conventional DSC measurement because the change of the specific heat capacity is not taken into account in the DSC measurement. When comparing the glass transition temperatures, T_g, it should be noted that in the DSC measurement a heating rate of 10K/min was used whereas with TOPEM™ only 2K/min.

Conclusions The determination of the softening point of uncured resins according to ASTM D6090 is very reproducible.
The glass transition temperature is much lower than the softening point due to slow viscous flow in the latter case. For this reason, there is no direct correlation between the two temperatures because the flow behavior has a large influence on the softening point. The real cause of softening is however the glass transition.
In addition to this result, DSC is able to detect possible curing or degradation of the materials above the glass transition temperature.

6.4 两种不同的填充三聚氰胺甲醛/酚醛树脂模塑料
Two different filled MF/PF molding compounds

目的 说明如何通过测试玻璃化转变来表征新进树脂—填料体系的质量。典型的 DSC 测试，只需用 15min 左右就能完成。
如果树脂的玻璃化转变温度超过指标，则模塑料就不能正常加工。

Purpose To show how the quality of an incoming resin-filler system can be characterized by measuring the glass transition. Typically this can be done in a DSC measurement that takes only about 15min.
If the glass transition temperature of the resin is above specification, the molding compound cannot be properly processed.

样品 MF/PF 1 和 MF/PF 2—两种不同的适用于电气应用（电绝缘）注塑成型制件制造的木材和矿物填料填充的粉末状三聚氰胺/苯酚—甲醛模塑料。

Sample MF/PF 1 and MF/PF 2-two different melamine-/phenol-formaldehyde molding compounds in powder form filled with wood and mineral fillers for the manufacture of injection-molded parts for electrotechnical applications (electrical insulation).

Conditions

Measuring cell: DSC

Pan: Aluminum 40μL with pierced lid

Sample preparation: Samples weighing 6 to 8mg

DSC measurement: Cooling from 25℃ to 0℃ at 15K/min, 1 min at 0℃, Heating from 0℃ to 125℃ at 10K/min

Atmosphere: Dry air, 50mL/min

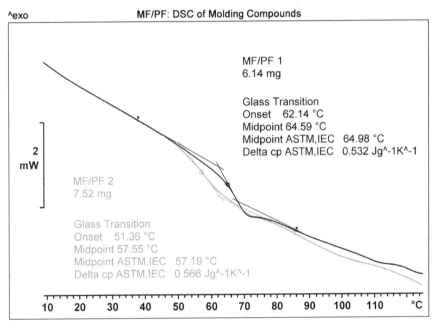

图 6.7 Fig. 6.7

Interpretation The measurement curves of the two MF/PF molding compounds differ significantly in the glass transition region. The effects above 100℃ may be due to postcuring.

Evaluation The tangents to determine the glass transition temperature should all be parallel if such thermosets have to be compared.

	中点温度, ℃ Midpoint temperature in ℃	ASTM 中点温度, ℃ Midpoint temperature ASTM in ℃	Δc_p ASTM J/gK
MF/PF 1	65	65	0.53
MF/PF 2	58	57	0.57

| 结论 | 用 DSC 可表征两种三聚氰胺/苯酚—甲醛模塑料,例如,将其用作电气开关的原材料。相应的玻璃化转变温度能用来设定交付指标和作质量保证。 | **Conclusions** The two melamine-/phenol-formaldehyde molding compounds, which are used for example as raw materials for electrical switches, can be characterized by DSC measurements. The corresponding glass transition temperatures can be used to define delivery specifications and for quality assurance. |

6.5 酚醛树脂:胶合板的纸预浸料 PF: Paper prepregs for plywood

目的	说明 DSC 测试和动力学计算如何能用于预测特定铸压温度下的反应时间。
	酚醛树脂是制造胶合板最常用的黏合剂。当在热的作用下层压时,了解反应时间和温度是重要的。本应用比较了三种不同的纸上涂层酚醛树脂。
Purpose	To show how DSC measurements and kinetic evaluations can be used to predict reaction times at particular press temperatures.
	Phenolic resins are the adhesives most frequently used for the manufacture of plywood. When the layers are compressed under the action of heat, it is important to know the reaction times and temperatures. In this application, three different phenolic resins coated on paper are compared.
样品	含三种不同树脂薄涂层的纸预浸料样品
Sample	Paper prepreg samples with thin coatings of three different resins.

条件	测试仪器:DSC	**Measuring cell**:DSC
Conditions	坩埚:	**Pan**:
	可重复使用的 30μL 高压坩埚	High pressure 30μL, reusable
	样品制备:	**Sample preparation**:
	用刀从预浸料刮下约 3mg 树脂。	Approx. 3mg of resin was scraped off from the prepreg using a knife.
	DSC 测试:	**DSC measurement**:
	以 10K/min 从 35℃升温至 220℃;一个样品还以 2 和 5K/min 测试。	Heating from 35℃ to 220℃ at 10K/min; one sample was also measured at 2 and 5K/min.
	气氛:空气,50mL/min	**Atmosphere**:Air, 50mL/min

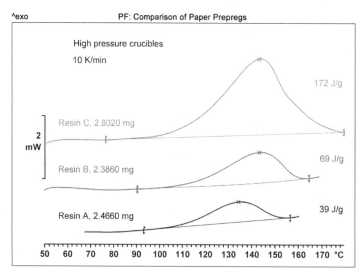

图 6.8 Fig. 6.8

解释 图 6.8 所示为在高压坩埚中测试样品所得到的 DSC 曲线。曲线反映的缩聚反应无任何由于蒸发产生的效应。三种树脂的反应焓显著不同。焓归一于刮去纸的树脂。假定在分析的树脂中不存在其他填料。

Interpretation Figure 6.8 displays the DSC curves obtained by measuring the samples in high pressure crucibles. The curves reflect the polycondensation reaction free of any effects due to evaporation. The reaction enthalpies of the three resins are noticeably different. The enthalpies refer to the resin scraped off the paper. It is assumed that no other fillers are present in the resins examined.

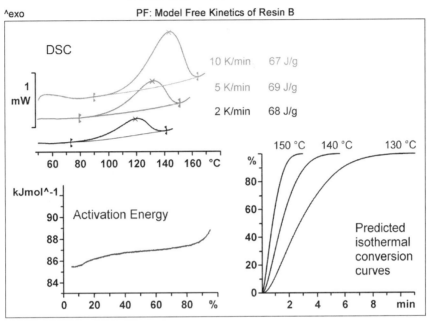

图 6.9　Fig. 6.9

计算 为了对在通常固化温度下的反应行为作出预测，以三个不同升温速率测试了树脂 B 并用非模型动力学 MFK 进行了计算（图 6.9）。三个升温热速率的反应焓都是相同的。活化能在 87kJ/mol 左右，且在较大的转化率范围内几乎是恒定的。在压铸温度 140℃ 下，反应约 3.1min 就达到 90% 转化率。

Evaluation To make predictions about the reaction behavior at the usual curing temperatures, Resin B was measured at three different heating rates and evaluated using model free kinetics, MFK (Fig. 6.9). The reaction enthalpies are the same for all three heating rates. The activation energy is of the order 87kJ/mol and is practically constant over a large conversion range. At the press temperature of 140℃, the reaction takes about 3.1 min to attain 90% conversion.

结论 为了防止缩聚反应中排除的水干扰动力学分析，使用了压力坩埚。只需几个 DSC 测试就可清晰地辨别三种酚醛树脂的反应焓和反应速率。对这些树脂在 140℃ 压铸温度下的反应时间也能通过非模型动力学预测。在这些高温下，这些时间是如此之短，因而它们无法用 DSC 测试。

Conclusions Pressure crucibles were used in order to prevent the water eliminated in the polycondensation reaction from interfering with the kinetic analysis. Just a few DSC measurements were needed to clearly distinguish between the three phenolic resins with regard to reaction enthalpy and reaction rate. The reaction times for these resins at a press temperature of 140℃ can also be predicted by model free kinetics. At these high temperatures, these times are so short that they can not be handled by DSC measurements.

6.6 酚醛树脂：缩聚反应的 TGA/SDTA 研究
PF: Condensation reaction investigated by TGA/SDTA

在生产胶合板和其他木材的热压过程中，缩聚反应中产生的水蒸汽能穿透进木材结构。因此问题是像这种"开放的"反应体系如何能用热分析来表征。

在上个实例中，使用了高压坩埚的 DSC 技术。在敞口的坩埚中，来自反应的水的蒸发是吸热的，与放热的缩合反应重叠，因而作为结果得到的 DSC 曲线是两个不同效应的加和。因此，为了只测试放热反应，使用了高压坩埚。

因为蒸发伴随着样品质量的损失，所以也能用 TGA 测试。而且，同步的 SDTA 提供由热效应产生的关于温度变化的信息。这两个测试的比较并结合蒸发焓的知识，能研究纯反应热流（见参考文献）。如果用 TGA/DSC 仪器，那么同步的 DSC 直接给出蒸发和缩合反应的热流变化。

In the hot press process for the production of plywood and other wood materials, the water vapor eliminated in the polycondensation reaction can penetrate into the structure of the wood. The question is therefore how an "open" reaction system like this can be thermoanalytically characterized.

In the previous example, the DSC technique was used with high pressure crucibles. In an open crucible, the vaporization of the water from the reaction is endothermic and overlaps the exothermic condensation reaction so that the resulting DSC curve is the sum of the two different effects. A high pressure crucible was therefore used in order to measure just the exothermic reaction.

Since vaporization is accompanied by a loss of sample mass, it can also be measured by TGA. Furthermore, simultaneous SDTA supplies information about the temperature changes caused by the thermal events. A comparison of these two measurements together with knowledge of the enthalpy of vaporization allows the net reaction heat flow to be investigated (see Reference). The simultaneous DSC directly gives a change in the heat flow of vaporization and condensation reaction, if a TGA/DSC instrument is used.

目的	本应用的目的是说明酚醛树脂的缩合反应如何能用 TGA/SDTA 在敞口坩埚中来定量研究。	
Purpose	The purpose of the application is to show that the condensation reaction of a phenolic resin can be quantitatively investigated in an open crucible by TGA/SDTA.	
样品	含酚醛树脂薄层的预浸料纸（样品 B，见 6.5 节）。	
Sample	Prepreg paper with a thin layer of phenolic resin (Sample B, see Section 6.5).	
条件	测试仪器：TGA/SDTA	**Measuring cell**：TGA/SDTA
Conditions	坩埚：	**Pan**：
	40μL 铝坩埚，盖钻孔	Aluminum 40μL with pierced lid
	样品制备：	**Sample preparation**：
	用刀从预浸料刮下约 3.565mg 树脂。	3.565 mg of resin was scraped off from the prepreg using a knife.
	TGA 测试：	**TGA measurement**：
	以 10K/min 从 30℃升温至 250℃	Heating from 30℃ to 250℃ at 10K/min
	气氛：氮气，50mL/min	**Atmosphere**：Nitrogen, 50mL/min

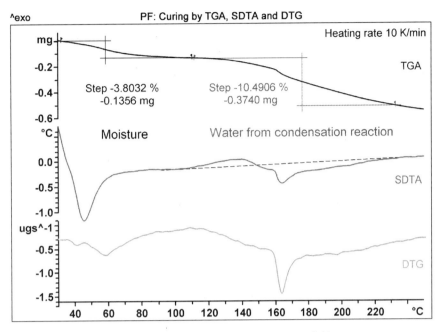

图 6.10　酚醛树脂缩合反应的 TGA/SDTA 曲线。
Fig. 6.10　TGA/SDTA curves of the condensation reaction of a phenolic resin.

解释　TGA 记录由湿气形式的水或来自缩合反应的水的蒸发产生的失重。DTG 曲线表示失重的速率而 SDTA 曲线表示测量的样品温度与参比温度之差，与 DSC 仪器类似。

至约 110℃，样品变干，失去约 3.8% 湿气。然后反应开始，表示为约 10.5% 的失重。在反应开始时，SDTA 曲线显示一个放热峰，在 150℃ 以上变为吸热峰。在 160℃，观察到一个导致 SDTA 和 DTG 曲线出峰的快速的额外的失重。对此可能的原因是树脂通过失水在样品表面形成薄膜。然后薄膜由于反应中排除的水的压力导致突然破裂。

缩合反应的净热流（$\Phi_{reaction}$）按下列方程从 DTG 和 SDTA 计算：

因此净热流为 SDTA 测得的和从蒸发算得的热流的加和（因为 DTG 是负的，所以方程中出现负号）。相应的热流示于图 6.11 中。

Interpretation　The TGA records the mass loss due to evaporation of water in the form of moisture or from the condensation reaction. The DTG curve shows the rate of loss of mass and the SDTA curve the difference between the measured sample temperature and a reference temperature, similar to in a DSC instrument.

Up to about 110℃, the sample dries and loses about 3.8% moisture. The reaction then begins and is shown by a mass loss of about 10.5%. At the beginning of the reaction, the SDTA curve exhibits an exothermic effect that becomes endothermic above 150℃. At 160℃, a rapid additional mass loss is observed that results in peaks on the SDTA and DTG curves. A possible reason for this is that the resin forms a film on the surface of the sample through the loss of water. The film is then ruptured suddenly by the pressure of the water eliminated in the reaction.

The net heat flow ($\Phi_{reaction}$) of the condensation reaction is calculated from the DTG and SDTA curves according to the following equation:

$$\phi_{reaction} = SDTA \cdot 4.81 - DTG \cdot 2400$$

The net heat flow is therefore the sum of the heat flows measured by SDTA and calculated from vaporization (since DTG is negative, a minus sign appears in the equation). The corresponding heat flows are shown in Figure 6.11.

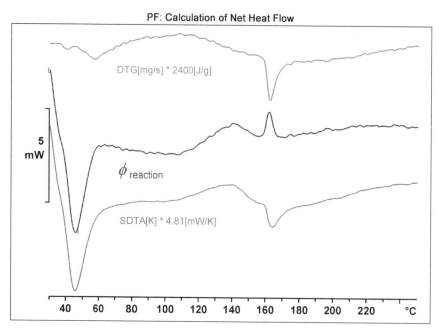

图 6.11 为测定净热流而从 DTG 和 SDTA 曲线计算的热流 $\Phi_{reaction}$。
Fig. 6.11 Heat flows calculated from the DTG and SDTA curves to determine the net heat flow, $\Phi_{reaction}$.

在图 6.11 中描述了从 SDTA 曲线上计算热流的方法。从 SDTA 曲线计算热流的校准因子 K 由不同金属的熔融焓测定,储存为单位为 mWK^{-1} 的校准曲线 $K(T)$。于是热流通过与 SDTA 曲线相乘计算。本例中,因为在狭窄温度范围的灵敏度几乎不变,所以 SDTA 曲线乘以 $4.81 mWK^{-1}$。

水的蒸发焓是 2400 J/g。

计算 图 6.12 表示在敞口坩埚中(TGA/SDTA 测试)和在高压坩埚中(6.5 节中的 DSC 应用)测试的反应热流的比较。两个反应条件下的反应焓大致相同。高至 140℃ 的反应进程是相似的。在该温度以上,不同的反应速率可解释如下。在密封的坩埚中,反应是与生成水的分压平衡的。在敞口坩埚中,形成扩散障碍(薄膜),这防止了水的释放并减慢了反应速率。随着水蒸气压力增加,障碍破裂使水能逸,而反应能继续进行。这由上边

The method used to calculate the heat flow from an SDTA curve is also described in Fig 6.11. The calibration factor, K, for calculating the heat flow from the SDTA curve is determined from the enthalpies of fusion of different metals and is stored as the calibration curve $K(T)$ with the unit mWK^{-1}. The heat flow is then calculated by multiplication with the SDTA curve. In this case, since the sensitivity over the narrow temperature range is practically constant, the SDTA curve was multiplied by $4.81 mWK^{-1}$.

The enthalpy of vaporization of water is 2400 J/g.

Evaluation Figure 6.12 shows the comparison of the reaction heat flows measured in an open crucible (TGA/SDTA measurement) and in a high pressure crucible (DSC, application in Section 6.5). The reaction enthalpy is about the same for both reaction conditions. The course of the reaction up to 140℃ is similar. Above this temperature, the different reaction rates can be explained as follows. In the sealed crucible, the reaction is in equilibrium with the partial pressure of the water produced. In the open pan, a diffusion barrier (film) is formed. This prevents the release of water and slows the reaction rate. With increasing vapor pressure, the barrier ruptures allowing the water to escape and the reaction can proceed. This is shown by the second peak in the upper DSC curve.

DSC 曲线的第二个峰所显示。

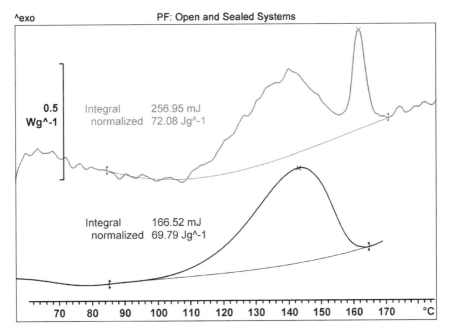

图 6.12　在敞口坩埚中(上面)和密封坩埚中(下面)酚醛树脂缩合反应的热流
Fig. 6.12 Heat flow of the condensation reaction of a phenolic resin in an open crucible (above) and in a sealed crucible (below).

结论　TGA/SDTA 能定量测定热流,这转而使在常压的敞口体系中研究缩合反应成为可能。在本酚醛树脂实例中,固化反应(即温度和反应速率)在敞口体系中与在密封体系中是大致相同的。在密封的高压坩埚中,由于水的蒸发在反应末产生了大约 2MPa 的内压。在敞口体系中,水的蒸发导致薄膜的形成。这使反应慢下来,特别是反应中排除的水的蒸发。

Conclusions　TGA/SDTA enables quantitative heat flows to be determined, which in turn allows the polycondensation reactions to be investigated in open systems at normal pressure. In this PF resin example, the curing reaction (i. e. the temperature and reaction rate) is about the same in the open system as in the sealed system. In a hermetically sealed high-pressure crucible, an internal pressure of about 2 MPa is produced at the end of the reaction due to the vaporization of the water. In the open system, the vaporization of the water leads to the formation of a film. This slows down the reaction, and in particular the vaporization of water eliminated during the reaction.

参考文献　R. Riesen, K. Vogel, M. Schubnell, 考虑质量变化的 TGA/SDTA851^e 的 DSC, Journal of Thermal Analysis and Calorimetry, Vol. 64 (2001) 243—252

Reference　R. Riesen, K. Vogel, M. Schubnell, DSC by the TGA/SDTA851^e considering mass changes, Journal of Thermal Analysis and Calorimetry, Vol. 64 (2001) 243—252

6.7 酚醛树脂：可溶性酚醛树脂的固化动力学
PF: Curing kinetics of Resol resins

目的　本应用的目的是说明 DSC 测试如何能用来预测树脂的反应行为。
了解温度对酚醛树脂(适合作黏合剂和胶水的可溶性酚醛树脂 Resol)的反应速率的影响可以预测缩聚反应速率。这对于例如木材复合材料的制造或树脂的贮存具有极大的实际意义。DSC 测试能够快速表征新生产的黏合剂的性能。最重要的是反应速率和转化程度，它们强烈影响力学性能和防水性。

Purpose　The purpose of the application is to show how DSC measurements can be used to predict the reaction behavior of a resin.
Knowledge of the influence of temperature on the reaction rate of phenol-formaldehyde resins (PF Resols for adhesives and glues) allows the polycondensation rate to be predicted. This is of great practical importance for example in the manufacture of wood composite materials or the storage of resins. DSC measurements can rapidly characterize the performance of newly produced adhesives. The most important quantities are the reaction rate and the degree of conversion, which strongly influence the mechanical properties and the resistance toward moisture.

样品　含高和低分子量分数的液态酚醛树脂 Resol。
Sample　Liquid phenol-formaldehyde Resol with high and low molecular mass fractions.

条件
Conditions

测试仪器：DSC　　　　　　　　　**Measuring cell**：DSC
坩埚：30 μL 高压铝坩埚　　　　　**Pan**：High-pressure crucible, 30 μL
样品制备：　　　　　　　　　　　**Sample preparation**：
称量约 3mg 放入坩埚，为了得到良好的密封，确保边缘和螺纹不弄湿。　　About 3 mg was weighed into the crucible, making sure that the rim and screw-threads were not wetted in order to achieve a good seal.
DSC 测试：以 5、10 和 20K/min 从 25℃升温至 250℃　　**DSC measurement**：Heating from 25℃ to 250℃ at 5, 10 and 20 K/min
气氛：静态空气　　　　　　　　　**Atmosphere**：Static air

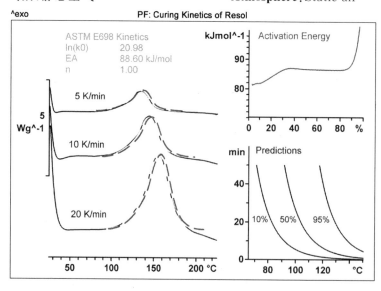

图 6.13　　Fig. 6.13

解释　图示为以不同加热速率测试的酚醛树脂的三条 DSC 曲线。由于动力学效应，放热反应峰在加热速率增大时向较高温度移动。200℃以上可见到吸热分解反应的开始。这与固化反应末尾重叠。因此用"样条"基线来积分 DSC 峰以测定转化率和反应速率。总反应焓对每个计算是相同的，因为可假定它是与升温速率无关的。按照 ASTM E698 计算了三条 DSC 曲线，求得 88kJ/mol 的活化能。MFK 计算表明，活化能不是常数。在反应开始，E_a 增加约 5%，然后在 40% 和 80% 之间几乎恒定为约 86kJ/mol 的值。90% 转化率以上 E_a 的快速增加表示反应机理的显著改变。

计算　模拟了 DSC 曲线（红色虚线）以表明动力学计算结果很好地描述了固化过程。在评估范围内，MFK 计算给出几乎完全相同的曲线。这意味着对时间—温度行为的预测（图右下）是可靠的。

结论　动力学分析能预测转化率—温度—时间行为，并提供技术问题的答案。ASTM E698 方法只在最高反应速率下（DSC 峰温）给出活化能。但 MFK 表明与转化率有关的活化能。除了给出可靠的模拟数据，还能提供有关固化机理解释的有用信息。
如之前讨论的，非模型动力学应该用于实际应用，因为假定活化能不变的动力学得到的结果通常可靠性很小。He 等人[1]已经表明，对于木材复合材料中的酚醛树脂，木材中树脂的固化度没有树脂以纯粹形态使用时那么高。由于木材的多孔结构，木材中树脂的分散因为黏合剂的凝胶化而受阻碍或甚至被阻止。

Interpretation　The figure shows the three DSC curves of the PF resin measured at different heating rates. The exothermic reaction peaks are shifted to higher temperatures at increasing heating rates due to kinetic effects. The beginning of an endothermic decomposition reaction can be seen from 200℃ onward. This overlaps the end of the curing reaction.
A "spline" baseline was therefore used to integrate the DSC peaks to determine the conversion and reaction rate. The total reaction enthalpy for each evaluation was the same because it can be assumed that it is independent of the heating rate. The three DSC curves were evaluated according to ASTM E698. This gave an activation energy of 88kJ/mol. As the MFK evaluation shows, the activation energy is not constant. At the beginning of the reaction, E_a increases by about 5% and is then practically constant between 40% and 80% conversion with a value of about 86 kJ/mol. The rapid increase of E_a above 90% conversion indicates a significant change in the reaction mechanism.

Evaluation　DSC curves were simulated (red dashed curves) to show how well the kinetic evaluation results describe the curing process. The MFK evaluation gave practically identical curves in the range of the evaluation. This means that the predictions (diagram bottom right) for time-temperature behavior are reliable.

Conclusion　Kinetic analyses allow the conversion-temperature-time behavior to be predicted and provide answers to technological questions. The ASTM E698 method yields the activation energy only at the highest reaction rate (DSC peak temperature). MFK however shows the activation energy curve as a function of conversion. Besides yielding reliable simulation data, this can also provide useful information relating to the interpretation of the curing mechanisms.
As previously discussed, model free kinetics should be used for practical applications because kinetics assuming a constant activation energy usually yield less dependable results.
He et. al. [1] have shown for PF resins in wood composite materials that the degree of cure of the resin in the wood is not so high as when the resin is used in the pure form. Due to the porous structure of the wood, diffusion of the resin in the wood is hindered or even prevented through gelation of the adhesive.

参考文献 [1] He, G., Riedl B., Aït-Kadi A., J. Applied Polymer Science, 87 (2003), p. 433.

References [1] He, G., Riedl B., Aït-Kadi A., J. Applied Polymer Science, 87 (2003), p. 433.

6.8 脲醛树脂模塑料：加工(模塑)的影响
UF molding compounds: Influence of processing (molding)

目的	本应用的目的是用DSC说明模塑工艺对材料性能的影响，例如比较模塑条与相应的未模塑粉。
Purpose	The purpose of this application is to demonstrate the influence of the molding process on material properties using DSC, for example to compare a molded bar with the corresponding unmolded powder.
样品	适用于生产电气应用模塑外壳制件的含纤维素填料的粉末状脲醛树脂(UF1型)模塑料。粉末状模塑料和在150℃模塑和固化的条。
Sample	UF (Type UF 1) molding compound with cellulose filler in powder form for the production of parts of molded housings for electrotechnical applications. Molding compound in powder form and a bar molded and cured at 150℃.

条件 / Conditions

测试仪器：DSC	Measuring cell：DSC
坩埚：	Pan：
40μL 铝坩埚，盖钻孔	Aluminum 40μL with pierced lid
样品制备：	Sample preparation：
模塑料：	Molding compound：
称重约5mg；	Approx. 5 mg was weighed in (one corn weighs about 1mg)
150℃模塑的条：	Bar molded 150℃：
用刀切出3至5mg的一片，称量后放入坩埚。	A piece weighing 3 to 5mg was cut off using a knife and weighed into the crucible.
DSC测试：	**DSC measurement：**
空气中的粉末和模塑条：	Powder and molded bar in air：
第一次测试：−40℃下恒温1min，以20K/min从−40℃升温至225℃，然后以20K/min降温至−40℃	1st run：1min at −40℃, heating from −40℃ to 225℃ at 20K/min, then cooling to −40℃ at 20 K/min and
第二次测试：以20K/min从−40℃升温至250℃	2nd run：Heating from −40℃ to 250℃ at 20K/min
在氮气气氛中的模塑条：	Molded bar in nitrogen atmosphere：
第一次测试：以20K/min从25℃升温至225℃，然后以20K/min降温至25℃	1st run：Heating from 25℃ to 225℃ at 20K/min, then cooling to 25℃ at 20 K/min and
第二轮：以20K/min升温至250℃	2nd run：Heating to 250℃ at 20K/min
气氛：干燥空气或氮气，50mL/min	**Atmosphere**：Dry air or nitrogen, 50mL/min

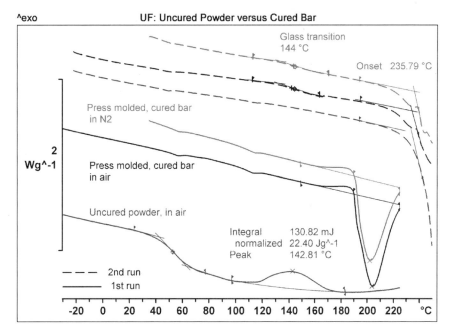

图 6.14　Fig 6.14

解释　未固化模塑料的第一次升温呈现玻璃化转变以及后固化峰。第二次测试中没有观察到玻璃化转变。分解反应显示起始值为约234℃,是吸热的(即未受空气中氧的明显影响)。

空气和氮气中模塑条样品的测试表明没有明显的不同。例如高至240℃没有氧的侵蚀。在180℃有小的放热后固化反应,跟随着一个显著的吸热效应。这大概是由于模塑过程中水的蒸发。假定蒸发焓为2400J/g,这等于约3%的水。

在第二次升温中,材料是干燥的,可观察到固化的模塑料的玻璃化转变。在粉末的固化过程中,反应排除的水能逃逸(通过钻孔的盖)。固化样品曲线呈现两个在50℃和115℃的小的、可重复的效应。

空气和氮气气氛中的分解温度是基本相同的,这证实了上面关于分解的吸热性质的结论。

Interpretation　The 1st heating run of the uncured molding compound shows the glass transition as well as the postcuring peak. A glass transition is not observed in the 2nd run. The decomposition reaction exhibits an onset value of about 234℃ and is endothermic (i. e. no significant influence of oxygen from the air).

Measurements of samples of the molded bar in air and in nitrogen show no significant difference. For example there is no attack of oxygen up to 240℃. There is small exothermic postcuring reaction at 180℃ followed by a pronounced endothermic effect. This is presumably due to vaporization of water that was trapped during the molding process. Assuming an enthalpy of vaporization of 2400 J/g, this corresponds to about 3% water.

In the 2nd heating run, the material is dry and the glass transition of the cured molding material is observed. In the curing process of the powder, the water eliminated by the reaction can escape (through the pierced lid). The curves of the cured samples show two small, reproducible effects at 50 and 115℃.

The decomposition temperatures in air and nitrogen atmospheres are to a large degree equivalent, which confirms the above conclusions about the endothermic nature of the decomposition.

计算 200℃下反应效应的表观焓只归一到吸热部分，见积分界限。

Evaluation The apparent enthalpy of the effect at 200℃ refers only to the endothermic part, see the integration limits.

	玻璃化转变 Glass transition	固化 Curing		吸热峰 Endothermic peak		分解 Decom position
	中点温度, ℃ Midpoint temperature in ℃	表观焓, J/g Apparent enthalpy in J/g	峰, ℃ Peak in ℃	表观焓, J/g Apparent enthalpy in J/g	峰, ℃ Peak in ℃	起始点, ℃ Onset in ℃
空气中粉末 Powder in air	53（第一次测试） 53 (1st run)	22	143	—	—	234
空气中固化条 Cured bar in air	144（第二次测试） 144 (2nd run)	—	—	60	203	234
氮气中固化条 Cured bar in nitrogen	145（第三次测试） 145 (2nd run)	—	—	55	202	236

结论 高压下模塑的部件在加工温度以上呈现吸热峰。如果在常压下进行固化，该峰就不存在，如果样品在峰温以上的温度退火，就消失。粉末状模塑料与成品之间的差异清晰可见。固化体系的玻璃化转变也可用 DMA 分析来测定。

Conclusions Components molded at high pressures exhibit endothermic peaks above the processing temperature. The peaks are not present if the curing is performed at normal pressure or disappear if the sample is annealed at temperatures above the peak temperature.
The differences between the molding compound in powder form and the finished product can be clearly seen. The glass transition of the cured system could also be determined by DMA analysis.

6.9 脲醛树脂：模塑料固化动力学
UF: Curing kinetics of molding compounds

目的 说明仅根据若干次 DSC 测试的动力学分析如何能给出固化时间的快速的全面评估。
在由填充模塑料的外壳制件的制造中，了解固化时间和反应温度之间的关系是重要的。这对工艺和循环周期的设计具有重要影响，不管对引进改性模塑料、对优化生产能力还是对重新设计模塑工艺。

Purpose To show how kinetic analyses based on just a few DSC measurements can give a rapid overview of curing times.
In the manufacture of parts for housings from filled molding compounds it is important to know the relationship between the curing time and the reaction temperature. This has an important influence on the design of the process and the cycle time, whether this is for introducing modified molding compounds, for optimizing throughput or for redesigning the molding process.

样品 适用于生产电气应用模塑外壳制件的含纤维素填料的粉末状脲醛树脂(UF1 型)模塑料。
Sample UF (Type UF 1) molding compound with cellulose filler in powder form for the production of parts of molded housings for electrotechnical applications.

条件	测试仪器：DSC	**Measuring cell**：DSC
Conditions	坩埚：	**Pan**：
	40μL 铝坩埚,盖钻孔	Aluminum 40μL with pierced lid
	样品制备：	**Sample preparation**：
	称量约 5mg 的粉末。	Approx. 5mg of the powder was weighed in.
	DSC 测试：	**DSC measurement**：
	在−20℃恒温 5min,以 5、10 和 20K/min 从−20℃升温至 190℃	Isothermal 5 min at −20℃, heating from −20℃ to 190℃ at 5, 10 and 20K/min
	气氛:干燥空气,50mL/min	**Atmosphere**：Dry air, 50mL/min

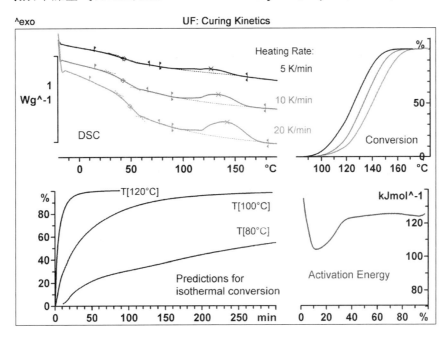

图 6.15 Fig. 6.15

解释 玻璃化转变和反应峰随着升温速率的增加向较高温度移动。比反应焓的值始终大约相同。因为这是一个填充体系,所以反应焓是填充模塑料而不是反应树脂体系的"表观"反应焓。

转化率曲线(图中右上图)是从反应峰积分计算的。用非模型动力学来计算表观活化能(在右下图中的曲线)和对三个等温转化率曲线的预测(左下)。

Interpretation The glass transition and the reaction peak shift to higher temperatures with increasing heating rates. The specific reaction enthalpy has always about the same value. Since this is a filled system, the reaction enthalpies are "apparent" reaction enthalpies that refer to the filled molding compound and not the reacting resin system.

The conversion curves (upper right diagram in the figure) were calculated from the reaction peaks by integration. Model free kinetics was used to calculate the apparent activation energy (the curve in the bottom right diagram) and the predictions for the conversion curves at three isothermal temperatures (bottom left).

计算 玻璃化转变和固化反应的计算结果汇总于下表：

Evaluation The results of the evaluation of the glass transitions and the curing reactions are summarized in the following table.

升温速率, K/min Heating rate in K/min	5	10	20
玻璃化转变 T_g (ASTM), ℃ Δc_p, J/gK Glass transition T_g (ASTM) in ℃ Δc_p in J/gK	41 0.56	42 0.53	47 0.53
峰温, ℃ Peak temperature in ℃	127	134	142
表观活化能, J/g Apparent reaction enthalpy in J/g	22.9	23.3	22.0

对质量 1 至 10mg 样品的其他测试表明,得到的反应焓几乎是相同的,平均值为 22.5J/g。不过对质量小于 5mg 的,出现较大的不确定性。未显示相应的曲线。

Other measurements with sample masses of 1 to 10 mg showed that practically the same reaction enthalpies resulted, with a mean value of 22.5 J/g. Larger uncertainties can however occur with masses less than 5 mg. The corresponding curves are not shown.

应用动力学:在三个等温反应温度下达到 20％、50％或 90％转化率的时间预测:

Applied kinetics: Prediction of the time to reach 20％, 50％ or 90％ conversion at the three isothermal reaction temperatures:

转化率 Conversion	80℃	100℃	120℃
20％	45min	6min	1min
50％	252min	26min	4min
90％	—	131min	17min

例如,如果固化在 80℃ 进行,那么 252min 后反应完成 50％(50％转化率)。这表明了必要的加工时间。MFK 计算表明,活化能只是在约 35％转化率后基本上是不变的。这表示固化反应初始与后来的进行是不同的。在 80℃ 下可最好地看到该效应对预测的影响。

For example, if curing is performed at 80℃, the reaction is 50％ complete (50％ conversion) after 252 min. This indicates the necessary processing times.

The MFK evaluation shows that the activation energy is only more or less constant after about 35％ conversion. This indicates that initially the curing reaction proceeds differently to afterward. The influence of this effect on the prediction is best seen at 80℃.

结论 对给定工艺条件的反应动力学行为可从仅仅几个 DSC 测试来评估。动力学分析能作出关于转化率、温度和时间的预测,甚至在实际上困难的甚至不可能达到的或费时太多的条件下。预测也可用来计划其他详细的测试。

预测的条件是优化生产的基础。结果应该用合适的测试来检验。

Conclusions The reaction kinetic behavior for given process conditions can be estimated from just a few DSC measurements. The kinetics analysis allows predictions concerning conversion, temperature and time to be made, even under conditions that are difficult or even impossible to achieve in practice or that take too much time. The predictions can also be used to plan other detailed measurements.

The predicted conditions are the basis for optimization of the production. The results should be verified using appropriate tests.

6.10 酚醛树脂：热导率的测定
PF: Determination of thermal conductivity

材料的热导率 λ 与热扩散系数 λ/(ρc_p) 是决定电子零件散热的重要性能。可用 DSC 测试来快速测定热固性树脂和其他聚合物的热导率，准确性约±10%。在此方法中，测量圆柱形样品即顶端盘片上纯金属的熔融。为了简化样品处理，金属被包含在坩埚中，放置在样品上。

The thermal conductivity, λ, together with the thermal diffusivity, $\lambda/(\rho c_p)$, of a material are important properties that determine the heat dissipation from electronic components. DSC measurements can be used to rapidly determine the thermal conductivity of thermosets and other polymers with an accuracy of about ±10%. In this method, the melting of a pure metal on top of a cylindrical sample or disk is measured. To simplify sample handling, the metal is contained in a crucible and placed on the sample.

目的	说明如何能用 DSC 来测定热导率 λ。	
Purpose	To show how the thermal conductivity, λ, can be determined by DSC.	
样品	酚醛树脂板。	
Sample	Phenolic resin plates.	
条件	测试仪器：DSC	**Measuring cell**：DSC
Conditions	坩埚：	**Pan**：
	含纯镓的无盖 20μL 铝坩埚。为了防止镓与铝坩埚形成合金，坩埚内表面涂了一油漆薄层。纯镓（熔点 29.8℃）对测试接近室温的热导率是理想的。镓（80mg）必须冷却至至少 10℃令其结晶。	Aluminum 20μL without lid to contain pure gallium. To prevent the gallium forming an alloy with the aluminum crucible, the inside surface of the crucible was coated with a thin film of lacquer. Pure gallium (melting point 29.8℃) is ideal for measuring thermal conductivity close to room temperature. The gallium (80 mg) has to be cooled down to at least 10℃ for it to crystallize.
	样品制备：	**Sample preparation**：
	从板冲出一个圆柱体（即盘片）并仔细磨光直至完全为圆柱形。用细砂纸细心磨光圆形末端的表面，因而高度测定至少可准确到 10μm 和直径到 20μm。样品的直径与坩埚底部的直径大致相同（6mm）。	A cylinder (i.e. a disk) was punched out of the plate and carefully ground until it was perfectly round. The circular end faces were carefully polished with fine emery paper so that the height could be accurately determined to at least 10μm and the diameter to 20μm. The diameter of the samples was about the same as that of the bottom of the crucible (6 mm).
	直径 6.40mm，高 1.432mm。	Diameter 6.40 mm, height 1.432 mm
	DSC 测试：	**DSC measurement**：
	以 0.5K/min 从 28℃升温至 38℃，然后以 10K/min 降温至 −5℃以使镓在测试后再次凝固。	Heating from 28℃ to 38℃ at 0.5K/min, then cooling to −5℃ at 10 K/min so that the gallium solidifies again after the measurement.
	气氛：氮气，50mL/min	**Atmosphere**：Nitrogen，50mL/min

背景 为了测定固体材料的热导率,将一个纯金属圆片(例如铟或镓)放在样品的上部圆形末端表面上(见参考文献)。样品不用坩埚直接放在 DSC 测试传感器上。在测试中,金属达到它的熔点,在金属熔化时温度保持不变。在该瞬间,圆柱体上端表面的温度由熔点测定。圆柱体下端表面的温度和进入圆片的热流由 DSC 测量。样品的热导率于是能从圆柱体上下端表面之间的温差和热流计算。

Background To determine the thermal conductivity of solid materials a disk of a pure metal (e.g. indium or gallium) is placed on the upper circular end surface of the sample (See Reference). The sample is placed directly on the DSC measuring sensor without using a crucible. During heating, the metal reaches its melting point and the temperature remains constant while the metal melts. At this instant, the temperature of the upper end surface of the cylinder is determined by the melting point. The temperature of the lower end surface of the cylinder and the heat flow into the disk are measured by the DSC. The thermal conductivity of the sample can then be calculated from the temperature difference between upper and lower end surfaces of the cylinder and the heat flow.

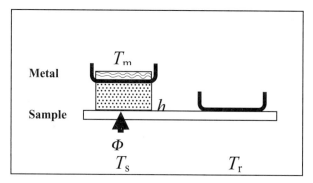

图 6.16 DSC 传感器上样品排列的示意图。h 是样品圆柱体的高度; Φ 是从传感器流入样品的热流;T_m 是熔融金属的温度; T_s 是样品下的传感器温度;T_r 是参比样品的温度。 参比坩埚是空的。含纯金属的同一型号的坩埚放在样品上。

Fig. 6.16 Schematic diagram of the sample arrangement on the DSC sensor. h is the height of the sample cylinder; Φ the heat flow that flows from the sensor into the sample; T_m the temperature of the metal melt, T_s the sensor temperature under the sample; T_r the temperature of the reference sample. The reference crucible is empty. A crucible of the same type containing the pure metal is placed on the sample.

在静止的条件下,通过热阻 R_s 的样品的热流 Φ 与温差 ΔT 成正比:

Under stationary conditions, the heat flow, Φ, through the sample with a thermal resistance, R_s, is proportional to the temperature difference, ΔT:

$$\Phi = \frac{1}{R_S + R_T} \Delta T$$

材料的热阻 R_s 由依赖于材料的热导率和样品的几何形状给出:

The thermal resistance, R_s, of the material is given by the material-dependent thermal conductivity and the geometry of the specimen:

$$R_S = \frac{h}{\lambda A}$$

式中 λ 是热导率,A 是截面积,h 是样品的高度。

Here λ is the thermal conductivity, A the cross-sectional area, and h the height of the specimen.

从传感器至纯金属的热流不仅取决于样品的热阻,而且取决于传感器—样品(R_1)和样品—金属(R_2)界面的热阻。

The heat flow from the sensor to the pure metal depends not only on the thermal resistance of the sample, but also on the thermal resistances at the sensor-sample (R_1) and sample-metal (R_2) interfaces.

为了确保热阻 R_1 和 R_2 是可重复的,界面处的空间用硅油充满。因此,如果始终使用同样的样品截面积 A,则可以假定 R_1 和 R_2 与样品无关。

如果已知 Φ、R_T、T_m 和 T_s,现在 R_s 和样品的热导率就能测定了。熔融时 T_m 值是已知的,因为使用了纯金属。Φ 和 T_s 的值从 DSC 测试得到,R_T 能通过进行几次测试测定。

熔融时 ΔT 是时间 t 时的温度 T_s 与金属熔点(分别是熔融开始的温度 T_{onset})之差。对应的热流 Φ 是相同时间 t 时的热流与熔融开始时的热流之差。

因此 S 是熔融峰线性边的斜率。
如果用同一材料但圆柱高度不同的两个样品测试,λ 能如下计算:

To ensure that the thermal resistances R_1 and R_2 were reproducible, the spaces at the interfaces were filled with silicone oil. It can therefore be assumed that R_1 and R_2 are independent of the sample if the same sample cross-sectional area, A, is always used.

R_s and hence the thermal conductivity of the sample can now be determined if Φ, R_T, T_m and T_s are known. The value of T_m during melting is known because a pure metal is used. The values of Φ and T_s are obtained from the DSC measurement, and R_T can be determined by performing several measurements.

During melting ΔT is the difference between the temperature T_s at a time t and the melting point of the metal (respectively the temperature of the beginning of melting, T_{onset}). The corresponding heat flow Φ is the difference between the heat flow at the same time t and the heat flow at the beginning of melting.

$$\frac{\Phi}{\Delta T} = \frac{\Phi_t - \Phi_{onset}}{T_t - T_{onset}} = S$$

S is therefore the slope of the linear side of the melting peak.
If two samples of the same material but with different cylinder heights are measured, λ can be calculated as follows:

$$\lambda = \frac{\Delta h}{A\left(\dfrac{1}{S_2} - \dfrac{1}{S_1}\right)}$$

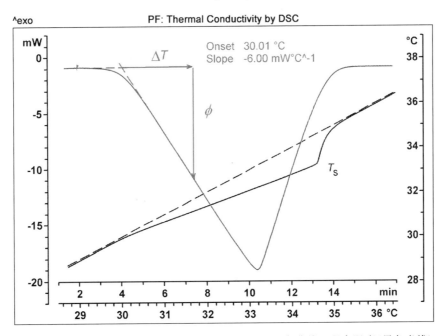

6.17 镓的熔融峰。热流(红色曲线)、样品底部温度(黑色曲线)、程序温度(黑色虚线)。
Fig. 6.17 Melting peak of gallium. Heat flow (red curve), sample bottom temperature (black curve), program temperature (black dashed curve).

式中 Δh 是圆柱体高度差(h_2-h_1)，S_1 是较小样品的 DSC 曲线的斜率，S_2 是较大样品的 DSC 曲线的斜率。

为了简化，可从无样品的测试一次测定 S_1。

where Δh is the difference of the cylinder heights (h_2-h_1), S_1 is the slope of the DSC curve of the smaller sample, and S_2 the slope of the DSC curve of the larger sample.

For simplicity, S_1 can be determined once from a measurement with no sample.

解释 红色曲线是与样品温度关系测试的在酚醛树脂样品上的镓的熔融 DSC 曲线。黑色曲线表示在样品下端的样品温度的进程。虚线是程序温度（参比温度）。起始点计算的结果给出斜率，用灰色箭头标记。

Interpretation The red curve is the DSC curve of the melting of gallium on the phenolic resin sample measured as a function of the sample temperature. The black curve shows the course of the sample temperature at the lower end of the sample. The dashed curve is the program temperature (reference temperature). The result of the onset evaluation gives the slope, which is marked by a gray arrow.

计算 通过样品温度(T_s)的 DSC 曲线，来测定斜率 S。在测定起始温度时，必须确保切线精确地拟合在 DSC 曲线上。斜率 S 可直接读出。

Evaluation To determine the slope, S, the DSC curve has to be plotted against the sample temperature (T_s). In the determination of the onset temperature, one must make sure that the tangents fit exactly on the DSC curve. The slope, S, can be read off directly in the result block.

样品 Sample	文献值 Literature value $Wm^{-1}K^{-1}$	h Mm	D mm	S $mW \cdot K^{-1}$	λ $Wm^{-1}K^{-1}$	与文献差 Difference to literature %
只含镓的坩埚 Only crucible with gallium		0	5.80	90.10		
PF 样品 PF sample	0.30	1.432	6.40	6.00	0.286	-4.6

结论 在特定温度下，用 DSC 测量法可测得已知尺寸的圆柱体样品的热导率。

11 个不同聚合物的结果与文献数据的比较表明，绝对值低于文献值 4%，标准偏差约为 10%。

样品圆柱体即盘片应该高约 1 至 3mm，必须与含纯金属的坩埚底部直径相同。在熔融状态，金属必须完全覆盖坩埚底部。如果镓用作参比金属，铝坩埚必须在内部用油漆涂层以防止合金形成。

DSC 应该只用坩埚和纯金属（即无圆柱体样品）以通常的方法校正，

Conclusions The thermal conductivity of cylindrical samples with accurately known dimensions can be determined at a particular temperature using DSC measurements.

Comparison of the results of 11 different polymers with literature data showed that the absolute values were 4% lower than the literature values with a standard deviation of about 10%.

The sample cylinder or disk should be about 1 to 3 mm in height and must have the same diameter as the bottom of the crucible containing the pure metal. In the molten state, the metal must completely cover the bottom of the crucible. If gallium is used as the reference metal, the aluminum crucible must be coated internally with a lacquer in order to prevent alloy formation.

The DSC should be adjusted in the usual way with just the crucible and the pure metal (i.e. without the cylindrical sample)

以使熔融焓和熔融温度与文献值吻合。

参考文献 G. Hakvoort, L. L. Van Reijen, Thermochimica Acta, 93 (1985) p. 317

so that the enthalpy of melting and the melting temperature agree with literature values.

References G. Hakvoort, L. L. Van Reijen, Thermochimica Acta, 93 (1985) p. 317

7 甲基丙烯酸类树脂 Methacrylate/Acrylic resins (MMA)

7.1 牙科复合材料的光固化 Light curing of a dental composite

光固化,也称作光引发聚合或光聚合,当今在牙科中广泛应用来固化复合填补材料和用于填补齿洞的密封填充物。牙科树脂中的主要吸收剂是光引发剂(经常是樟脑醌,一种青光光引发剂)。它启动光聚合过程。这意味着填补材料能如糊状物一样嵌入孔洞、成形,或以另外的方式储备,牙科医生没有时间压力。当一切准备好时,固化反应通过照光启动。

因为填补材料吸收光,所以光固化在技术上只对一定厚度的材料是可能的。对于厚度大于2mm的填补物,就需要使用所谓的三明治技术。这种填补物的质量与光的强度,特别在开始时,和曝光时间有关。太高的光强度会对所谓的边缘配合带来负面影响,而如果强度太低,则影响材料的物理性能。

Light curing, also known as photoinduced polymerization or photopolymerization, is nowadays widely employed in dentistry to cure composite filling materials and sealants used to fill cavities in teeth. The primary absorber in dental resins is the photoinitiator (very often camphorquinone, a blue light photoinitiator). This starts the photopolymerization process. This means that the filling material can be inserted as a paste into the cavity, shaped and otherwise prepared without the dental surgeon being under pressure of time. When everything is ready, the curing reaction is started by supplying light.

Since the filling material absorbs light, light curing is only technically possible to a certain thickness of material. With fillings thicker than 2 mm, the so-called sandwich technique is used. The quality of such fillings depends on the light intensity, especially at the beginning, and the exposure time. A light intensity that is too high has a negative effect on the so-called margin fit, whereas if the intensity is too low, the physical properties are negatively affected.

目的 / **Purpose**	说明牙科复合材料固化速度快和用DSC-光量热仪系统来测定需要的曝光时间。 To show how rapidly a dental composite cures and to determine the required exposure time using a DSC-photocalorimeter system.	
样品 / **Sample**	糊状牙科复合材料。 Dental composite in paste form.	
条件 / **Conditions**	测试仪器: 带光量热附件的DSC	**Measuring cell**: DSC with photocalorimeter accessory
	坩埚: 无盖40μL铝坩埚	**Pan**: Aluminum 40μL, no lid
	样品制备: 23.831mg糊状物均匀地涂在坩埚底上。	**Sample preparation**: 23.831 mg of the paste was spread uniformly over the bottom of the crucible.
	DSC测试: 在37℃等温,曝光50s,强度为220mWcm^{-1}(在400至500nm之间高光谱输出的白光)	**DSC measurement**: Isothermally at 37℃, 50s exposure time with an intensity of 220 mWcm^{-1} (white light with a high spectral output between 400 and 500nm)
	气氛: 干燥空气,40mL/min	**Atmosphere**: Dry air, 40mL/min

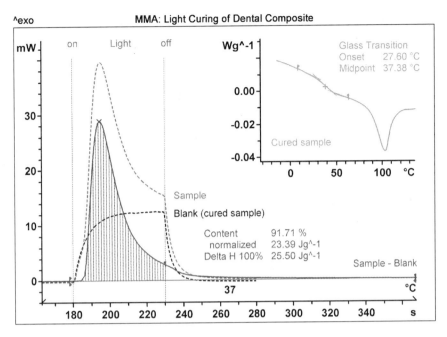

图 6.18 Fig. 6.18

解释 图示为曝光前短时间、曝光中和曝光后的 DSC 曲线。为了测定净反应功率(红色曲线),再次测试了固化样品(Blank)并被第一次测试(Sample)减去。总反应焓为 25.5J/g。在切断光时,已达到总转化率的 92%,即之后反应 8%。本例中的曝光时间有点太短。不过材料继续反应因而达到了足够的固化度。

插入图表示在等温测试后的样品的升温曲线。玻璃化转变温度为 37℃。这意味着在体温下固化材料部分处于橡胶弹性态。玻璃化转变后的吸热峰是由于复合材料另一组分的熔融引起的。

结论 DSC-光量热法能用来定量跟踪光固化反应的进程。牙科应用的最佳条件可通过改变光强度和曝光时间来快速测定。于是,就能表征获得最佳材料质量的光类型和曝光时间与每类复合材料的填补寿命。

Interpretation The figure shows the DSC curves shortly before, during and after exposure to light. To determine the net reaction power (red curve), the cured sample was measured again (Blank) and subtracted from the first measurement (Sample). The total reaction enthalpy is 25.5 J/g. At the moment the light is switched off, 92% of the total conversion has been reached, i.e. 8% reacts afterward. The exposure time in this case was a bit too short. Nevertheless the material continues to react so that a sufficient degree of cure is achieved.

The inserted diagram shows the heating curve of the sample after the isothermal measurements. The glass transition temperature is 37℃. This means that at body temperature the cured material is partially in the rubbery-elastic state. The endothermic peak after the glass transition is due to the melting of another component of the composite.

Conclusions DSC-photocalorimetry can be used to quantitatively follow the course of light curing reactions. The optimum conditions for dental applications can be rapidly determined by varying the light intensity and exposure time. It would then be possible to characterize the type of light and exposure time for optimum material quality and longevity of the filling for each type of composite.

8 聚氨酯体系 PUR systems

8.1 聚氨酯：含溶剂的双组分体系
PUR: Two-component system with solvent

目的 Purpose	说明用 DSC 测试含溶剂的样品。 为了改善加工性，溶剂（例如甲基乙基酮）被加到聚氨酯树脂中以调节黏度。这就有必要研究溶剂如何影响了反应的进程和是否在工艺控制中必须予以考虑。 To demonstrate the use of DSC measurements of samples with solvent. Solvents (e.g. methylethylketone) are added to polyurethane resins to adjust the viscosity in order to improve processability. This makes it necessary to investigate how the solvent affects the course of the reaction and whether this must be taken into account in process control.
样品 Sample	适合于耐侯聚合物薄膜制品的由多元醇和异氰酸酯组分组成的含溶剂的双组分聚氨酯体系。 Solvent-containing two-component PUR system consisting of a polyol and an isocyanate component for weather-proof polymer film laminates.
条件 Conditions	测试仪器：DSC 坩埚： 40μL 铝坩埚，盖钻孔 样品制备： 混合含溶剂的多元醇异氰酸酯组分，直接加入坩埚。 在混合组分后约 30min 开始测试。 DSC 测试： 第一轮：以 5K/min 从 25℃ 升温至 205℃，然后以 20K/min 降温至 25℃，在 25℃ 等温 2min。 第二轮：以 20K/min 从 25℃ 升温至 205℃。 气氛： 干燥空气，50 mL/min	**Measuring cell**：DSC **Pan**： Aluminum 40μL with pierced lid **Sample preparation**：： The solvent-containing polyol isocyanate components were mixed and filled directly into the crucibles. The measurement was started about 30 min after mixing the components. **DSC measurement**： 1st run：Heating from 25℃ to 205℃ at 5 K/min, 2 min at 205℃ followed by cooling to 25℃ at 20 K/min, 2 min at 25℃ and 2nd run：Heating from 25℃ to 205℃ at 20 K/min. **Atmosphere**： Dry air, 50 mL/min

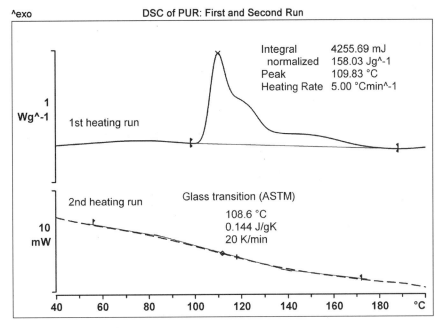

图 8.1　Fig. 8.1

解释　DSC 曲线表示固化行为。加入坩埚时含溶剂的样品,不呈现吸热的蒸发峰。这间接地表明实际上所有溶剂在早于测试开始前的时间里已经蒸发了。

在第二次测试中,测试曲线显示宽大扁平的玻璃化转变。

Interpretation　The DSC curve shows the curing behavior. The sample that contained solvent when it was filled into the crucible does not exhibit an endothermic vaporization peak. This suggests that practically all the solvent has evaporated in the time prior to starting the measurement.

In the 2nd run, the measurement curve shows a broad glass transition.

计算 **Evaluation**	峰,℃ Peak in ℃	反应焓,J/g Reaction enthalpy in J/g	玻璃化转变,℃ Glass transition in ℃
含已蒸发溶剂的聚氨酯 PUR with evaporated solvent	110	158.0	108

结论　DSC 测试能用来检查溶剂是否影响反应进程。在本特例中,溶剂在测试前已完全蒸发。

Conclusions　DSC measurements can be used to check whether a solvent influences the course of a reaction. In this particular case, the solvent was completely evaporated before measurement.

8.2　聚氨酯:在不同温度下加成聚合
PUR: Polyaddition at different temperatures

目的　用DSC测试来说明加聚反应进行速度较快。该信息能够直接检查和优化固化温度和时间。

Purpose　To use DSC measurements to show how fast the polyaddition reaction proceeds. This information allows curing temperatures and times to be directly checked and optimized.

样品　适合于耐侯聚合物薄膜制品的由多元醇和异氰酸酯组分组成的双组分聚氨酯体系。在生产中,混合后的组分应用于薄膜,然后在 120℃固化 1min。

Sample Two-component PUR system consisting of a polyol and an isocyanate component for weather-proof polymer film laminates. In production, the mixed components are applied to a film and then cured at 120℃ for 1min.

Conditions

Measuring cell: DSC

Pan:
Aluminum 40μL with pierced lid

Sample preparation:
The polyol and isocyanate components were mixed and the solvent evaporated in a vacuum drying cabinet at room temperature for about 1h. The crucibles were then filled.

DSC measurement:
Isothermally at 100℃, 110℃ and 120℃ for up to 60 min

Atmosphere:
Dry air, 50mL/min

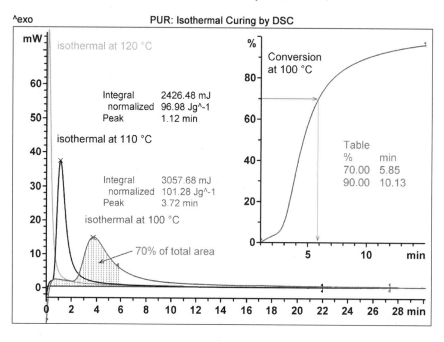

Fig. 8.2

Interpretation At higher temperatures, the polyaddition reaction begins more rapidly, the reaction peak is higher and the reaction time is shorter. At 120℃, the reaction is so fast that the DSC curve does not cover the whole enthalpy of reaction. The DSC curve at 100℃ clearly shows that an initial reaction occurs first during a so-called induction period of 2min. The main reaction that follows has still not quite finished even after 30min.

Evaluation The time for a certain conversion is determined by

分积分测定；图中表示对于 100℃和 70%的例子。从转化率曲线可直接读出值或表示为表格。90%的转化率在 100℃需要至少 10min 的反应时间。

partial integration; an example is shown in the diagram for 100℃ and 70%. Values can be read off from the conversion curve or presented as a table. A conversion of 90% requires a reaction time of at least 10 min at 100℃.

反应温度,℃ Reaction temperature in℃	反应焓,J/g Reaction enthalpy in J/g	峰,min Peak in min	70%转化率的时间,min Time for 70% conversion in min	90%转化率的时间,min Time for 90% conversion in min
100	101.3	3.7	5.9	10.1
110	97.0	1.1	1.8	3.8

结论 加聚反应的进程能用等温 DSC 测试及其测定的最佳固化温度和时间来表征。
在高温，反应很快，只可能通过使用动力学研究和预测来优化过程条件。

Conclusions The course of the polyaddition reaction can be characterized by isothermal DSC measurements and the optimum curing temperature and time determined.
At high temperatures, the reactions are very fast and the process conditions could only be optimized through the use of kinetic studies and predictions.

8.3 聚氨酯漆涂层的软化温度 Softening temperature of PUR lacquer coatings

目的 固化的质量控制和软化温度。

Purpose Quality control of curing and the softening temperature.

样品 胶合板上的聚氨酯涂层(镶木地板漆)，样品 1 至 3。

Sample PUR coatings on plywood (lacquer for parquet flooring), Samples 1 to 3.

条件

测试仪器：TMA
球点探头

样品制备：
用刀分离油漆涂层。样品厚度约 0.5mm。

TMA 测试：
以 5K/min 从 25℃升温至 150℃；
探头末端上的力为 0.05N。

气氛：静态空气

Conditions

Measuring cell: TMA
Ball-point probe

Sample preparation:
The lacquer coating was separated using a knife. Sample thickness was about 0.5 mm.

TMA measurement:
25℃ to 150℃ at 5 K/min;
Force of 0.05 N on the probe tip.

Atmosphere: Static air

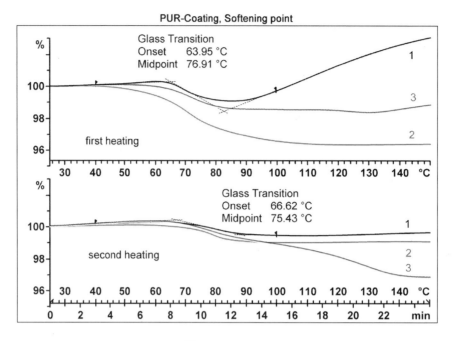

图 8.3 Fig. 8.3

解释 图示为纵坐标表示为原始样品厚度百分数的第一次和第二次 TMA 升温曲线。对每种情况,都用同一样品进行第一次和第二次升温。在第二次测试中,冷却后探头在样品上更换位置(即在不同部位上)。

第一次和第二次升温的软化行为是非常不同的,因为预期在较高温度下有后固化反应。与样品 1 和 2 不同,样品 3 始终显示两次软化行为,它表明存在两层。样品 1 与样品 2 不同最明显的是在第一次升温,其中样品 1 在软化点后膨胀相当快。

计算 测定作为软化特征值的起始温度(玻璃化转变温度)。对所有曲线,40℃和100℃用作切线的界限。

Interpretation The figure shows the first and second TMA heating curves with the ordinate expressed as a percent of the original sample thickness. In each case, the 1st and 2nd heating runs were performed with the same sample. In the 2nd run, the probe was repositioned on the sample (i.e. on a different part) after cooling.

The softening behavior of the 1st and 2nd heating runs is very different because at higher temperatures a postcuring reaction is expected. In contrast to Samples 1 and 2, Sample 3 always shows two-step softening behavior, which indicates the presence of two layers. Sample 1 differs from Sample 2 most noticeably in the 1st heating run where Sample 1 expands relatively rapidly after the softening point.

Evaluation Determination of the onset temperature as characteristic value for softening (glass transition temperature). For all curves, 40℃ and 100℃ were used as limits for the tangents.

样品 Sample	第一次升温 1st heating run 起始点,℃ Onset in ℃	第二次升温 2nd heating run 起始点,℃ Onset in ℃
1	64	67
2	58	69
3	61	64

样品 2 第一次和第二次升温的起始温度的差异表现出最大,因此此例的固化最差。

结论 热机械分析能够辨别不同涂层,特别是软化行为。在第一次和第二次升温之间有显著的不同,最可能的原因是由于后固化。

注:因为木材膨胀,所以对于评估不同涂层的质量来说,涂层连同木材基板一起测试是不合适的。因此涂层必须分开,而可能有少量木材附着在涂层上。不过,与刮下油漆涂层并在坩埚中压实的 DSC 测试相比,这类样品制备是更简单的。

Sample 2 exhibits the largest difference in the onset temperatures of the 1st and 2nd heating runs. Curing is therefore poorest in this case.

Conclusions Thermomechanical analysis can distinguish between different coatings, especially with respect to softening behavior. There is a significant difference between the 1st and 2nd heating runs, most probably due to postcuring.

Note: Measurement of the coating together with the wood substrate is not suitable for assessing the quality of different coatings because of the expansion of the wood. The coating must therefore be separated, whereby it is possible that a small amount of wood remains attached to the coating. This type of sample preparation is however simpler than scraping off the lacquer coating and compacting it in a crucible for DSC measurements.

8.4 聚氨酯模塑料:作为质量标准的玻璃化转变
PUR casting compounds: Glass transition as a quality criterion

聚氨酯浇铸树脂用于中压变压器的制造。为了达到良好的绝缘性能和低介电损耗因子,使通常为大制件的每一部分都彻底硬化是必需的。测定玻璃化转变温度作为固化的质量标准。为此目的在一些特定部位取样保存,用 DSC 进行测试。这种测试是按常规进行的,以检查在大炉体中的固化过程是否已经真正完成。

在随机样品检查中,变压器客户自己进行 DSC 测试,发现得到的玻璃化转变温度(84℃)不满足指标值 90℃。而变压器的供应商已经测得的玻璃化转变温度为 94℃ 确认为正确,因而交货。
这产生了下面的重要问题:
1. 如何正确地测试玻璃化转变温度才能使比较有意义?
2. 固化度不同以及因此导致的玻璃化转变温度的不同发生在哪里?

Polyurethane casting resins are used in the manufacture of medium voltage transformers. To achieve good insulating properties and a low dielectric loss factor, thorough hardening throughout the often large components is necessary. The glass transition temperature is determined as a quality criterion for curing. This is done by taking samples at certain places reserved for this purpose and measuring them by DSC. Such measurements are routinely performed to check whether the curing process in the large furnaces has been properly done.

In a random sample check, the transformer customer performed DSC measurements himself and discovered that the glass transition temperature (84℃) he obtained did not satisfy the specified value of 90℃. The supplier of the transformer had however measured the glass transition temperatures correctly at 94℃ and had therefore proceeded with the delivery.
This leads to the following important questions:
1. How can the glass transition temperature be correctly measured for comparison purposes?
2. Where do the differences in the degree of cure and hence in the glass transition temperature occur?

目的	说明如何测试玻璃化转变温度才具有比较意义。 此外,讨论导言里问题的答案和关于结果不同的原因。
Purpose	To show how glass transition temperatures can be measured so that they can be meaningfully compared. Furthermore, the answers to the questions in the introduction are discussed and the reasons for the different results.
样品	三片固化的聚氨酯树脂,第1至第3号,由树脂生产商提供,进行比较测试。样品含61%填料。
Sample	Three cured pieces of polyurethane resin, No. 1 to 3, supplied by the resing producer to perform comparison measurements. The samples contained 61% filler material.
条件	测试仪器:DSC
Conditions	

坩埚:	**Measuring cell**:DSC
40μL 铝坩埚,盖钻孔	**Pan**: Aluminum 40μL with pierced lid
样品制备:	**Sample preparation**:
用侧剪钳剪下约25mg。	Approx. 25 mg were cut off using side-cutting pliers.
DSC 测试:	**DSC measurement**:
以 10K/min 从 40℃升温至 150℃	Heating from 40℃ to 150℃ at 10K/min
气氛:氮气,50mL/min.	**Atmosphere**:Nitrogen, 50mL/min.
步骤:	**Procedure**:

1) 为了处理第一点,所有有关当事人被请求在相同条件下校准他们的 DSC 系统,用铟作基准物(熔点 156.6℃)和 10K/min 的升温速率(用于测定玻璃化转变温度的相同速率)。来自同一块板的聚氨酯片发送给有关当事人让他们测试和计算玻璃化转变。结果的比较表明,所有的实验室得到了误差在±1K 内的相同聚氨酯玻璃化转变值。因而这保证了有关实验室提供的可比较的结果。

1) To deal with the first point, all the parties involved were requested to calibrate their DSC systems under the same conditions, using indium as reference substance (melting point 156.6℃) and a heating rate of 10 K/min (the same rate used for the determination of the glass transition). Pieces of polystyrene from the same plate were sent to the parties involved for them to measure and evaluate the glass transition. A comparison of the results showed that all the laboratories obtained the same value for the glass transition of polystyrene within a scatter of ±1 K. This thereby ensured that the laboratories involved supplied comparable results.

2) 由各实验室进行的 DSC 测试用来检查固化的聚氨酯树脂样品1至3号是否呈现必需的玻璃化转变温度。为此所有实验室均采用下面给出的测试条件,由下面描述的测试进行比较。

2) DSC measurements performed by the laboratories were used to check whether the cured PUR resin Samples 1 to 3 showed the required glass transition temperatures. To do this the measurement conditions given below were used in all the laboratories. The measurements described below show this comparison.

3) 第三步,在供应商设施处检查了变压器的生产工艺(炉温、时间等)。结果可在本文末找到。

3) In a third step, the manufacturing process of the transformer was checked at the supplier's facility (furnace temperatures, time, etc.). The results can be found at the end.

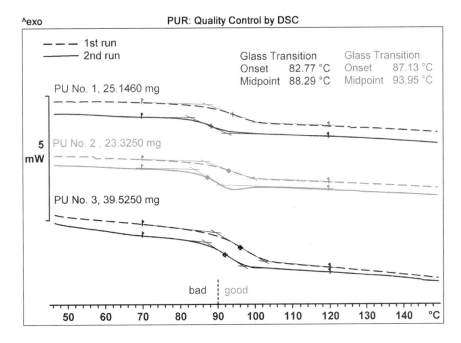

图 8.4　Fig. 8.4

解释　DSC 曲线清晰地呈现玻璃化转变,尽管填料含量相当高。样品 2 在第一次升温中显示小的松弛效应,可能是由于它比其他样品冷却慢。在第二次升温中,由于在第一次升温至 150℃ 时的后固化,导致玻璃化转变温度的提高非常明显,虽然没有观察到放热反应峰。属于作为质量标准极限的 90℃ 接近测得的玻璃化转变温度范围。因此为了达到质量目标,需要更长时间的固化。

Interpretation　The DSC curves show the glass transitions clearly despite the relatively high filler content. Sample 2 exhibits a small enthalpy relaxation effect in the 1st heating run, which is possibly due to it being cooled more slowly than the other samples. In the second run, the increase in the glass transition temperature due to postcuring on heating to 150℃ in the first run is very noticeable, although an exothermic reaction peak is not observed. The limit of 90℃ as quality criterion is almost in the range of the measured glass transition temperatures. Longer curing is therefore needed to achieve the quality target.

计算　玻璃化转变温度的中点温度 (T_g,℃)汇总于表:

Evaluation　The midpoint temperatures of the glass transitions (T_g in ℃) are summarized in the table:

样品/加热测试 Sample/Heating run	供应商 Supplier	客户 Customer	梅特勒-托利多实验室 MT laboratory	ΔT_g,K ΔT_g in K
1/1	89.8	90.8	88.3	
1/2	95.3		94.0	5.7
2/1	93.2	94.7	87.4	
2/2	97.0		93.0	5.6
3/1	94.4	93.9	91.8	
3/2	98.6		95.7	3.9

当比较各个温度时,应考虑到与温度有关的切线如何取,变化可以在 0.5K 左右。供应商和客户实验室的

When comparing the individual temperatures, one should take into account that they can vary by about 0.5K depending on how the tangents are drawn. The results of the supplier and customer

结果比梅特勒－托利多实验室的高约3K。

在两个实验室的结果中，由于后固化导致的玻璃化转变温度的提高可清晰地观察到，且是显著的。

结论 比较测试表明如下：
1. 从各实验室得到的结果在通常的重复性范围内是相同的。样品2由于焓松弛呈现最大的偏差。
2. 玻璃化转变温度的测定要求所有有关实验室严格使用相同的实验条件（升温速率、样品质量、坩埚等）。
3. 必须非常细心地取用于测定 T_g 的切线。
4. 质量极限90℃低于通常的校准物铟的熔点66K。为了避免外推误差，DSC须用另外的物质如萘（80.3℃）或硬脂酸（69.4℃）校准。

因此，对客户最初得到的、评价为不足的84℃玻璃化转变温度，不能够用不同的测试技术来解释。

供应商和客户之间的进一步研究表明，必须改进工艺：
- 用于评估固化度的样品取自变压器上不同的位置。因此新位置必须以取得代表性样品来确定。
- 生产炉中的温度梯度太大。在相当短的几个小时的加热和固化时间中，不是所有的变压器部件达到了必需的最终温度。这导致了来自同一变压器的样品的玻璃化转变温度差异大。

因此，对大部件必须给予足够的时间加热和达到树脂块内的温度平衡，这对产量当然有负面影响。

laboratories are about 3K higher than those of the MT laboratory.

The increase of the glass transition temperature through postcuring in the results of both the laboratories can be clearly seen and is significant.

Conclusions The comparative measurements showed the following:
1. The results obtained from the laboratories were the same within the usual limits of reproducibility. Sample 2 shows the largest deviations due to enthalpy relaxation.
2. The determination of the glass transition temperatures requires all the laboratories involved to use exactly the same experimental conditions (heating rate, sample size, crucible, etc.).
3. The tangents used for the determination of T_g must be very carefully drawn.
4. The quality limit of 90℃ is 66 K below the melting point of usual calibration substance indium. To avoid extrapolation errors the DSC's have to be calibrated with an additional substance such as naphthalene (80.3℃) or stearic acid (69.4℃).

The glass transition temperature of 84℃ initially obtained by the customer and criticized as insufficient cannot therefore be explained by different measurement techniques.

Further investigations between the supplier and the customer showed that the process must be improved:
- The samples to assess the extent of cure were taken from different places on the transformer. It follows that a new place must be defined for taking a representative sample.
- The temperature gradients in the production furnace were too large. In the relatively short heating and curing times of a few hours, not all the parts of the transformer reached the necessary final temperature. This led to the large differences in the glass transition temperatures of samples from the same transformer.

With large components, sufficient time must therefore be allowed for heating and temperature equilibration within the resin mass, which of course has a negative effect on throughput.

9 其他树脂体系 Other resin systems

9.1 双马来酰亚胺树脂—碳纤维：贮存温度对预浸料黏性的影响
BMI-CF: Influence of storage temperature on tackiness of prepregs

目的 例如适用于航空器制造的双马来酰亚胺预浸料的加工要求各层在固化前有足够的黏性。黏性由于树脂的交联而降低。

本应用的目的是找到贮存后的"表观"剩余焓与预浸料的黏性的关系。为此,预浸料在不同温度贮存几个月。这能找到贮存条件(温度和时间)的标准,在这些条件下材料仍可被加工(所谓的外置寿命)。

Purpose The processing of bismaleimid prepregs, for example for aircraft construction, requires layers to be sufficiently tacky (sticky) before curing. The tackiness decreases due to crosslinking of the resin.

The purpose of the application is to relate the "apparent" residual enthalpies after storage with the tackiness of the prepreg. To do this, the prepregs were stored at different temperatures for several months. This allows a criterion for storage conditions (temperature and time) to be found, under which the material can still be processed (the so-called outlife).

样品 由碳纤维和双马来酰亚胺树脂制成的预浸料 BMI—CF。
Sample BMI-CF, prepreg made of carbon fibers and BMI resin matrix.

条件 测试仪器:DSC   **Measuring cell**:DSC
Conditions 坩埚: **Pan**:
40μL 铝坩埚,盖钻孔   Aluminum 40μL with pierced lid
样品制备: **Sample preparation**:
预浸料在 −20℃、5℃ 和 23℃ 下贮   The prepregs were stored at −20℃, 5℃ and 23℃
存超过 15 个月的时间。贮存后,用   over a period of 15 months. After storage, the
刀从预浸料切出样品。   samples were cut out of the preregs using a knife.
DSC 测试: **DSC measurement**:
以 10K/min 从 25℃ 升温至 350℃   Heating from 25℃ to 350℃ at 10 K/min
气氛:干燥空气,50 mL/min   **Atmosphere**:Dry air, 50 mL/min

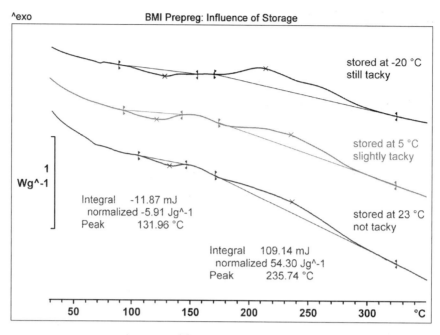

图 9.1　Fig. 9.1

解释　图示为三个树脂的第一次升温。样品的固化以两个阶段发生：第一是在其中排除挥发性产物的缩合反应（90 至 155℃），然后第二是在较高温度（高于 170℃）的实际固化阶段。

贮存在－20℃的预浸料仍是黏性的，而 5℃贮存的预浸料的黏性对适当的加工已不足。在 23℃贮存温度下，预浸料不再是黏性的，树脂在形变和机械应力下有破裂的倾向。

这些测试能对该预浸料体系的足够加工性（黏性）的极限值作出规定：例如对预浸料归一化的在较低温度范围内大于约 11J/g（吸热）的表观剩余焓和在较高温度范围内 80J/g（放热）的表观剩余反应焓。不过应该明白，这些不是每单位树脂值而与碳纤维含量有关。

Interpretation　The figure shows the 1st heating runs of the three resins. The curing of the samples occurred in two steps: first the condensation reaction (90 to 155℃) in which the volatile products were eliminated, and then second the actual curing step at higher temperature (above 170℃).

While the prepreg stored at －20℃ was still tacky, the tackiness of the prepreg stored at 5℃ was already insufficient for adequate processing. At 23℃ storage temperature, the prepreg was no longer tacky and the resin tended to break apart on deformation and mechanical stress.

These measurements enable limiting values to be set for sufficient processability (tackiness) for this prepreg system: for example an apparent residual enthalpy in the lower temperature range greater than about 11 J/g (endothermic) and an apparent residual reaction enthalpy in the upper temperature range of 80 J/g (exothermic) referred to the prepreg. One should be aware these values are however not resin specific but depend on the carbon fiber content.

计算
Evaluation

	－20	5	23
贮存温度，℃ Storage temperature in ℃	－20	5	23
第一峰温，℃ Temperature of first peak in ℃	128	121	132

续表

表观吸热缩合焓,J/g Apparent endothermic condensation enthalpy in J/g	12.3	12.9	−5.9
第二峰温,℃ Temperature of second peak in ℃	213	235	236
表观反应焓,J/g Apparent reaction enthalpy in J/g	84.7	59.2	54.3
贮存后黏性 Tackiness after storage	仍然良好 still good	不好 bad	不能使用 unusable

结论 加工实验及其与 DSC 结果的关系,能为贮存后预浸料体系的允许的加工性设定定量标准。

注:上面的双马来酰亚胺树脂体系大概是所谓的 PMR(单体反应物聚合)树脂。可惜得不到关于所用单体的性质和有关化学的更多信息。

Conclusions Processing experiments and their correlation with DSC results allow quantitative criteria to be defined for acceptable processibility of prepreg systems after storage.

Note: The above BMI system is presumably a so-called PMR (polymerization of monomeric reactants) resin. Unfortunately no further information is available on the nature of the monomers used and the chemistry involved.

9.2 黏合剂的光固化 Light curing of adhesives

现代电子元件由许多小型化的部件和连接件组成。在生产过程中,它们必须被准确地固定在适当位置。实行它的一个方法是使用在几秒内固化的光固化黏合剂。该技术有若干重要的优势:元件不受热或机械应力的影响,实际的固定过程发生非常快。

Modern electronic components consist of a multitude of miniaturized parts and connections. In production processes, these have to be fixed accurately in place. One way to do this is to use light-curing adhesives that cure within a few seconds. The technique has several important advantages: the components are not subjected to thermal or mechanical stresses and the actual fixing process takes place very quickly.

目的 本应用描述如何研究一种用于制造硬盘储存设备的读/写头存储臂的工业黏合剂的固化行为(见图)。所选择的步骤是在 DSC—光量热仪系统中模拟生产方法,分析以此方法固化的样品的后固化行为。
然后将结果与取自不同存储臂的实际样品的测试作比较。

Purpose This application describes how the curing behavior of a technical adhesive used in the manufacture of access arms for the read/write heads of hard disk storage devices was investigated (see Figure). The approach chosen was to simulate the production methods in a DSC-photocalorimeter system, and to analyze the samples cured in this way with respect to their postcuring behavior.
The results were then compared with measurements of real samples taken from different access arms.

样品 液态黏合剂由巯基酯、聚氨酯低聚物和四氢化糠基异丁烯酸酯组成。
在用于存储臂的实际生产过程中,将黏合剂固定在适当位置的部件用 $1W/cm^2$ 强度的光

Samples	Liquid adhesive consisting of a mercapto-ester, polyurethane oligomers and tetrahydrofurfurylmethacrylate. In the actual production process used for the access arms, the part to be fixed in place with the adhesive was exposed to a light intensity of 1 W/cm² for 1 s and then heat treated (annealed) at 120℃ for 10 minutes. This process was simulated in the DSC-photocalorimeter using samples of pure adhesive in three different processes (Samples A, B and C). Sample D: The comparison measurements were performed using small samples (approx. 0.4 mg) cut out from the region marked "Sample D" from the cured access arm of the read/write head of a hard disk storage device. The width of the arm was about 1 mm and the total length 51 mm.

Conditions	**Measuring cell**: DSC with photocalorimetry accessory. A Hamamatsu "Lightning Cure 200" with a xenon-mercury lamp was used as the UV light source. **Pan**: Aluminum 40 μL without lid **Sample preparation**: Samples A to C: the viscous adhesive was used as received, weighed into the crucible and spread evenly over the bottom. See notes on Sample D. **DSC measurement**: No light exposure. The sample was inserted at 120℃ and annealed at 120℃ for 10 min. The sample was exposed to a light intensity of 1 W/cm² for just 1 s at 30℃. The sample was exposed to a light intensity of 1 W/cm² for 1 s at 30℃, and then heated to 120℃ at 50 K/min and annealed at 120℃ for 10 min. A data point interval of 0.1s was used to obtain DSC curves with the best possible time resolution. The postcuring measurements (Samples A to D) were performed in the temperature range －50℃

进行。

气氛:氮气,30mL/min

to 230℃ at 20 K/min.

Atmosphere:Nitrogen,30mL/min

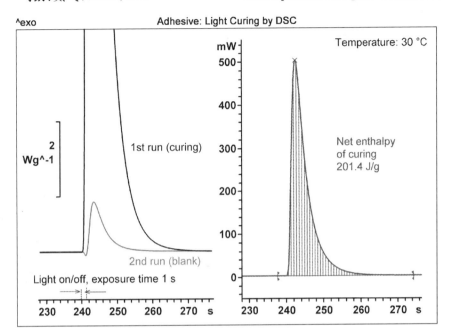

图 9.2 对紫外光曝光 1s 的样品 B 的固化

Fig.9.2 Curing of Sample B by exposure to UV light for one second.

光照 图 9.2 所示为 DSC 测试的样品 B 用 $1W/cm^2$ 光强照射 1s 后的固化反应。实际的固化反应仅发生在样品的第一次光照中(第一次测试)。在相同条件下同一样品对光的第二次曝光生成空白曲线(第二次测试)。然后该曲线被第一条曲线减去得到差分曲线(图 9.2 右边图中的净固化焓)。此曲线表示固化反应的真实进程。应当注意正好超过 500mW 的最大热流对 DSC 测试异乎寻常地大,且在几秒内达到。这是当样品用很高光强照射时发生的极快固化反应的直接结果。

Light exposure Figure 9.2 shows the curing reaction measured by DSC after exposing Sample B to a light intensity of 1 W/cm^2 for 1s. The actual curing reaction takes place only during the first exposure of the sample to light (1st run). Exposing the same sample to light a second time under the same conditions yields a blank curve (2nd run). This curve is then subtracted from the first curve to yield a difference curve (Net enthalpy of curing in the diagram on the right of Fig. 9.2). This curve shows the true course of the curing reaction. It should be noted that the maximum heat flow of just over 500 mW is unusually large for DSC measurements and is reached within a few seconds. It is a direct consequence of the extremely fast curing reaction that takes place on exposing the sample to a very high light intensity.

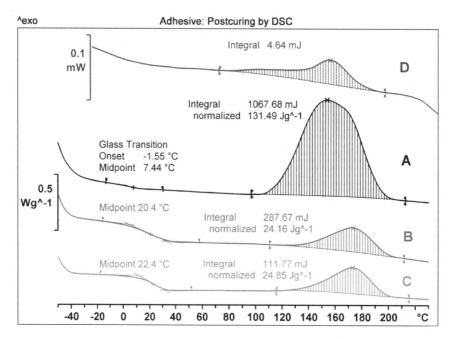

图 9.3　样品 A、B、C 和 D 的 DSC 后固化曲线的比较。
（用于样品 D 的纵坐标刻度是不同的，因为用的样品质量小。）
Fig. 9.3　A comparison of the DSC postcuring curves of Samples A, B, C and D. The ordinate scale used for Sample D is different because of the small sample mass used.

解释　就如预期的，仅是退火的样品 A 呈现大得多的后固化峰（图 9.3）。样品 A 的玻璃化转变温度也是显著低于样品 B（光照）或样品 C（光照加退火）的。

样品 B 和 C 最重要的不同在于玻璃化转变。虽然玻璃化转变温度（中点）大致相同（20℃ 和 22℃），但玻璃化转变的宽度明显不同：对样品 B（仅光照），起始点和终止点之间的差别约 30K，而对样品 C（光照加退火）只有 20K。

样品 B 和 C 后固化焓之间的小差别可解释如下。退火样品在退火过程中除去气体，而未退火样品在升温 DSC 测试过程中排除气体。除气过程是吸热的，稍稍降低了后固化焓（放热）的测量值。热重测试表明，在 120℃ 退火产生 1.0% 的失重。

存储臂小部件（样品 D）的 DSC 分析也呈现一个后固化小峰。因为样品量很少，所以无法看到玻璃化转变。

Interpretation　As expected, Sample A, which was only annealed, exhibits a much larger postcuring peak (Figure 9.3). The glass transition temperature of Sample A is also significantly lower than that of Sample B (exposed to light) or Sample C (exposed to light and annealed).

Samples B and C differ above all with respect to the glass transition. Although the glass transition temperatures (midpoint) are about the same (20℃ and 22℃), the widths of the glass transition are noticeably different: for Sample B (exposed only to light) the difference between the onset and endset is about 30 K, while for Sample C (exposed to light and annealed) it is only 20 K.

The small difference between the postcuring enthalpies of Samples B and C can be explained as follows. The annealed sample outgases during the annealing period, whereas the non-annealed sample outgases during the DSC measurement while it is being heated. The outgassing process is endothermic and slightly lowers the value measured for the postcuring enthalpy (exothermic). Thermogravimetric measurements showed that annealing at 120℃ results in a mass loss of 1.0%.

The DSC analysis of a small part of the access arm (Sample D) also shows a small postcuring peak. The glass transition cannot be seen because of the very low sample mass. Endothermic

样品的吸热热解在约220℃开始。
除了聚合物,样品D也含有不同的金属作为导线通道。为了计算相对于聚合物质量的后固化焓,样品中的金属含量用TGA作为完全热解后的残留物(47%)来测定。这意味着这时测试的样品含有0.214mg的聚合物。这给出约为22J/g的后固化焓值,一个与纯黏合剂的结果近似一致的值。

pyrolysis of the sample begins at about 220℃.
Besides the polymer, Sample D also contains different metals as conduction tracks. To calculate the postcuring enthalpy relative to the polymer mass, the metal content in the sample was determined by TGA as the residue (47%) after complete pyrolysis. This means that the sample measured here contained 0.214 mg polymer. This gives a value of about 22 J/g for the enthalpy of postcuring, a value which agrees approximately with the results for the pure adhesive.

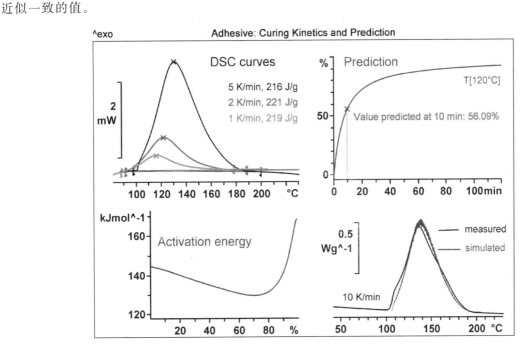

图 9.4　用非模型动力学分析的黏合剂的固化。

Fig. 9.4　Curing of the adhesive, analyzed by model free kinetics.

为了更好地深入了解退火的影响,用非模型动力学(MFK)研究了热固化过程。这是通过测试在1K/min、2K/min和5K/min三个不同升温速率下的热固化(即无光照)进行的。DSC曲线表示在图9.4中的左上图。然后从相应的转化率曲线计算了表观活化能。活化能首先是连续下降的,但是从转化率80%往上显示显著的增大。对10K/min升温速率的DSC曲线用活化能作了模拟,并与测试曲线比较(右下)。曲线表明良好的一致。因此,对120℃等温固化计算得到的转化

To gain a better insight into the influence of annealing, the thermal curing process was investigated using model free kinetics (MFK). This was done by measuring the thermal curing at three different heating rates of 1 K/min, 2 K/min, and 5 K/min (i.e. without exposure to light). The DSC curves are displayed in the upper left diagram of Figure 9.4. The apparent activation energy was then calculated from the corresponding conversion curves. The activation energy first of all decreases continuously but then shows a marked increase from a conversion of 80% onward. A DSC curve for a heating rate of 10K/min was simulated using the activation energy and compared with a measured curve (bottom right). The curves show good agreement. The calculated conversion curve for isothermal curing at 120℃ is therefore reliable. It shows that, at the temperature of interest of 120℃,

率曲线是可靠的。它显示,在所关心的120℃温度下,甚至在2h后还达不到最大转化率。根据非模型动力学的结果,在120℃退火10min后,固化转化率为56%。这与从测试得到的估算的42%固化度吻合相当好。这意味着热固化过程能用非模型动力学恰当地描述。

the maximum degree of cure is not reached even after two hours. According to the results from the model free kinetics, the degree of cure is 56% after annealing for 10 min at 120℃. This agrees reasonably well with the estimated degree of cure of 42% obtained from the measurements. This means that the thermal curing process can be adequately described using model free kinetics.

计算 完全固化材料的反应焓(样品B)是226J/g(22.4J/g + 201.4J/g)。因此,退火后(无光照)样品A有大约42%的固化度。结果汇总结在下表。从动态DSC曲线得到的热固化反应焓(图9.4)比在包括光照加退火(后固化)中得到的稍低(约220J/g)。

Evaluation The reaction enthalpy for the completely cured material (Sample B) is 226 J/g (24.2 J/g + 201.4 J/g). After annealing (no light exposure) Sample A therefore has a degree of cure of about 42%. The results are summarized in the following table. The thermal curing reaction enthalpies obtained from the dynamic DSC curves (Fig. 9.4) are slightly lower (about 220 J/g) than those obtained in the process involving light exposure and annealing (postcuring).

样品 Sample	30℃下 UV 光强度 1W/cm² 1s UV light intensity 1 W/cm² for 1s at 30℃	120℃下等温 10min Thermally at 120℃ for 10 min	T_g, ℃ T_g in ℃	固化度,% Degree of cure in %	ΔH 后固化, J/g ΔH postcuring in J/g
总值 Total				0	226
A		√	7.4	42	131.5
B	√		20.4	89	24.2
C	√	√	22.4	89	24.8
D	√	√		90	22

结论 测试表明,当前进行的过程产生90%的固化度。退火过程加速黏合剂中气体的排除,这影响了玻璃化转变的宽度。不同的是,退火对固化度几乎无影响。这为用非模型动力学(MFK)对等温固化的动力学分析所确认。用DSC得到的结果,与对直接来自生产过程的存储臂进行的测试吻合很好。因此,推荐DSC-光量热法作为控

Conclusions The measurements show that the process as it is currently operated results in a degree of cure of 90%. The annealing process accelerates the outgassing of the adhesive, which influences the width of the glass transition. In contrast, annealing has practically no influence on the degree of cure. This is confirmed by a kinetic analysis of isothermal curing using model free kinetics (MFK). The results obtained by DSC agree well with measurements performed on access arms that originated directly from the production process.
DSC-photocalorimetry can therefore be recommended as a straig-

制、模拟、了解和必要时优化技术上苛求的过程的一种直接和经济的方法。

htforward and economical method to control, simulate, understand and if necessary optimize technologically demanding processes.

附录：缩写和首字母缩拼词
Appendix: Abbreviations and acronyms

α	固化度,转化率;线膨胀系数,CTE	Degree of cure, conversion; linear expansion coefficient, CTE
$\bar{\alpha}$	平均膨胀系数	Mean expansion coefficient
ADSC	调制 DSC	Alternating DSC
AMFK	高级非模型动力学	Advanced model free kinetics
ASTM	美国试验材料协会	American Society for Testing Materials
BMC	团状模塑料	Bulk molding compound
BMI	双马来酰亚胺	Bismaleimide
CF	碳纤维增强的	Carbon-fiber reinforced
CTE	热膨胀系数	Coefficient of thermal expansion
CTT	转化率－温度转换	conversion-temperature transformation
DAIP	间苯二甲酸二烯丙酯	Diallyl isophthalate
DAP	邻苯二甲酸二烯丙酯	Diallyl phthalate
DDM	二氨基二苯甲烷	Diaminodiphenylmethane
DDS	二胺基二苯砜	diaminodiphenyl sulfone
DEA	介电分析	dielectric analysis
DETA	介电热分析	dielectric thermal analysis
DGEBA	双酚 A 二缩水甘油醚（双酚 A 环氧树脂）	diglycidyl ether of bisphenol A
DDA	双氰胺	Dicyanodiamide
DLTMA	动态负载 TMA	dynamic load TMA
DMA	动态热机械分析(动态力学分析)	dynamic mechanical analysis
DSC	差示扫描量热法	differential scanning calorimetry
DTA	差热分析	differential thermal analysis
Ea	活化能	activation energy
EDA	乙二胺	Ethylenediamine
EGA	逸出气体分析	evolved gas analysis
EP	环氧树脂	epoxy resin
FTIR	傅立叶变换红外光谱法	Fourier transform infrared spectrometry
GC	气相色谱法	gas chromatography
GF	玻璃纤维增强的	glass-fiber reinforced
GPC	凝胶渗透色谱法	gel permeation chromatography
IPN	互穿聚合物网状结构	interpenetrating polymer network
IR	红外光谱法	infrared spectroscopy
ISO	国际标准化组织	International Organization for Standardization

		续表
IUPAC	国际纯粹与应用化学联合会	International Union of Pure and Applied Chemistry
LC	液相色谱法	liquid chromatography
MF	甲基甲酰胺	methyl-formamide
MFK	非模型动力学	model free kinetics
MS	质谱法	mass spectrometry
PAA	聚丙烯酸	poly(acrylic acid)
PDA	苯二胺	Phenylenediamine
PF	苯酚甲醛	phenol-formaldehyde
phr	份/百份树脂	parts per 100 parts resin
PMMA	聚甲基丙烯酸甲酯	poly(methyl methacrylate)
PVC	聚氯乙烯	poly(vinyl chloride)
PI	聚酰亚胺	Polyimide
RT	室温(约20至30℃)	room temperature (about 20 to 30℃)
RIM	反应注射成型	reaction injection molding
SDTA	单式差热分析	single differential thermal analysis
SMC	片状模塑料,预浸料(半固化片)	sheet molding compound, prepreg
SMD	表面贴装器件	surface mounted device
T_g	玻璃化转变温度	glass transition temperature
TGA	热重分析	thermogravimetric analysis
TGA-MS	TGA 联用 MS	TGA coupled with MS
TMA	热机械分析	thermomechanical analysis
TMDSC	温度调制 DSC	temperature-modulated DSC
TTS	时间—温度叠加	time-temperature superposition
TTT	时间—温度—转换	time-temperature-transformation
UF	脲醛树脂	urea-formaldehyde resin
WLF	Williams、Landel 和 Ferry	Williams, Landel and Ferry

与热固性树脂有关的所用术语
Terms used in connection with thermosets

在本书中使用的以及常用的与热固性树脂有关的许多术语汇总于下。也可在相关标准或推荐(IUPAC,见文献)中部分查到这些术语。实际上,使用的是以往一般公认的术语和名称。

A number of terms used in this publication and in general usage in connection with thermosets are summarized below. These can also be found in part in the corresponding standards or recommendations (IUPAC, see Literature). In practice, historically accepted terms and names are used.

术语 Term	解释 Explanation
冷固化 Cold curing	在室温下或低于室温的交联 Crosslinking at or below room temperature
缩合树脂 Condensation resins	当缩合树脂固化时,(通常)形成低分子量的挥发性物质如水、氨等。 When condensation resins cure, (usually) volatile substances of low molecular mass such as water, ammonia, etc., are formed.
交联键 Crosslinks	高分子间形成的共价键 Covalent bonds that occur between macromolecules
交联密度 Crosslink density	交联密度表示为交联键之间聚合物摩尔单体的平均数,或更简单地,聚合物中化学交联键的密度。 Crosslink density is expressed as the average number of mol monomer of polymer between crosslinks; or more simply, the density of chemical crosslinks in a polymer.
交联 Crosslinking	1. 涉及几对聚合物链的反应,导致聚合物中小区域的形成,从该区域至少生出四个链。 2. 在聚合物或预聚物的不同分子链之间形成的化学键或键接的形成。这导致聚合物性能的变化,例如对于热固性塑料或弹性体(硫化)。 1. A reaction involving pairs of polymer chains that results in the formation of small regions in a polymer from which at least four chains emanate. 2. The formation of chemical bonds or links between different molecular chains of a polymer or prepolymer. This leads to a change in the properties of the polymer, e. g. with thermosetting plastics, or elastomers (vulcanization).
固化 Curing	1. 将黏性或固态(树脂)的预聚物或聚合物转变为更高摩尔质量或网状结构聚合物(热固性树脂)的化学交联过程。注意:固化通常通过加热(热固化)或对紫外光曝光(光固化)或电子束辐射(EB固化)来完成。 2. 热固性塑料经历的液态树脂交联形成固体的化学过程。固化一般发生在成型过程中,可能需要几秒钟至几个小时来完成。 1. A chemical crosslinking process of converting a prepolymer or a polymer in a viscous or solid state (resin) into a product containing a polymer with higher molar mass or network (thermoset). Note: Curing is usually accomplished by heating (thermal curing), or exposure to UV light (photocuring), or electron-beam irradiation (EB curing). 2. The chemical process undergone by a thermosetting plastic by which the liquid resin crosslinks to form a solid. Curing generally takes place during the molding operation and may require from a few seconds to several hours for its completion.

术语 Term	解释 Explanation
固化时间 Cure time	材料在特定温度下达到最佳固化度所需的时间。 The time required for a material to reach an optimum degree of cure at a particular temperature.
固化度 Degree of cure or Degree of curing	热固性树脂固化或硬化已经进行到的程度。 The extent to which curing or hardening of a thermosetting resin has progressed.
凝胶时间 Gelation time	凝胶时间是树脂在等温反应中交联成为如此强烈因而不能发生进一步成型所用的时间。按照 BS 3900:J3、DIN 55990 和 ISO 8130-6,粉末涂料的凝胶时间根据手动热板带方法测定。 The gelation time is the time taken for the resin in an isothermal reaction to become so strongly crosslinked that no further molding can occur. The gelation time of powder coatings is determined according to the manual hot plate and string method as per BS 3900:J3, DIN 55990 and ISO 8130-6.
凝胶点 Gel point	凝胶点是一个极限值,此时交联密度变得如此之大以致形成了初始网状结构。黏度达到极高值(趋向无穷大)。在凝胶点前,材料是液态的,之后是固体,不再能进行通常的加工(浇注、涂层、注入)。 The gel point is a limiting value at which the crosslinking density becomes so large that an incipient network is formed. The viscosity reaches extremely high values (tends toward infinity). Prior to the gel point, the material is liquid, afterward solid and can no longer be normally processed (cast, coated, pumped).
玻璃化转变 Glass transition	从固体玻璃态至液态或橡胶弹性态的转变。 Transition from a solid glassy state to the liquid or rubbery elastic state
硬化剂 Hardener	固化剂。辅助组分,交联反应的反应助剂 Curing agent. Second component, reaction partner for the crosslinking reaction
抑制剂 Inhibitors	抑制剂抑制活性基或其他活性中心的形成,因而延迟交联反应。 An inhibitor suppresses the formation of radicals or other activated centers and thus delays the crosslinking reaction.
成型、模塑 Molding	将热固性模塑料在模子中受热受压而固化成特定形状的固体材料的过程。 The process of subjecting thermosetting molding compounds to heat and pressure in a die so that they cure to solid material of a particular shape.
模塑料 Molding compound; Molding material; Molding mass 模塑树脂 Molding resin 模塑粉 Molding powder	准备好用于模塑加工的树脂、硬化剂、填料、颜料、增塑剂和其他成分的配方或混合物的同义词。 Synonyms for the formulation or mixture of the resin, hardener, filler, pigments, plasticizers and other ingredients ready for the molding process.

术语 Term	解释 Explanation
脱模剂 Mold release agent	经常应用于模具表面的物质,在成型周期结束时帮助聚合物产物脱模。 Material often applied to mold surfaces to aid release of polymer product at the end of molding cycle.
适用期 Pot life	催化树脂体系保持足够低的黏度用于加工的时间长度。 The length of time that a catalyzed resin system retains a viscosity low enough to be used in processing.
预聚物 Prepolymer	能够通过活性基团参加进一步聚合的聚合物或低聚物,因此其对最终聚合物的至少一个链贡献一个以上的单体单元。 A polymer or oligomer capable of entering, via reactive groups, into further polymerization, which thereby contributes more than one monomer unit to at least one chain of the final polymer.
预浸料 Prepregs	预浸料是增强织物,用部分固化的专门树脂预浸,由专门公司(称为预浸料商)按客户的技术规格制造。 Prepregs are reinforcing fabrics, preimpregnated with a special partially cured resin and manufactured by specialized firms (so-called prepregers) to customer specification.
反应树脂 Reaction resins 反应体系 Reaction systems	通过聚合或加聚作用自身固化或借助于反应介质固化的液态树脂或可液化树脂。无挥发性反应产物产生。 Liquid resins or liquefiable resins that cure by themselves or with the aid of reaction media through polymerization or polyaddition. No volatile reaction products occur.
反应介质 Reaction media	硬化剂、促进剂。固化反应也能通过紫外光或红外光的作用引发或进行。 Hardeners, accelerators. The curing reaction can also be initiated or performed through the action of UV or IR light.
树脂 Resins	1. 一般带有两个或多个进行化学交联官能团的未固化或轻微固化的聚合物(预聚物)。热固性树脂生产的原料。 2. 在历史上由天然树脂类推而命名的用于柔软、固态或高黏性物质的术语,常包含带活性基团的预聚物。 1. Usually uncrosslinked or lightly crosslinked polymers (prepolymers) with two or more functional groups for chemical crosslinking. Starting material for the production of thermosets. 2. A term used for soft, solid, or highly viscous substances, usually containing prepolymers with reactive groups that are named historically by analogy with natural resins.
树脂料 Resin mass	含溶剂、添加剂、填料或增强纤维的可加工预备树酯 Processible ready-prepared resins with solvents, additives, fillers or reinfor-cing fibers

术语 Term	解释 Explanation
树脂传递成型 Resin Transfer Molding 树脂注射成型 Resin Injection Molding	RTM(树脂传递成型)是工艺名称,在该工艺中,树脂/硬化剂混合物从储料罐转移至模具中。在RIM(树脂注射成型)中,高活性组分就在注射前被混合。 RTM (Resin Transfer Molding) is the name given to the process in which the resin/hardener mixture is transferred into the mold from a storage tank. In RIM (Resin Injection Molding), the highly reactive components are mixed immediately prior to injection.
抗热变形 Resistance to thermal deformation	根据标准测试方法定义的升温下的机械强度。 According to standard test methods, defined mechanical strength at increased temperature.
热固性树脂 Thermoset	高度交联的聚合物。热固性树脂是通过加热或其他方法由预聚物的不可逆固化制成的不熔不溶的聚合物网状结构。 注意:1. 固化前的软塑性或黏性状态材料常称为热固性树脂的预聚物或热固性聚合物。 2. 热固性树脂不能通过熔融或在溶剂中溶解进行再加工。 Heavily crosslinked polymer. A thermoset is an infusible insoluble polymer network prepared by the irreversible curing of a prepolymer by heat or other means. Notes: 1. The material in the soft plastic or viscous state before curing is usually called the prepolymer of a thermoset, or a thermosetting polymer. 2. A thermoset cannot be reprocessed by melting or by dissolving in a solvent.
热塑性塑料 Thermoplastic	未交联的聚合物。能重复熔融和冷却而且性能没有明显变化的任何聚合物,例如聚苯乙烯、聚酰胺等。 Uncrosslinked polymer. Any polymer that can be repeatedly melted and cooled without appreciable change in properties, e.g. polystyrenes, polyamides, etc.
热固性聚合物 Thermosetting polymer 热固性塑料 Thermosetting plastic 热固性树脂 Thermosetting resins	常处于未固化阶段但也作为固化材料的热固性树脂的同义词。 Synonyms for thermosets often in the uncured stage but also as cured materials.

文献 Literature

热分析 Thermal analysis

Michael E. Brown, *Introduction to Thermal Analysis, Techniques and Applications*, Chapman and Hall, London, 1988.
此书适合想获得不同技术、动力学和纯度测定概况的热分析初学者。
The book is for beginners to thermal analysis who want to obtain an overview of the different techniques, kinetics, and purity determination.

C. Darribère, *Collected Applications Thermal Analysis: EVOLVED GAS ANALYSIS*, METTLER TOLEDO, 2001, (ME-51725056).
在介绍了 TGA 仪器与 MS 和 FTIR 光谱仪耦联后,借助合适的实例讨论了应用的可能性。
Following an introduction to the coupling of a TGA instrument to MS and FTIR spectrometers, the application possibilities are discussed with the aid of suitable examples.

G. W. H. Höhne, W. Hemminger and H.-J. Flammersheim: *Differential Scanning Calorimetry*, Springer Verlag, 1996.
详细描述了技术和物理背景并从数学上进行理论探讨。
The technique and the physical background is described in detail and the theory mathematically treated.

塑料 Plastics

H. Domininghaus, *Die Kunststoffe and ihre Properties (VDI-Buch)*, Springer-Verlag, Heidelberg, 1998
附带许多材料性能表格的关于市场上可得到的塑料的物理和化学性能、加工和应用的信息。也探讨了近年来的高性能聚合物包括它们的改性、聚合物共混物和热塑性弹性体。
Information about physical and chemical properties, processing and application of the commercially available plastics with many material properties tables. Also covered are high performance polymers of recent years including their modifications, polymer blends and thermoplastic elastomers.

J. D. Ferry, *Viscoelastic Properties of Polymers*, John Wiley & Sons, New York, 1980.
IUPAC Glossary of Basic Terms in Polymer Science (Recommendations 1996), *Pure Appl. Chem.*, 68, 2287-2311, 1996.

V. B. F. Mathot, *Calorimetry and Thermal Analysis of Polymers*, Hanser Publishers, 1994.

N. G. McCrum, B. E. Read and G. Williams, *Anelastic and Dielectric Effects in Polymeric Solids*, Dover Publications, 1991.

L. H. Sperling, *Introduction to Physical Polymer Science*, John Wiley & Sons, New York, 1992.

热固性树脂 Thermosets

G. W. Ehrenstein, E. Bittmann, *Duroplaste*; *Aushärtung*, *Prüfung*, *Eigenschaften*, Carl Hanser Verlag, München Wien, 1997 (text in German).
介绍了最重要的树脂和它们的应用以及表征它们的许多测试曲线。用具体实例讨论了三种热分析方法以及力学测试方法和计算步骤。
The most important resins and their application are presented together with many measurement curves for their characterization. Three TA methods as well as mechanical test methods and evaluation procedures are discussed using concrete examples.

S. H. Goodman (Ed.): *Handbook of thermoset plastics*, 2nd ed., Noyes Publications, Westwood, N. J., USA, 1998.
详细讨论了化学和配方。主要重点放在高性能热固性树脂上。给出了各自的和非常特殊的树脂的加工、性能和应用的实例。
Chemistry and formulations are discussed in detail. Great emphasis is placed on high perfomance thermosets. Examples on processing, properties and application of individual and very special resins are given.

A. Hale, *Thermosets*, Chapter 9 in S. Z. D. Cheng (Ed.), *Handbook of Thermal Analysis and Calorimetry*, Volume 3, Elesevier, Amsterdam, 2002, p. 295-354
讨论了最重要的树脂体系和它们的性能。另外的主题是反应转化率的测定、玻璃化转变、反应动力学、光引发聚合和 TMDSC;96 篇参考文献。
The most important resin systems and their properties are discussed. Further topics are: the determination of reaction conversion, the glass transition, reaction kinetics, photoinduced polymerization and TMDSC; 96 Literature references.

J.-P. Pascault, H. Sautereau, J. Verdu, R. Williams, *Thermosetting Polymers*, Marcel Dekker Inc., New York, 2002.
此书讨论了热固性聚合物、共混物和网状结构的分析、合成和化学,描述了工业中用的专门热固性聚合物材料的实用方法、工艺和配方。有章节关于:交联聚合物合成化学;凝胶化和网状结构的形成;玻璃化转变和转换图;网状结构形成动力学;网状结构形成的流变学和介电监测;固化热固性树脂的多相性;改性热固性树脂的制备;加工中的温度和转化率曲线;网状结构的基本物理性能;交联密度对弹性和黏弹性能的影响;增韧网状结构的屈服和断裂;材料的耐久性。
The book covers the analysis, synthesis and chemistry of thermosetting polymers, blends and networks and describes practical methods, processes and formulations for specialized thermosetting polymer materials for use in industry. There are chapters on: the chemistry of crosslinked polymer synthesis; gelation and network formation; glass transition and transformation diagrams; the kinetics of network formation; rheological and dielectric monitoring of network formation; the inhomogeneity of cured thermosets; preparation of modified thermosets; temperature and conversion profiles during processing; basic physical properties of networks; the effect of crosslink density on elastic and viscoelectric properties; the yielding and fracture of toughened networks; and the durability of a material.

R. B. Prime, "Thermosets" in E. A. Turi (Ed.): *Thermal Characterization of Polymeric Materials*,

Academic Press, Chapter 6, p. 1380-1766, 1997.

在热固性树脂领域中的热分析应用给予了详细评述,援引超过 800 篇参考文献。此书出色地介绍了各个测试技术、所讨论的物理和化学模型的专门用途。

A detailed overview of the TA applications in the field of thermosets is given and backed up with over 800 literature references. The book gives an excellent presentation of the specific use of the individual measurement techniques, and physical and chemical models discussed.

E. Schindel-Bidinelli, *Strukturelles Kleben and Dichten*, Bd. 1: *Grundlagen des strukturellen Klebens and Dichtens*, *Klebstoffarten*, *Kleb-and Dichttechnik*, Hinterwaldner Verlag München, 1988

评述包括黏合剂的设计、黏合剂的类型和它们的加工、黏合剂技术和直至安全问题和产品责任的测试。

The overview covers the design of adhesives, types of adhesive and their processing, adhesive technology and the testing through to safety issues and product liability.